W9-CIC-089

Algae as Ecological Indicators

Algae as Ecological Indicators

edited by

L. Elliot Shubert

Department of Biology
University of North Dakota
Grand Forks, ND 58202
USA

ACADEMIC PRESS, INC.
Harcourt Brace Jovanovich, Publishers
London   Orlando   San Diego
New York   Austin   Boston   Sydney
Tokyo   Toronto

ACADEMIC PRESS INC. (LONDON) LTD
24/28 Oval Road, London NW1 7DX

United States Edition published by
ACADEMIC PRESS INC.
Harcourt Brace Jovanovich, Inc.
Orlando, Florida 32887

British Library Cataloguing in Publication Data
Algae as ecological indicators
1. Algae—Ecology
I. Shubert, L. E.
589.3′5   QK565
ISBN 0-12-640620-0

LCCCN 83-70584

PRINTED IN THE UNITED STATES OF AMERICA

87 88   9 8 7 6 5 4 3

Contributors

D. J. Bonin
Université d'Aix-Marseille
Station Marine d'Endoume et Centre
 d'Oceanographie
Chemin de la Batterie-des-Lions
13007 Marseille
France

T. B. Boyle
Columbia National Fisheries Research
 Laboratory
US Department of the Interior
Fish and Wildlife Service
Columbia, MO 65201
USA

V. H. Dale
Center for Quantitative Science
University of Washington
Seattle, WA 98195
USA

M. R. Droop
Dunstaffnage Marine Research
 Laboratory
PO Box 3
Oban
Argyll, PA34 4AD
Scotland

M. T. Elnabarawy
Environmental Laboratory
3M Corporation
PO Box 33331
St Paul, MN 55133
USA

H. G. Levine
Marine Sciences Research Center
State University of New York
Stony Brook, NY 11794
USA

S. Y. Maestrini
Centre de Recherche en Ecologie
 Marine et Aquaculture de
 l'Houmeau
Case 5
17137 Nieul-sur-Mer
France

R. G. Merrill
Industrial Environmental Research
 Laboratory
Process Measurements Branch
US Environmental Protection Agency
Research Triangle Park, NC 27711
USA

Contributors

A. E. Pipe
Department of Plant Sciences
University of Western Ontario
London, Ontario
Canada

C. L. Schelske
Great Lakes Research Division
University of Michigan
Ann Arbor, MI 48109
USA

L. E. Shubert
Department of Biology
University of North Dakota
Grand Forks, ND 58202
USA

E. F. Stoermer
Great Lakes Research Division
University of Michigan
Ann Arbor, MI 48109
USA

G. L. Swartzman
Center for Quantitative Science
University of Washington
Seattle, WA 98195
USA

D. G. Swift
Graduate School of Oceanography
University of Rhode Island
Kingston, RI 02881
USA

F. B. Taub
College of Fisheries
University of Washington
Seattle, WA 98195
USA

F. R. Trainor
Botany Section, U-42
University of Connecticut
Storrs, CT 06268
USA

G. E. Walsh
Environmental Research Laboratory
Sabine Island
US Environmental Protection Agency
Gulf Breeze, FL 32561
USA

A. N. Welter
Environmental Laboratory
3M Corporation
PO Box 33331
St Paul, MN 55133
USA

B. A. Whitton
Department of Botany
University of Durham
Durham DH1 3LE
England

Foreword

> Things are so strictly related, that
> according to the skill of the eye, from
> any one object the parts and properties
> of any other may be predicted.
>
> Ralph Waldo Emerson (1803–1882)

Writing this foreword affords a welcome opportunity to express admiration and gratitude to the scientists who have contributed to the common endeavour of this book. Their genius and steadfast pursuit of the aim have resulted in this integrated account of the research involved in studies on algae as ecological indicators.

As the science of biology explores deeper into the problems of the preservation and improvement of the environment on the global scale, purer and more specialized knowledge is needed. The use of algal indicators and bioassay methods is indispensable in obtaining necessary information about the different factors that act together in various environments and cause displacement effects in the ecosystem.

Management of natural resources on a sound ecological basis confronts us with problems that are biological. We are dealing with processes which determine the lives of organisms and which are themselves subject to the action of living things, so it is inevitable that biological methods provide advantageous possibilities for the assessment of environmental impacts. Ecological imbalance must be identified and measured biologically.

Ecology is the biological science that considers the environmental relations of organisms. Autecology is concerned with the relations of individual organisms to the various factors of their environment. Synecology is concerned with communities of organisms, and its essential feature is that many organisms are interacting, with each other and with the environment. The algae occupy an unique position among the primary producers: they are an important link in the food chain and are essential to the economy of marine and freshwater environments as food organisms. The factors influencing the fluctuations of algal populations, the interactions with other organisms, their fundamental role in food webs, their importance as primary producers, and the effects of man are prominent ecological problems. Algal assays are exceptionally suitable in analysing autecological as well as synecological problems relating to the management of living resources and the evaluation of environmental qualities.

As in ecology generally, understanding of the growth of algae will only be achieved by synthesis of laboratory investigations and field studies. Biological indicators and bioassay methods are aids for the purpose. Algal assays contribute to the efficient analysis of biological water quality and are necessary to obtain appropriate quantitative data expressing the relationship between the pollution load and the biological response of the receiving water. Chemical analysis provides information on the concentrations of substances present but supplies no direct knowledge of their influence on the water as a growth medium for organisms. Algal assays—in the laboratory and *in situ* experiments—open up the possibility of a fruitful combination of physical, chemical and biological measurements resulting in relevant information.

Algal assays are the source of relevant and quantitative information about the availability of chemical substances to algae and their different stimulative and inhibitory effects. Evaluation of toxic materials is unthinkable without biological assays, but this also holds for algal assays related to the fertility of waters. The overall information obtained from algal assays surpasses the significance of the sum of all partial data about the relative abundance of particular water components.

Algae as Ecological Indicators is a timely discussion of current research and the deficiencies that exist in our knowledge. Let it guide our imagination to further research.

Olav M. Skulberg
Blindern, Norway

Preface

This book was first conceived in 1979 and is an outgrowth of a symposium, "Algae as Ecological Indicators", which I organized and chaired at the annual meeting of the American Institute of Biological Sciences (AIBS) at Oklahoma State University–Stillwater. The symposium was co-sponsored by the Phycological Society of America and the Ecological Society of America.

The concept of algal bioassays provides a methodology for determining limiting or toxic factors for growth and productivity. Professor Wilhelm Rodhe (1978) quoted Goethe to describe the utility of bioassays: "Gott gibt die Nüsse, aber er beißt sie nicht auf." ("Nature provides the nuts, and we have to crack them.") Rodhe explained that "the experimental use of algal cultures offers excellent implements for cracking problems concerning algae as well as the interplay between biota and environment in natural and disturbed waters". This book provides valuable data and ideas on how we can crack "Nature's nuts".

The origin of algal bioassays can be traced to the work of Professor Martinus Beijerinck (1890), who was the first person to obtain a pure (axenic) culture of algae. Axenic cultures are an important component to algal bioassay culture methods. The *in situ* methods rely primarily on naturally occurring algal communities. The culturing of algae was advanced by the development of zoospore release methods for pure culture by G. Klebs (1896), the introduction of agar–agar by N. Tischutkin (1897), and the first culture solution specifically designed for algae by M. Beijerinck (1898). Since 1900, tremendous advances have been made in the development of defined inorganic media, isolation and culturing methods, primary productivity and radioisotope methods, *in situ* enclosures, industrial applications, and modelling.

This book focuses on the recent developments in the use of algal bioassays as indicators of ecological changes. The application of algae as ecological indicators is diverse, both in habitat type and ecological parameter. The various chapters provide examples of the different approaches to using algae as bioassay organisms. The reader will find some redundancies between chapters and some differences of opinions. Experience has shown that no single bioassay method can be used as a standard, since each experimental situation poses its own unique problems, such as marine v. freshwater, culture v. *in situ*, etc. It is apparent that increased human activities and the resulting environmental problems are providing a lot of "nuts to crack". Scientifically sound experimental approaches explored in this book demonstrate that the use of algae as ecological indicators are efficient and effective "nutcrackers".

This book could not have been completed without the help of many people. I am very grateful to William Miller and Joe Greene of the US Environmental Protection Agency, for their advice, encouragement and assistance in procuring a grant to support the development of this publication, to Academic Press for providing continued support and encouragement, especially Ruth Gadsby, Emily Wilkinson, Rosemarie Ginty and Chris Adamson, to the authors of the chapters for their cooperation, time and ideas, to the following reviewers for their helpful suggestions and ideas: Dr J. R. Cain, Dr K. M. Duke, Dr P. Eloranta, Dr V. Eloranta, Dr M. Hohn, Dr S. Huntsman, Dr J. W. Leffler, Dr J. Lewin, Dr R. L. Lowe, Dr A. G. Payne, Dr H. E. Schlicting, Jr, Dr T. L. Starks, Dr P. M. Stokes, Dr J. Verduin and Dr C. Yarish, to Linette Erickson, Rhonda Hughey and Suzanne Poe for assisting with the correspondence and the retyping of the manuscripts, to Chuck Crummy of the Academic Media Center at the University of North Dakota for the graphics, to Thomas L. Starks for assistance with the subject index, and most of all to my family, especially my wife, Marilyn, for their constant support and moral encouragement.

L. Elliot Shubert

September, 1983

Contents

Part One

Freshwater Ecosystems

1 Indicator Algal Assays: Laboratory and Field Approaches

F. R. Trainor

Biological Science Group–Botany
University of Connecticut
Storrs, Connecticut, USA

I. Introduction

For a number of years there has been a series of proposals indicating that one or more algae could be used as organisms indicative of water quality. A variety of suggestions has been proposed, for instance Palmer (1959) demonstrated that algal assemblages could be used as indicators of clean water or polluted water. A great number of instructors, students and researchers have become familiar with the very useful series of colour plates first published by Palmer in 1959, which have now become an integral part of Standard Methods.

In 1969 Palmer published a composite rating of organisms such as *Euglena*, *Oscillatoria*, *Chlamydomonas*, *Scenedesmus*, *Chlorella*, *Stigeoclonium*, *Nitzschia* and *Navicula* which could be used to indicate that the water was polluted. On the other hand, different organisms, such as *Lemanea*, *Stigeoclonium*, and certain species of *Micrasterias*, *Staurastrum*, *Pinnularia*, *Meridion* and *Surirella* would indicate that one was sampling from a body of water considered clean water. Cairns (1974) pointed out that enumeration of these kinds of indicator species, i.e. a species list, might not give as much useful information as a study of community structure.

Patrick (1971) also described a technique which could be useful in determining water quality. She proposed that an examination of the diatom flora would give us a good idea of the quality of a body of water. Using glass slides as artificial substrates, she examined a number of streams and on the basis of the data obtained suggested that we might use the frequency distribution of the algae as an indicator of the water quality. Clean water would support a wide variety of organisms, whereas polluted water would yield just a few organisms, with one or two being the dominant forms. The graphs showing the normal distribution of diatoms in these types of waters are now well known. Patrick's studies indicated that we could use a numerical approach, rather than merely a list of species.

ALGAE AS ECOLOGICAL INDICATORS
ISBN 0 12 640620 0

When we use the term assay, other types of manipulations might also come to mind, e.g. using algae to determine the level of a vitamin, using light and dark bottles, and the organisms in them, to determine the primary productivity in a body of water, or using growth of selected organisms to aid us in measuring the carrying capacity of a body of water. It is in this last sense that the term is used here. At times mixed microalgal populations in membrane diffusion chambers (Schlichting, 1975) are used as assay organisms.

Some years ago a series of investigators from government, industry and academia joined together to determine if algal growth in water samples could be used to aid in finding out more concerning the chemistry of a body of water and therefore its carrying capacity. The rationale was to determine if more information concerning the pollutants, both inorganic and organic, which were then known to be serious and troublesome additions to our lakes, streams and rivers, could be gathered by the use of one or more assays, involving photosynthetic microorganisms. After a series of studies and a number of publications, several approaches involving a variety of laboratories were suggested. Further experimentation followed, but eventually the several types of assays, whether field incubation studies, experiments using continuous cultures, or approaches using several different types of organisms, were found unnecessary. Soon, the so-called provisional algal assay procedure (PAAP) became a standardized test, and one organism, the green alga, *Selenastrum*, became the national test organism (Bartsch, 1971). Much interesting data have been assembled since that time. This procedure is discussed in more detail in the Introduction to this volume.

Most algal assays are very similar in theory. After elimination of the indigenous organisms by mechanical removal, usually sterile filtration, one inoculates *Selenastrum* into the water sample, which now serves as a culture medium. Incubation is for days under standard conditions. Growth, or lack of it, is the index of available nutrients or toxic elements.

II. Related Research

In working with synchronized cultures of microalgae, populations are *transferred each day* and the organisms, in a constant temperature and light–dark regime, are automatically entrained to grow in the light period and both divide and release new progeny in the dark period (Trainor and Rowland, 1968). Only two doublings of the population are usually possible, because in most systems light intensity, temperature or available carbon dioxide are limiting. Since fresh nutrients are supplied each day, as a typical laboratory medium, nutrient supply is more than adequate. With these points in mind while considering algal assays, it became clear that with a slight modification of the procedures used in cell synchrony experiments, an algal assay for determining the carrying capacity of a body of water could be developed. We felt that this should be a quick, reliable and uncomplicated measurement of algal growth in the sample water. The key differences in the approach would be that river water would be used as a medium and as a result cell numbers would be reduced.

With the daily transfer system it is easy to determine the cell number for the inoculum and daily transfer. In initial experiments one merely inoculates the culture adequately. After a number of days in standard conditions of growth, the cell number will come into equilibrium with the conditions used, that is, the light–dark cycle, temperature and available nutrients in the sample water. Typically one works with 10^4–10^5 cells ml^{-1}.

In developing two new assays, the algal bioassay compromise (ABC) and the sensitive algal assay procedure (SAAP), we had in mind some recycling of nutrients and the five-day work week, two quite different, but important facts of life (Cain and Trainor, 1973; Klotz *et al.*, 1975). We worked with unspiked natural water, initially with an inoculum of 4×10^5 cells ml^{-1} and standard conditions of light and temperature. We measured the amount of growth at the end of a 24-hour period, and then diluted the population (with sterile-filtered river water) back to the cell density in the original inoculum. This procedure was followed each day of the week and growth was reported as doublings per day. In this way cells were examined while they were in the active stage of growth (Tables 1 and 2). Both algal assays (ABC and SAAP) dealt with procedures which could be carried out in five days. One of the difficulties with the PAAP, and also our bioassay compromise, was the problem of nutrient carry-over. This is especially critical when dealing with oligotrophic waters. As seen in Table 1, the growth rate (doublings per day) at site B decreases each day. This is due to nutrient carry-over in the inoculum. Thus, in the first day of the assay, the measurement is not carrying capacity of the river water but growth from luxury consumption of nutrients within cells of the inoculum. But the sensitive assay, in which a quite weak nutrient solution was used, enabled us to

Table 1 Growth of *Selenastrum* over a 5-day period, using the bioassay compromise, in water collected from 4 sites.[a]

	Doublings per day			
Day	Site A	Site B	Site C	Site D
1	1·243	0·899	1·433	1·272
2	1·0	0·547	1·469	1·359
3	0·841	0·532	1·273	1·058
4	0·826	0·222	1·392	1·282
5	0·807	0·140	1·495	1·261
5-day mean	0·943	0·468	1·412	1·246

[a] Site A was in the river as it flowed through the city of Willimantic; site B was in a quite clean river; site C was just below the confluence of the latter two rivers at the point where sewage was discharged; site D was some distance from the point of sewage outfall. (After Cain and Trainor, 1973.)

Table 2 A comparison of growth rates, presented as doublings per day (for a 5-day average) of *Selenastrum*, using the algal bioassay compromise (ABC) and the sensitive algal assay procedure (SAAP). (After Klotz *et al.*, 1975.)

Site	Doublings per day (average)	
	ABC	SAAP
A	0·746	1·652
B	0·586	1·492
C	0·796	1·688
D	0·736	1·554

avoid nutrient carry-over, especially when assaying water from a relatively clean river.

The growth rates in the sensitive assay (SAAP) were typically double those in the bioassay compromise, mainly because of the lower cell number used as an inoculum, 10^4 cells ml^{-1}. Thus individual cells were in contact with more river water nutrients each day than in the bioassay compromise. The advantages to the SAAP are that the procedure can be carried out during the work week, cell counts are needed only at the beginning and the end of the experimental period, and one can assay a relatively clean body of water. The disadvantages are few, but mainly involve the need for daily transfer and the problems associated with working with small populations.

A few years ago we thought that more information could be gained from an assay by selective use of a particular organism. For example, it had been reported for years that nitrogen depletion was necessary for sexual reproduction in a variety of microalgae. If an appropriate organism was selected for an assay, standard curves would be produced relating degree of sexual reproduction and level of available nitrogen. Thus not only could one determine the carrying capacity of the body of water, but by a measurement of the degree of sexual reproduction there could also be a calculation of the amount of nitrogen which was available for organisms in the body of water. Several species of *Chlamydomonas*, or closely related colonial green algae would appear to be excellent organisms for such a combination assay, or for what might be called an indicator assay. (The organism could be an *indicator* of the carrying capacity of the river water and also indicate the level of available nitrogen.) Species of *Chlamydomonas* such as *C. eugametos* are a good choice because they produce persistent tandem pairs soon after mating. Shortly after plus and minus are mixed, if there is to be a mating reaction, cells of the opposite mating type join together at their anterior poles, forming the tandem pairs.

When plus and minus *C. eugametos* mating types were used for assay organisms (ABC) of Willimantic River water, a growth rate of one doubling per day was achieved (Trainor, 1975). Both organisms were then grown in water samples from 6 sites in the Willimantic–Natchaug–Shetucket River system. Sites 1–3 were near the city of Willimantic or established in the clean Natchaug River (site 3). The remaining sites were in the Shetucket River below the discharge of secondary treated effluent from the city of Willimantic. The data in Table 3 are incompatible with the nitrogen starvation theory: rather than showing an inverse relationship with nitrogen, the percentage of mating increased when the level of available nitrogen was elevated. Compare the percentage pairing at sites below the discharge of sewage effluent, sites 4–6, with those from cleaner water. Contrary to previous theory, there was good mating, even with elevated nitrogen (Table 3). In addition, when the level of nitrogen was artificially increased in culture tenfold, matings were still excellent; the percent pairing was identical in the control and in flasks with a tenfold increase of nitrogen. Although chlamydomonads are useful organisms for our algal assay, because of growth characteristics and ease of handling, they could not be used as organisms indicating nitrogen levels.

Table 3   The mating of *Chlamydomonas* in water collected from 6 sites in the Willimantic River system. The percentages are 4-day averages. Only nitrate nitrogen was measured. (After Trainor, 1975.)

Site	Percentage pairs	Nitrate nitrogen $(mg\ l^{-1})$
1	26	0·17
2	22	0·18
3	12	0·16
4	35	0·31
5	31	0·21
6	26	0·29

The reason for the lack of success is that the nitrogen starvation theory was based on results obtained with highly artificial laboratory media. In order to support dense populations and keep cultures healthy for weeks or months, algae have traditionally been grown in media with artificially high nitrogen levels (Table 4). Obviously there is little or no nutrient cycling in actively growing algal cultures. Thus it is not at all unusual to have nitrogen levels two orders of magnitude more than what is in nature, usually less than 1 mg l^{-1}. Apparently mating could be blocked by extraordinarily high laboratory nitrogen levels; in the laboratory this block was removed when a 10% nitrogen level was utilized. But even with 10% Bristol's medium there is much more nitrogen than in nature. With this information it should not be surprising to learn that chlamydomonads will mate at almost any nitrogen level found in nature. Seldom, if ever, is the organism in a river or pond overwhelmed sufficiently to block mating; this level is probably around 10 mg l^{-1}.

Inasmuch as laboratory findings could not be duplicated in the field, C. *eugametos* could not be used as an indicator organism.

Table 4   The levels of nitrogen found in a variety of freshwater media, and in nature. (After Trainor, 1978.)

Medium	Nitrogen $(mg\ l^{-1})$
Tamiya *et al.*	693
Sager and Granick	105
Ishiura and Iwasa	42
Bristol	41
Chu	7
PAAP	4
Nature	~1

A second attempt to utilize a combination assay involved the use of the phenomenon of polymorphism in *Scenedesmus*. These colonial four-celled algae are interesting because they do not consistently form colonies, but with specific nutrient stimuli they may form a unicellular stage. Previously we had demonstrated with *Scenedesmus* that certain phenomena discovered in the laboratory could be duplicated in the field. For example, a low temperature was needed for sexual reproduction in certain species. In field incubation studies in a local pond in autumn, we were able to produce gametes in dialysis sac cultures incubated in nature (Trainor, 1965b). Also we found that a spine-bearing *Scenedesmus*, which became spineless in older culture deficient in Fe and/or Ca, would become spineless in dialysis sacs placed in a local pond. Encouraged by these results we felt that we might use the unicellular stage of one of a variety of spine-bearing *Scenedesmus* species as an indicator of the nutrient status of the body of water. Unfortunately the place of origin of many species available in culture collections is not known. With other strains it is often difficult to do field studies at the isolation site. Also certain heavy traffic areas, such as the Willimantic River system, are not suitable for such work. Damage and loss run high. Accordingly we isolated a new organism from a small, somewhat isolated body of water, Agronomy Pond, so that we could examine the colony-unicell phenomenon in nature, especially precisely where the organism originally grew (Trainor, 1979).

The new isolate, *Scenedesmus* AP 1, grew well in culture, both in defined media and in water collected from the pond the day it was first observed and collected. It formed both colonies (with four corner spines) and unicells (each bearing four spines) in defined media (Trainor, 1979). However, in cultures grown in sterile-filtered pond water the organism was always colonial. This included water sterilized at the time of isolation as well as samples collected many times over a two-year period. In addition, cell density and several physical parameters were not involved in the colony-unicell transformation.

Scenedesmus AP 1 was then grown in dialysis sac cultures incubated in Agronomy Pond (Table 5). A completely unicellular population was used for the inoculum. Each day sacs were sacrificed and brought back into the laboratory for examination. The unicell population, produced for the inoculum in a laboratory medium, was not sustained; cultures in the field sacs, as they divided and reproduced, became

Table 5   Incubation of dialysis sacs in nature in Agronomy Pond. A 100% unicell population was used as an inoculum. (After Trainor, 1979.)

Day	Percentage unicells
1	63
3	23
5	1

colonial. *Scenedesmus* AP 1, will not form unicells in Agronomy Pond water. With several species we know that formation of the unicell stage is subject to nutritional control (Trainor, 1971; Shubert and Trainor, 1974), even involving a variety of individual factors (Siver and Trainor, 1981). Although some strains might work as indicator assay organisms, *Scenedesmus* AP 1 is not to be included among them.

An additional problem concerning algal cultures which is not frequently discussed is that of strain reliability. Cain (1979) recently reported what many workers with chlamydomonads have encountered, that is, organisms losing sexual potential over a period of months. We have noted some similar problems with a *Scenedesmus* isolate. Can we rely on comparative data accumulated with certain of these organisms which are apparently rather fine tuned?

Several workers have written about certain of these problems; I shall quote only from Alexander (1971) in *Microbial Ecology*.

The science of microbiology (and I include the algae here) has to a large extent blossomed because of careful investigations of isolates obtained from nature and nurtured in pure culture, in which microbial behaviour could be studied in carefully controlled, albeit highly artificial, conditions. The nutrition, genetics and metabolism of at least selected species are, as a result, now reasonably well understood. But almost all the information has come from axenic cultures maintained in artificial media ... In nature, however, strangers are nearly always present, and the environmental conditions are frequently highly variable.

To what extent can *in vitro* nutritional, genetic and biochemical information be applied *in vivo*, in natural ecosystems where microorganisms live, proliferate, and die? The metabolic process so prominent in a culture pampered and nourished with choice nutrients designed to maximize growth may never be expressed in the reality of a harsh environment.

No one would dispute the fact that biology has come a long way in understanding the functioning of organisms by examination of distinct and discrete processes and components. Although the interactions among microorganisms are complex, it should be possible to establish, under controlled or recorded conditions, ecological models that represent or mimic nature. In this way extrapolations from the prototype to nature become increasingly less difficult or tenuous.

With some microalgae, especially certain species of *Chlamydomonas* and *Scenedesmus*, there is still quite a challenge to be faced. The conditions of culture commonly used in the laboratory for studies of growth and reproduction might not always allow us to apply laboratory data to the field.

III. Methods

A. Field Incubation Studies

"Incubation of cultures *in situ* in the sea or lakes, inconvenient though it is, is perhaps a method that ought to be used more frequently." (Fogg, 1975.) The advantage of field incubation or, *in situ*, study if that one does not use "standard conditions" of growth but relies on available nutrient supplies as well as the temperature and light conditions prevailing at the time. The disadvantage is the infinite range of variables found in nature. Field incubations of algae can be carried out using any one of a variety of chambers (e.g. Trainor, 1965a). A very useful and much less expensive approach involves incubation in dialysis sacs (Trainor, 1965b). The cultures are grown in nature within dialysis tubing, a material commonly used in chemistry laboratories for purifying complex mixtures. The sacs are prepared in the following manner.

One end of a 15-cm length of cellophane dialysis tubing is folded and tied, and the resulting sac is filled with distilled water. A short length of 5-mm glass tubing is then inserted into the other end and secured. The end of the glass tubing is plugged with cotton. The sac is sterilized in the autoclave, submersed in a beaker of water. Later it is inoculated by way of the tube, the tube is removed, and the sac is tied securely. In nature, sacs can be incubated at various depths by fastening them firmly to a vertical rod. Sacs can become covered with silt and are a popular site for epiphytes. One can expect some damage and loss, but they retain axenic cultures for an average of 7–10 days. With high siltation they become useless in a few days, but they can last for 2–3 weeks in cleaner and colder waters.

There is considerable support for conducting assays in an *in situ* open system, i.e. dialysis sacs, or apparatus with a membrane filter, rather than to work with assays in closed systems, i.e. incubated in bottles suspended in the environment (Jensen *et al.*, 1972; Schlichting, 1975). Some types of apparatus are available commercially (Schlichting, 1975); others are fabricated for a particular set of circumstances (Jensen *et al.*, 1972). The possibilities for monitoring either the environment or an elaborate laboratory system appear almost endless, and the individual investigator must choose which technique works for the problem at hand.

B. The Sensitive Algal Assay Procedure

In the sensitive algal assay procedure (Klotz *et al.*, 1975) cultures are grown in medium 3·6 (Table 6) at $20°C$ under 4300 lux. Six days before starting the test, 30 ml medium were placed in a 125 ml Erlenmeyer flask and inoculated with sufficient cells from a week-old stock culture to establish an initial cell density of 10^4 cells ml^{-1}. The culture was then incubated under standard conditions. On the fourth and second days prior to the assay, 30 ml sterile medium were added to the flask. At the end of the 6-day growth period the inoculum cultures contained approximately 5×10^5 cells ml^{-1}. Cells were harvested by centrifugation and washed by suspension, centrifugation and resuspension in aliquots of the sterile river water sample being tested. Cultures were established in 18×150 mm test tubes by inoculating 10 ml of water sample with an initial cell density of 10^4 cells ml^{-1}. Cultures were allowed to grow for 24 h, and then cell counts were taken. Daily, cultures were transferred 1:1 by removing half of the culture and replacing it with an equal volume of sterile water sample. Growth determinations were carried out in

Table 6 The composition of medium 3.6

Component	Concentration $(mg\ l^{-1})$
$NaNO_3$	2·0
K_2HPO_4	0·2
$MgSO_4.7H_2O$	1·0
Tris	40·0
Trace mix (0·1 ml)	
The trace mixture contains:	
$FeCl_3.6H_2O$	540
EDTA	500
$CaCl_2.2H_2O$	2655
$MnCl_2.4H_2O$	30
$CoCl_2.6H_2O$	2
$CuSO_4$	1
$ZnSO_4.7H_2O$	4
$NaMoO_4.2H_2O$	2

duplicate for 4 or 5 successive days, and growth was expressed as the number of population doublings per day.

IV. Future Research

One line of experimentation which would enable us to gather considerably more useful data with algal assays would be the development of a continuous culture procedure. Rather than providing the organism with a generous supply of nutrients, i.e. a small inoculum in a large flask of assay water (Bartsch, 1971), the growth conditions could mimic nature, especially with regard to cell densities. There is literature dealing with continuous culture, but little employing the algal assay procedure (Ukeles, 1973).

Although *Selenastrum* has now been designated as the standard assay organism, any one of a number of microalgae might be used along with the latter, or as a replacement assay organism. However, the choice would have to be thoroughly tested to be certain that it has the durability needed for daily use. If a suitable alga is selected, the organism could be used in the manner described in this chapter. The concept of indicator assays appears to be sound; the critical factor appears to be in the selection of the organism.

A basic problem with using an indicator assay concept is that we have to question whether we can use the abundant laboratory data to understand the functioning or organisms in nature. In some instances the conditions of culture which have been used for decades preclude any reasonable understanding of a field situation. Assumptions that whatever is reported in the test tube will apply in the field are often unfounded and usually have not been tested, even in a preliminary fashion. In certain cases the laboratory investigator has actually had little interest in processes taking place in a local stream or pond and makes no attempt to extrapolate from one area to the other. In addition, one might believe in the unwritten maxim that whatever is reported in the laboratory will be precisely duplicated in the field.

We can always improve the techniques followed in algal assays. However, with potential indicator assays, the problems are more fundamental. With these organisms we still need to know much about their basic biology, even prior to selection for a particular assay.

V. Conclusions

After noting the success of investigators using the provisional algal assay procedure to determine the carrying capacity of rivers and lakes, we attempted first to improve this type of assay and then to expand the concept. We had apparently achieved sufficient success in maintaining nutritional control of various structural and reproductive phenomena in some algae that we could use this information in a combination assay. Specifically we believed that we could utilize selected data on reproduction to develop a combination assay. For example, if species of *Chlamydomonas* reproduce sexually only when there is "nitrogen starvation" or if a particular *Scenedesmus* will form a developmental stage at a time when there is a precise nutrient level, then one could experiment with either of these algae not only as assay organisms for the carrying capacity of the river, as in PAAP, but also to obtain additional nutritional data each time the assay was run. Theoretically this information could be obtained with little additional effort. Based on data collected during this study, we know that neither a chlamydomonad nor a *Scenedesmus* would be practical for an indicator assay of this type.

It is well known that with laboratory media, most with rather high levels of nitrogen, one has to reduce that concentration at least 90% in order to stimulate sexual reproduction. However, in nature we found that chlamydomonads mated in water collected from a variety of sites, even when the concentration was artificially increased. After we examined a morphological manipulation with a colonial green alga in defined media in the laboratory, we could not duplicate this response in field incubation studies at the site where the organism originated, or even in growth studies in the laboratory in pond water. These problems are now compounded by the fact that some strains of the microalgae which we might select for indicator assays have not always shown the reliability needed for algal assays.

Although the concept of combination assays might some day be a useful one, we have to carefully select the organism and the phenomenon we choose to manipulate. Certainly we need more concern for field conditions when we routinely study algal strains in culture. Otherwise, for some time in the future we will find extrapolation from one area to the other increasingly difficult.

Acknowledgements

I would like to thank several former graduate students for assembling much of the data reported here and for useful discussions concerning algal assays. H. Schlichting provided a constructive and useful review of the manuscript. Ellie DeCarli aided considerably in preparing the manuscript copy.

References

Alexander, M. (1971). "Microbial Ecology". J. Wiley, New York.
Bartsch, A. (1971). "Algal Assay Procedure Bottle Test". National Eutrophication Research Program, US EPA.
Cain, J. (1979). *Phycologia* 18, 24–29.
Cain, J. and F. Trainor (1973). *Phycologia* 12, 227–232.
Cairns, J. (1974). *Wat. Res. Bull.* 10, 338–347.

F. R. Trainor

Fogg, G. (1975). "Algal cultures and phytoplankton ecology" (2nd edn). University of Wisconsin Press, Madison.

Jensen, A., B. Rystad and L. Skoglund (1972). *J. Exp. Mar. Biol. Ecol.* 8, 241–248.

Klotz, R., J. Cain and F. Trainor (1975). *J. Phycol.* 11, 411–414.

Palmer, C. M. (1959). "Algae in Water Supplies". US Department of Health, Education and Welfare, Public Health Service, Cincinnati.

Palmer, C. M. (1969). *J. Phycol.* 5, 78–82.

Patrick, R. (1971). Diatom communities. *In* "The Structure and Function of Fresh-Water Microbial Communities" (Cairns, J., ed.), pp. 151–164. Res. Div. Mon. 3, Virginia Polytechnic Institute and State University, Blacksburg.

Schlichting, H. (1975). *Biocontrol Techniques* 1, 4–5.

Shubert, L. and F. Trainor (1974). *Brit. Phycol. J.* 8, 1–7.

Siver, P. and F. Trainor (1981). *Phycologia* 20, 1–11.

Trainor, F. (1965a) *Canad. J. Bot.* 43, 701–706.

Trainor, F. (1965b). *Bull. Torrey Bot. Club* 92, 329–332.

Trainor, F. (1971). Development of form in *Scenedesmus*. *In* "Contributions in Phycology" (Parker, B. and Brown M., eds), pp. 81–92. Allen Press, Lawrence.

Trainor, F. (1975). *Phycologia* 14, 167–170.

Trainor, F. (1978). "Introductory Phycology". John Wiley, New York.

Trainor, F. (1979). *Phycologia* 18, 273–277.

Trainor, F. and H. Rowland (1968). *J. Phycol.* 4, 310–317.

Ukeles, R. (1973). Continuous culture—a method for the production of unicellular algal foods. "Phycological Methods" Vol. 1. (Stein, J., ed.), pp. 233–254. Cambridge University Press, Cambridge.

2 *In Situ* and Natural Phytoplankton Assemblage Bioassays

C. L. Schelske

Great Lakes Research Division
and Department of Oceanic Science
University of Michigan
Ann Arbor, Michigan, USA

I. Introduction

A large amount of literature exists under the general heading of algal assays or bioassays of algal responses to different environmental perturbations. This chapter is mainly concerned with literature that deals with bioassay type experiments on natural phytoplankton assemblages and includes experiments conducted in the laboratory as well as *in situ*. The reason for including laboratory studies is that natural assemblages used in laboratory experiments have undergone succession in nature. These successional changes are the product of a broad spectrum of factors that cannot be replicated in the laboratory (see Smayda, 1980), whereas the physical characteristics of light and temperature that exist in the water column can be controlled in the laboratory. It is of course impossible to simulate natural turbulence and its associated patterns of irradiance in the laboratory, nor does a natural field of turbulence exist in *in situ* enclosures.

One of the reasons commonly cited for employing *in situ* experiments is that natural light and temperature conditions are simulated for the enclosed phytoplankton. However, because of reduced turbulence and probably different patterns of irradiance, organisms in enclosures may experience different light and temperature conditions than those in the natural body of water. These characteristics of experiments with *in situ* enclosures must be recognized and considered in interpreting results, and together with other factors involved in designing and evaluating bioassay experiments, will be discussed in this chapter.

The emphasis of this chapter will be to point out that natural phytoplankton assemblages can be used in different types of bioassay experiments and that these experiments can be conducted *in situ* or in the laboratory. Much of the literature that will be cited deals with freshwater phytoplankton. A large amount of marine literature exists on experiments that deal with food chains or more than one trophic

ALGAE AS ECOLOGICAL INDICATORS
ISBN 0 12 640620 0

level (see Grice and Reeve, 1981; Boyd, 1981), and because large enclosures are required, these experiments have generally been conducted *in situ*. Other sources of information on marine work are studies by Steele (1979) and Davies and Gamble (1979).

A diverse array of enclosures has been used for *in situ* experiments with phytoplankton. These range from static systems to devices designed for flow-through studies and in volume by a factor of 10^9 from 16 ml (Owens *et al.*, 1977) to 16 000 m^3 (Lund, 1972). Some experiments are terminated after a few hours, others were still in progress seven years after initiation (Lund, 1978). Steele (1979) makes the point that experimental units can range in size from small containers in the laboratory to ponds (Hall *et al.*, 1970) or to whole lakes (Schindler, 1974, 1980).

The experimental approach to studying phytoplankton dynamics has at least three advantages over descriptive approaches. First, the same population in the statistical sense is being sampled with time because this population is isolated in the enclosure at the beginning of the experiment. Second, a range of treatments or perturbations can be established. Third, replicates of these treatments also can be established so that results can be evaluated statistically. It is these advantages combined with incubation under "natural conditions" that has stimulated research with *in situ* bioassays.

II. Experimental Approaches

A. Types of Enclosures

Physically, *in situ* enclosures can be divided into five types: bottles, bags or spheres, tubes, curtains, and specialized types for flow-through studies. Different types with the exception of specialized flow-through devices are illustrated in Fig. 1.

Bottles are distinguishable from the other types in that they are readily available as shelf items from commercial sources for laboratory supplies. Sizes range from 250-ml bottles to 20-litre or larger "carboys". This type of *in situ* container can be suspended in the water column in much the same manner as bottles used for primary productivity experiments. This approach to *in situ* experiments is probably the oldest and originated with light and dark bottle studies in which oxygen production was utilized as a measure of phytoplankton photosynthesis (see Wetzel, 1965). Bottles can be used in the laboratory as well as for *in situ* experiments.

Spheres or bags were first used to study the dynamics of phytoplankton by Strickland and Terhune (1961). The plastic sphere utilized by Strickland and co-workers was large, enclosing 120 m^3. In small lakes, much smaller, less expensive and more simply designed bags can be employed, because wave action, currents and turbulence are less than in large lake and oceanic environments. Bags as small as 1 or

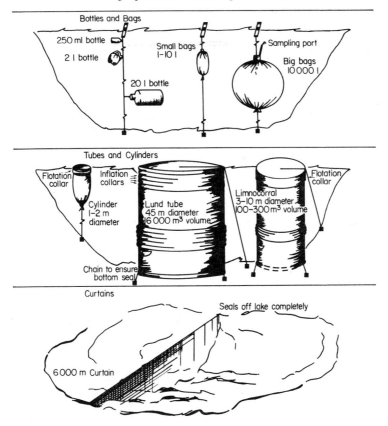

Fig. 1 Diagrams of different types of *in situ* enclosures

2 litres can be deployed in small lakes with relatively simple schemes for anchoring and sampling (Soeder and Talling, 1969).

Tubes differ from spheres in that they have a fixed diameter or other fixed dimensions over the entire length if the geometric form is not a cylinder. However, the most important difference is that tubes can be open at the bottom or both the top and bottom and do not necessarily isolate the enclosed water mass from the atmosphere or bottom sediments. A large range in size among different types of tubes has been employed for ecological studies. The largest is the Lund tube which is 45 m in diameter and contains $16\,000$ m^3 (Lund, 1972; Lack and Lund, 1974), and the smallest tubes have a diameter of about 1 m and are likely to be less than 10 m long. "Plankton-test-lots", small diameter (5·5 cm) Plexiglass tubes (6 to 8 m long) open at the bottom, were the first *in situ* enclosures that were used extensively for nutrient enrichment experiments (see Thomas, 1964). Marine enclosures (Figs 2 and 3) are generally larger than those in freshwater, the largest used in CEPEX (Controlled Ecosystem Pollution Experiment) for example contained 1700 m³ (Menzel and Case, 1977).

Placing tubes in lakes requires structural designs that ensure a minimum amount of water exchange at the surface and an effective seal at the bottom if the tube extends to the sediment surface. Some type of flotation device is needed to maintain the upper end of the tube at some distance above the water level and to minimize transfer of water between the tube and the surrounding water column. The tube must therefore be constructed of flexible material and be long enough to compensate for changes in water level in the environment in which it is placed (Landers, 1979). Hydrostatic pressure will maintain equal water levels within and outside the tube. The bottom must be sealed by weights incorporated into the end of the tube or by other means that bury it in the sediments.

Limnocorral, a term apparently coined by Canadian workers (Lean *et al.*, 1975), differs from the Lund tube mainly on the basis of size. The Canadian limnocorrals, however, were shaped as equilateral triangles 7·6 metres on a side and 4 metres deep and contained approximately 100 m³ of water (Charlton, 1975). The term limnocorral was given to cylindrical enclosures used during MELIMEX (MEtal LIMnological EXperiment), a long-term study of heavy metal pollution (Gächter, 1979) which is a freshwater analogue of CEPEX (Controlled Ecosystem Pollution

Fig. 2a Flotation module 0·25– size "controlled experimental ecosystem" (CEE). From Menzel and Case (1977).

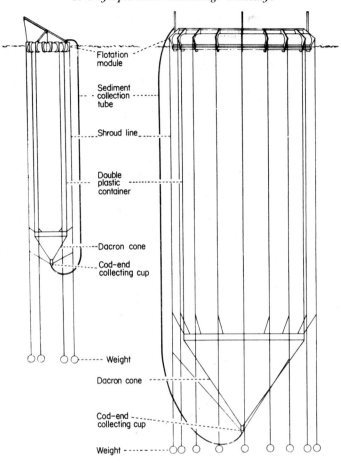

Fig. 2b Two-dimensional view of 0·25– and full-size "controlled experimental ecosystems" (CEE). From Menzel and Case (1977).

Dimensions (m)	Small	Large
Total length	16	29
Length of Dacron cone	2	7·3
Inside diameter of flotation module	2·5	10

Experiment) that was conducted at Saanich Inlet in a marine environment (Menzel and Case, 1977).

Curtains are used to separate one water mass from another within an aquatic system. Curtains therefore extend from the water surface to the bottom sediments and must effectively restrict the exchange of water between the two parts of the

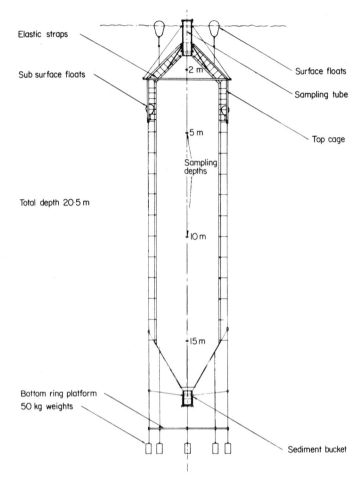

Fig. 3 The general design of the bags and outer support frame, side view. From Gamble *et al.* (1977).

system. Curtains have been used to separate parts of lakes where a narrow channel connects the two parts (Schindler, 1974) and could be used to divide a relatively small lake or pond in two or more parts or to enclose an embayment.

The specialized types of *in situ* enclosures that will be considered in this chapter are flow-through devices and dialysis tubes. Two flow-through devices are the *in situ* chemostat (deNoyelles and O'Brien, 1974) which is designed so that raw lake water can be passed through a culture chamber at a constant rate (Fig. 4) and phytoplankton cages which are designed so that filtered water can be added to the enclosure (Owens *et al.*, 1977). The *in situ* chemostat has a 5-litre volume, and

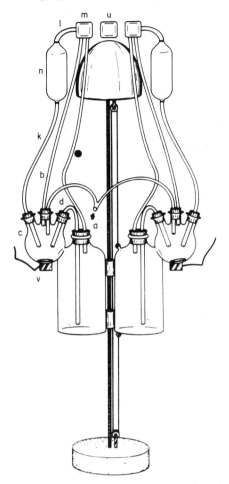

Fig. 4 Diagram of an *in situ* continuous culturing system with two culture chambers (*c*) each connected by *b* directly to the surface and by *d* to a collection bottle (*r*) which is in turn connected by *e* to an air valve at *m*. Each culture receives lake water inflow through *a* with regulation at *m* and liquid perturbation inflow through *k* from *n* with regulation at *m* through *l*. Magnetic stirrers (*v*) regulated at *u* provide culture mixing. Cultures are positioned along a cable which is attached to an anchor and float. From deNoyelles *et al.* (1980).

the *in situ* cages have volumes of 16 or 32 ml. In *in situ* chemostats raw water containing phytoplankton is added continuously, whereas in phytoplankton cages water is filtered so phytoplankton populations are affected only by conditions that exist within the enclosure. In both types the volume in the experimental chamber remains constant so that water collected from the outlet of the chamber can be used to determine changes in the system with time. Results from either type, however, will be a function of the flow rate.

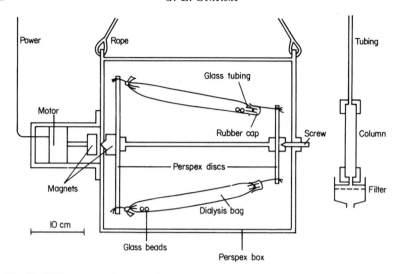

Fig. 5   Field apparatus for cage cultures and the chelating resin column. From Eide (1979).

Small tubes ($c.$ 30 ml) constructed from dialysis membranes have been utilized for *in situ* experiments with unialgal phytoplankton cultures. This general type of experiment has been reviewed recently by Sakshaug and Jensen (1978). Their use of the term "cage culture" may be somewhat confusing because some of the cages that are discussed only utilize dialysis sacs to enclose the phytoplankton sample (Fig. 5) and therefore differ from the phytoplankton cages described above (Owens *et al.*, 1977). According to Sakshaug and Jensen (1978), the term cage culture should be used because in some experiments with phytoplankton, nylon nets with perforations as large as 7 µm have been used instead of dialysis or other types of membranes with smaller porosities. With larger perforations, the rate of transport of materials across the membrane will not limit growth of phytoplankton, but with smaller perforations size will determine the rate of nutrient transport and related rate of nutrient limited photoplankton growth. The rate of nutrient transport under such conditions is controlled by molecular diffusion and must be determined for each experimental condition as follows:

$$N = P_m A_m \Delta S$$

where N is the rate of permeation (g h^{-1})
A_m is the total membrane area (cm^2)
ΔS is the concentration gradient across the membrane (g cm^{-3})
P_m is the permeability coefficient (cm h^{-1}) which must be determined for each experimental set-up.

The supply of nutrients for experiments (N) can be controlled by either altering the concentration gradient (ΔS) or the membrane area (A_m). The time course of N will also be a function of the ratio of volume within the cage to that of the surrounding medium.

Instead of sacs constructed from dialysis membranes other materials such as different types of membrane filters could be used to separate cultures from the surrounding water. This approach, however, apparently has not been utilized for *in situ* experiments (Sakshaug and Jensen, 1978) where dialysis bags have been suspended between two discs of different diameters and rotated within a perforated Perspex box suspended in the water (Fig. 5). The perforated transparent box allows exchange of water between the box and surrounding water. Glass beads placed within each dialysis bag roll back and forth along the length of the bags to maintain phytoplankton cultures in suspension. Beads roll as the bags are rotated because the bags are attached to discs of unequal diameter.

The dialysis technique also has been used in studies of the effects of toxic organics on natural phytoplankton assemblages (O'Connors *et al.*, 1978). In these experiments the toxicant was placed in the membrane bag with the natural phytoplankton assemblage and the bags were then incubated *in situ*. The opposite approach can also be employed with unialgal cultures suspended in water to monitor the uptake of nutrients and heavy metals and to study their effects on growth (Jensen and Rystad, 1973; Eide *et al.*, 1979). These experiments are more complex and difficult to interpret than completely isolated enclosures because exchange of materials with the surrounding water occurs as long as the membranes are suspended in the water and because growth may be affected by more than one substance in the water.

Materials No attempt will be made to present a complete discussion of different materials or of the suitability of different materials for *in situ* enclosures. Few experiments have been conducted with different materials, so no routine or standard techniques can be presented. However, it is important to include a few general comments about materials.

Enclosures must be constructred of non-toxic materials. Whether specific materials are toxic or growth–inhibiting under different conditions can be determined experimentally. Containers should also be checked to determine whether nutrients are leached from surfaces: soft glass bottles should be avoided because appreciable quantities of silica can be leached from glass. Problems related to leaching of toxic materials, other compounds, or nutrients are minimized with increasing sizes of containers which increases the ratio of volume to surface area. Materials also can be "preconditioned" in the environment prior to use (Lean *et al.*, 1975), although this may introduce an artefact resulting from fouling.

Some materials may sorb substances that are added to enclosures. This may be a problem especially for some of the potentially toxic organic compounds that may be added in concentrations of ng l^{-1} or less. Some materials also may sorb nutrients. In long-term experiments periphyton growing on surfaces may also accumulate significant fractions of substances added to the water. These effects would also be reduced as the sizes of containers are increased.

Fouling is a common problem when apparatus or instrumentation is maintained *in situ* for long periods of time. For example, Gächter (1979) stated that transparent material maintained *in situ* for a year or more became non-transparent because of fouling by periphyton. In experiments conducted on offshore waters in Lake Michigan, we observed no noticeable growth of periphyton (Schelske and Stoermer, 1972) even on plastic bags left in the lake for four or five weeks. Lack of fouling was possibly related to deployment in oligotrophic waters at relatively low light intensities. Fouling is not controllable except mechanically and should be expected with all materials, although it undoubtedly will be less of a problem in oligotrophic environments than in more eutrophic ones or in short-term experiments than in longer term exposures.

Metallic materials should be avoided whenever possible: there is increasing evidence that small amounts or concentrations of trace metals may affect phytoplankton growth rates (Carpenter and Lively, 1980; Huntsman and Sunda, 1980). Although most of the cited evidence is from marine environments these effects should also be expected in freshwater systems.

Transparent or translucent materials must be used for bags and bottles and for tubes of relatively small diameter so irradiance is not reduced significantly. Large tubes or cylinders can be constructed of opaque materials because light penetration will not be affected significantly if the diameter is large enough so that shading does not occur in the cylinder. However, Gächter (1979) indicated that shading was a problem in the MELIMEX limnocorrals, and Takahashi and Whitney (1977) found that floats and plastics in CEPEX enclosures reduced direct solar radiation as much as 50% and diffused incident light. Enclosures maintained at a fixed position in the water column will be shaded differentially throughout the day in relation to the angle of the sun. Shading can therefore affect the distribution of phototactic organisms within such enclosures. The amount of shading will vary with depth and will be determined by the diameter and depth of cylinders. Some plastic containers are not transparent but translucent and therefore will probably act as a neutral density filter, and the resulting reduced irradiance within the enclosure can be used to simulate depth in the water column (Schelske *et al.*, 1978). The amount of reduction can be determined experimentally with a quantum meter or other light-measuring instruments.

Durability of materials is a consideration for experiments that will last for long periods of time. Because of greater durability, materials constructed of rubber rather than plastic were used for Lund tubes and limnocorrals that were to be maintained in lakes for more than a year (Lack and Lund, 1974; Gächter, 1979). Lund (1978) stated that Lund tubes made of butyl rubber had been maintained *in situ* for seven years and probably would last seven years more. Plywood enclosures were used by Dickman and Efford (1972).

Experiments can be run to determine the durability and potential toxicity to organisms of materials used for enclosures. The material should be checked to determine whether nutrients or other materials of interest are leached from it or sorbed by it. Experiments should be run for each application, because characteristics of materials may vary depending on the type of water or the type of substances being used in the experiments.

B. Types of Experiments

Nutrients or other substances may be tested for effects on phytoplankton in five different types of experiments.

Spike One or more substances are added at one time, usually at the beginning of the experiment. Concentrations of the spikes decrease with time if the added substance is assimilated by phytoplankton or degraded under the experimental conditions. Responses are then a function of the initial concentration(s) of spike and interactions among the spikes if more than one substance is added.

Fixed level or fixed supply Concentrations of substances used are adjusted at specified intervals to a predetermined level and therefore can be maintained at the specified level so that supplies are not depleted in the medium. The fixed level approach has been used for major nutrients (Schelske *et al.*, 1974; Stoermer *et al.*, 1978), but it is not practical for substances that cannot be measured readily or for relatively low levels that cannot be measured precisely. If the substance of interest cannot be measured readily, it may be possible to utilize a fixed supply rather than a fixed level treatment. In the fixed supply treatment, a specified quantity is added at specified periods of time. If several levels are utilized in this experimental design, a known range of supply rates but not necessarily a measurable range of concentrations can be employed. The fixed supply approach therefore has the advantage that levels or concentrations of substances that are too small to measure can be used in experiments.

Dilution A specified fraction of the volume is removed at specified periods of time, resulting in a known dilution rate. New medium can be added at the same rate so that chemostat or steady-state conditions can be stimulated (Jones *et al.*, 1978a,b).

Chemostat This type of experiment has been utilized in many studies with unialgal populations and in some experiments with natural phytoplankton (deNoyelles and O'Brien, 1974; Barlow *et al.*, 1973; Peterson *et al.*, 1974; deNoyelles *et al.*, 1980; Thomas *et al.*, 1980b). This is a continuous flow-through system designed to produce steady-state conditions because the inflow and outflow rates are equal. Dilution rates for chemostats are determined by the flow rates and volume of the culture vessel.

Turbidostat   In essence, the turbidostat is a chemostat whose dilution rate is regulated to obtain a predetermined density of phytoplankton within the culture vessel. At steady state, the turbidostat is operated like a chemostat to maintain the desired phytoplankton density in the culture vessel. Turbidostat and chemostat experiments were used by Peterson *et al.* (1974).

The five types of treatments discussed above can be divided into those in which the initial water is not diluted appreciably (spike and fixed level or fixed supply) and those in which the initial water sample is diluted with "new" medium at a predetermined rate (dilution, chemostat, and turbidostat). A known flushing rate simplifies analysis of kinetics, but conditions simulated with a chemostat are not attained in nature (Jannasch, 1974).

C. Design of Experiments

Levels of Nutrients   In designing enrichment experiments, levels of added nutrients should be selected that are realistic in comparison to concentrations that exist in the natural environment. Adding $50-100\ \mu g\ P\ l^{-1}$ is an unrealistic enrichment, especially if the untreated water has a total phosphorus concentration of $10-20\ \mu g\ P\ l^{-1}$ or less. Large concentrations of added substances can in fact be toxic, inhibitory, or instrumental in determining species succession. A rule of thumb might be that added concentrations should no more than double or triple the lake concentration or the expected concentration if the substance has not been measured. For some applications the desired objective may be to determine the smallest quantity necessary to elicit a response. Phosphorus added at levels of $5\ \mu g\ P\ l^{-1}$ or less has stimulated the growth of phosphorus-limited phytoplankton in the Great Lakes when total phosphorus concentrations in untreated lake water ranged from $5-10\ \mu g\ P\ l^{-1}$ (Lin and Schelske, 1981; Schelske *et al.*, 1978). Magnitude of responses will not necessarily increase with increasing level of the added limiting nutrient, because at some point another nutrient will become secondarily limiting.

Complete treatment   A complete treatment is recommended for enrichment experiments and may prove to be very valuable in preliminary work on evaluating limiting nutrients. The complete treatment should in theory include spikes of all substances known to be required for phytoplankton growth, and unless it includes some toxic or inhibitory substance or combination of toxic substances it should produce the maximum growth response from the phytoplankton. In practice the composition of the complete treatment can probably be restricted to the major nutrients (phosphorus, nitrogen, and silica), to trace metals (iron, manganese, molybdenum, cobalt, zinc, and possibly others) and to vitamins (thiamine, cobalamine, and biotin). In waters with low dissolved inorganic carbon it may be necessary to include a carbon source in the spike. The rationale for the complete treatment is that it will identify potential limiting factors that might not be apparent if only single spikes or combinations of spikes were used. Responses from other treatments, including deletions from the complete treatment, can be evaluated in comparison to the complete treatment (Lin and Schelske, 1981; Smayda, 1974).

Control A control should be employed in all experimental designs. Only distilled water equal to the volume of liquid added to all other treatments will be added to the control.

Lake Water When experiments are maintained *in situ* samples of lake water should be obtained from the original sampling point whenever enclosures used in the experiment are sampled. If samples of lake water can be obtained for periods of time before and after the duration of the experiment they can be used to compare and interpret results from the experiments. Lake water samples also should be obtained if experiments are conducted in the laboratory. The same variables that are being used as response measures in the experiments should be measured on the lake water samples.

Other factors Results of experiments conducted with natural phytoplankton assemblages may be affected by factors other than those employed as treatments in the experimental design. These include depth and time of sampling and presence of zooplankton.

In most lakes the vertical distribution of phytoplankton will not be homogeneous throughout the year, so depth of sampling must be considered in the experiment. Samples collected should be maintained at the light and thermal conditions of the point of collection, unless either or both of these conditions are used as a variable. Light quality and intensity vary with depth in the water column which may affect the distribution of phytoplankton. Wall and Briand (1979) enclosed natural phytoplankton assemblages in different coloured plexiglass cubes to study effects of light intensity and quality on population growth.

Zooplankton cannot be eliminated completely from experiments with natural waters even though larger zooplankton can be screened with plankton nets. However, Venrick *et al.* (1977) found that this type of screening increased instead of reduced variability in experiments. It also may be possible to obtain samples from depths at which zooplankton are least abundant, possibly by taking advantage of vertical migration patterns. Smaller zooplankton, those which cannot be screened because of similarity in size to phytoplankton, will be included in the samples that are collected. Lehman (1980) used several concentrations of zooplankton to study the role of grazers in nutrient cycling; his enclosure approach with natural communities could be used to quantify zooplankton effects in experiments.

When samples are being collected for experiments, the sampling scheme should be designed to minimize light and temperature changes or shock to the natural communities. To minimize light shock we have collected samples at dusk or at night to avoid possible exposure of populations, even for a short time, to full sunlight. If samples are to be transported to a laboratory, insulated opaque boxes must be provided.

In adding substances to containers it is important to add the same volume of liquid to the entire series of experiments. For example if the nutrients being added are dissolved in distilled water the same volume of distilled water with or without nutrients should be added to each container including the controls. This procedure ensures that any dilution effects are controlled among the treatments and that any contaminants present in distilled water are added in equal concentrations to all treatments.

Chemicals with the highest possible purity should be used for treatments. Only small amounts of materials relative to the dissolved salts in moderately hard waters will be added with the treatments. However, it is possible that the contaminants may supply essential trace elements under some conditions. In addition appreciable quantities of cations may be added with nutrient additions, especially nitrate nitrogen and silica which may be added in concentrations that can be measured in mg l^{-1}.

Containers should be carefully cleaned to ensure they are free of chemical contamination. If containers are reused they should be identified, so that a container used for a high level treatment will not be used as a control because it may not be possible to completely remove all traces of the added substance.

D. Measurement of Responses

It is highly desirable to include redundancy in measurements of phytoplankton responses to experimental perturbations. At least two independent measurements should be employed for most purposes. Responses can be evaluated from different estimates of standing crops, from growth rates of phytoplankton and from short- or long-term rate changes of different chemical properties. The volume of sample required for a particular analysis and thus container size may be important in selecting variables that will be used to measure responses. If repetitive sampling is required and small containers or enclosures are used, sampling schemes must be planned judiciously and response measures must be selected carefully to avoid serious depletion of the original volume.

Chlorophyll a   Changes in standing crop can be estimated conveniently from measurements of chlorophyll *a*. This variable is widely used because it can be measured relatively simply with fluorometry either *in vivo* (Thomas *et al.*, 1974; Schelske *et al.*, 1978) or on extracted samples (Strickland and Parsons, 1972). Fluorometric analysis is economical in terms of numbers of samples that can be processed in a given amount of time and small sample sizes can be used because the method is extremely sensitive compared to spectrophotometric methods (Strickland and Parsons, 1972).

Cell counts   Changes in standing crop can be estimated from microscopic cell counts or from particle counts obtained with electronic particle counters. Microscopic counting is preferable because it also provides valuable information on individual populations within the phytoplankton assemblage.

Species composition For many applications, data on species composition can be the most important and useful information obtained from bioassay experiments with natural phytoplankton assemblages. One, and possibly the only, test of whether experimental conditions simulate those in nature can be obtained from analysis and comparison of data on species composition in experiments with those from the natural environment.

Population biomass Data on quantitative abundance of phytoplankton can be used for estimates of biomass at the population level. Methods of estimating cytoplasmic volume and carbon content (see Strathmann, 1967; Sicko-Goad *et al.*, 1977) should be used rather than estimates of cell volume.

Other biomass measurements Responses can be evaluated from estimates of assemblage biomass in terms of carbon, nitrogen and phosphorus (Jones *et al.*, 1978a) or in terms of different protoplasmic fractions, including protein, carbohydrate and fat. Measurements of protoplasmic fractions apparently have not been utilized to any great extent in nutrient enrichment experiments.

Indirect measurements Nutrients utilized by phytoplankton for uptake and growth can be measured indirectly from changes with time of nutrient concentrations in the water. Uptake rates can be calculated from short-term sampling and the phytoplankton community chemical composition can be estimated from these chemical changes in the water.

Chemical composition Chemical composition can be determined indirectly from chemical measurements of water or measured directly from samples of phytoplankton. Chemical composition can be used to determine whether plant tissues contain a critical level of an essential nutrient (Gerloff, 1969) or to determine whether the proportion of different nutrients in phytoplankton differs from an expected norm. In oceanography the Redfield ratio is widely used as the expected atomic ratio of 106:16:1 for C:N:P in marine phytoplankton (Redfield *et al.*, 1963).

^{14}C *uptake* Short-term ^{14}C uptake experiments have been widely used to measure photosynthetic rates of phytoplankton and to determine with bioassays whether nutrients limit primary productivity (Goldman and Carter, 1965). This response measure should not be employed independently for reasons presented in the discussion below.

Physiological indicators It may be important to have information on the physiological state of phytoplankton in determining responses to nutrient enrichment. For example enzymatic analyses can be used to determine the phosphorus status of cells (Fitzgerald, 1969; Fitzgerald and Nelson, 1966). Other physiological indicators possibly could be applied in bioassay studies (see Healey, 1978; Healey and Hendzel, 1980).

III. Discussion

Data on variables obtained during bioassay experiments can be used in several different ways to determine responses to treatments. Values for any variable on any given day, and differences or ratios of increase or decrease for any variable between two time periods, can be used to determine responses. The preferred method of measuring responses is to estimate rates of change for each treatment. For example, if adequate points are available, growth rates of populations can be estimated with linear regression of data that have been transformed to natural logarithms. Variances of mean growth rates that are obtained can be used in tests of statistical significance among or between treatments either at the community (Schelske *et al.*, 1974, 1978) or population level (Stoermer *et al.*, 1978).

It is very important to establish initial conditions for all experimental treatments, especially if analyses are to be made on phytoplankton species composition. It is virtually impossible to end up with homogeneous phytoplankton assemblages in the large series of containers or enclosures that will be used in an experiment. Problems in interpreting data for populations that clump and may not be distributed randomly among the replicated treatments and for rare populations that may not be counted initially in all replicated treatments can be obviated if these facts are known from samples taken at time zero from every replicate. Another reason for making initial measurements is that even though standing crops measured as population counts or as a community biomass measure may vary from replicate to replicate these differences can be normalized by different methods of data reduction. For example, analysis of slopes of growth curves is independent of initial standing crops as are ratios of increase and decrease or differences in values between two sampling periods.

Choosing which variables to measure depends largely on the purposes of the study and questions of interest. In studies on nutrient limitation of algal growth it is important to select variables that measure phytoplankton growth. Included in these standing crop measures are cell counts, cell carbon or biomass, chlorophyll *a*, and particulate carbon, phosphorus, and nitrogen. As discussed below, it may not be possible to predict this growth response from short-term ^{14}C uptake measurements of community photosynthesis.

Short-term ^{14}C uptake rates obtained several hours after nutrient perturbation may not be a valid measure of phytoplankton community response to experimental treatment (Golterman, 1975). Results from a number of studies indicate that this response measure cannot be used to determine growth limiting nutrients. Results obtained by Barlow *et al.* (1973) showed that only phosphorus limited the growth of natural phytoplankton in chemostats, whereas short-term ^{14}C uptake results only rarely indicated phosphorus limitation but did indicate limitation by silica and trace metals. Peterson *et al.* (1974) reported erratic results on nutrient limitation from short-term ^{14}C uptake but only phosphorus limitation for experiments with natural phytoplankton from Cayuga Lake run either in chemostats or turbidostats. In our work on the Great Lakes we found no difference among treatments in short-term ^{14}C uptake at the beginning of experiments (Schelske *et al.*, 1972, 1974) but did find differences later in the experiments which could also be determined from other response measures (Schelske *et al.*, 1978).

Workers who have compared methods for determining nutrient limitation have also raised questions about using short-term ^{14}C uptake to evaluate treatments in bioassay experiments. Gerhart and Likens (1975) found that nitrogen and phosphorus additions stimulated phytoplankton growth in experiments with Mirror Lake phytoplankton either when the nutrients were added to polyethylene tubes or to chemostats. However, results from ^{14}C uptake experiments showed no effect of nitrogen and phosphorus enrichments. O'Brien and deNoyelles (1976) showed that additions of nitrogen and phosphorus fertilizer increased both phytoplankton biomass and production in ponds. Experiments with the same pond water in batch and chemostat bioassays showed that nitrogen and possibly phosphorus were limiting factors but these enrichments did not increase ^{14}C uptake at the beginning of the experiments. These results and others cited above suggest that there is a lag between nutrient perturbation and phytoplankton growth response which flaws the universal application of the short-term ^{14}C uptake approach. Knoechel and deNoyelles (1980) have concluded that nutrient enrichment bioassays must last longer than 24 hours so that "initial shock effects can dissipate and adaptation or selection can occur". Their conclusion is based on ^{14}C uptake at the population level that was quantified with track autoradiography.

Three types of responses can be expected from nutrient enrichment experiments with natural phytoplankton assemblages. First is the typical lag response discussed above. Second is immediate exponential growth (Stoermer *et al.*, 1971; Schelske *et al.*, 1975) that apparently occurs when exponentially growing populations are sampled. Third is no response, which possibly may be due to the presence of blue–green toxins in the water which inhibit growth of other algae (Stoermer *et al.*, 1971; Murphy *et al.*, 1976) or to the lack of enrichment with critical nutrients or other substances required for growth. Similar types of responses can be expected at the population level. In addition it has been shown that nutrient interactions had a significant effect on population growth (Stoermer *et al.*, 1978; Jordan and Bender, 1973b).

IV. Applications and Advantages

The focus of interest will largely dictate the type and size of *in situ* enclosures that will be used in experiments (see Lund, 1978). For example, it is obvious that either transparent or translucent materials must be utilized for bags or bottles; whereas opaque materials could be utilized to construct curtains or large tubes and cylinders.

There appears to be no obvious advantages of one type of an *in situ* enclosure in relation to another (Lund, 1978). All types have shortcomings, some of which will be covered in this chapter and others which are fairly obvious. Which type or types can be applied for particular experiments may become obvious if the potential user considers the following factors in a proposed design of experiments.

Number of variables The number of enclosures that must be utilized in an experiment will be determined by the number of independent variables that are employed. The number of enclosures will be equal to $N+1$ in the simplest type of applications where N is the number of treatments.

Number of replicates For some experiments it will be necessary to include replicate treatments which may more than double the requirements for enclosures.

Analysis of data A statistical or another type of model must be available for the analysis of data. It is highly advisable to consult with a statistician if the outcome of the results hinge on statistical tests so an appropriate and efficient statistical design can be employed which also could greatly reduce requirements for replicates.

Temporal scale is best illustrated by an example. If experiments are conducted on factors that control seasonal succession of phytoplankton, new experiments must be initiated on time scales of days to one or two weeks to sample the stages in phytoplankton succession (Fournier, 1966). Initiating a number of experiments over a season precludes the use of curtains, Lund tubes, and limnocorrals because many experiments and enclosures must be employed over a season or an annual cycle to obtain samples of temporal changes in the phytoplankton community. However, if the effects of different loading rates on phytoplankton production and dynamics are being studied (Lean *et al.*, 1975; Twinch and Breen, 1981), then curtains, Lund tubes, and limnocorrals would be better suited than smaller containers or enclosures. Enclosures can be used in studies that allow simulation of particulate settling (Charlton, 1975; Gamble *et al.*, 1977) and regeneration of substances in sediments (Baccini and Suter, 1979; Davies and Gamble, 1979).

Sampling The size and volume of the enclosure must obviously be large in comparison to the total quantity of sample that will be required for repetitive sampling. However, in some experiments water that has been sampled can be replaced with an equal quantity of water, and this dilution can be employed as part of the experimental design. Generally speaking, there are physical and logistic constraints on experimental design such that increasing container size reduces the number of treatments and replicates that can be employed. Only large enclosures, such as curtains and limnocorrals, can be used if experiments are to be conducted for long periods of time.

Spatial scale For experiments with phytoplankton, problems with spatial scale are less troublesome than they would be for experiments with larger organisms. A large body of laboratory research on phytoplankton has been obtained over the years from experiments that were conducted in relatively small volumes of water so this should be possible also with *in situ* experiments. Container size must be increased as larger organisms and additional trophic levels are included in the experimental design (Boyd, 1981). Ordinarily when setting up phytoplankton experiments in small containers, effort is expended to eliminate spatial hetero-geneity that existed in nature because analysis of results is simplified by minimizing the initial variability among containers or treatments (Takahashi *et al.*, 1975).

Practical considerations Economic factors are certainly a part of determining whether enclosures or which type of enclosures should be employed. Costs for construction and deployment of different types of enclosures illustrated in Fig. 1 can be expected to range over several orders of magnitude. The most expensive *in situ* enclosures used to date have without a doubt been those used in marine experiments. Costs for construction, deployment and maintenance were greater because big enclosures were used and because deployment and maintenance are more expensive in large bodies of water which have tides, currents and waves. Likewise, costs of construction and deployment of a Lund tube in fresh water will be much greater than costs of obtaining and deploying small bottles and bags even though large numbers of the latter may be used only once in a series of replicated experiments over a season.

The advantage of tubes over closed containers is that tube construction allows studies of sediment–water and air–water interactions during the experiment. Water in tubes open to the atmosphere, however, can be disturbed by vandals, contaminated by birds (Moss, 1981) or insects and diluted by rainfall. Transparent covers can obviate these problems (Kuiper, 1977). Another problem is that enclosures that extend to the surface can be damaged by ice in the winter (Lund, 1978).

Deployment of enclosures in large bodies of water is also more complicated than in small lakes or ponds. If enclosures are to be used in large systems, they will usually have to be anchored and maintained in place during the period of the experiment. Methods of filling and sampling are also more complex for plastic spheres maintained in place during filling or sampling than if an entire sphere could be sampled or if the container could be sampled and then returned to the lake. We found that spheres of 1000 or 4000 litres could not be filled at the surface of Lake Michigan without rupturing and could only be filled successfully when the sphere was anchored at depth (Schelske and Stoermer, 1972). Therefore SCUBA divers were needed for filling and sampling operations. This system of filling and sampling has the advantage that containers can be filled at depth without subjecting organisms to changes in light or temperature that can occur if water is brought to the surface for extended periods of time.

Much attention must be given to engineering problems if large-scale enclosures are to be deployed, especially in the initial design. Also how these enclosures will be placed in the experimental site must be given adequate consideration. SCUBA divers may be required for placement or sampling (Strickland and Terhune, 1961; Schelske and Stoermer, 1972; Menzel and Case, 1977) and should definitely be employed initially to determine whether the structure is in place as designed and periodically to determine whether maintenance is necessary. Even small holes in tubes and spheres can lose significant amounts of water through pumping and squeezing actions induced by physical forces on the enclosures.

Problems in deploying large or small plastic spheres is one reason why researchers have considered other types of enclosures. Plastic carboys (8 l) and plastic bottles (2 l) have been employed successfully in experiments conducted on the Great Lakes (Marshall and Mellinger, 1980; Schelske *et al.*, 1975). An advantage of bottles of this general size is that all the bottles required for one day's sampling can be deployed on one anchor line (Fig. 6). Bottles with all the different treatments can be arranged on the line randomly, and these strings of bottles also

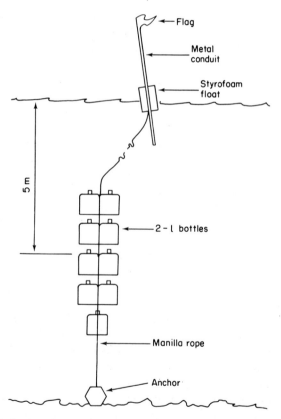

Fig. 6 Schematic drawing of *in situ* placement of bottles in Lake Michigan. Water depth was 17 m. From Schelske *et al.* (1975).

can be sampled randomly. In this type of an experiment only the string of bottles that is sampled on any given day is disturbed, so repetitive sampling affects only the samples that are removed each day.

Some of the problems of employing *in situ* enclosures were reviewed by Soeder and Talling (1969). They state that "great skill" and a "well-equipped team of experts" are needed for studies with plastic bags and cylinders. They also discuss advantages of employing different types of enclosures and different types of experiments.

V. Containment Effects

A relatively large amount of literature exists on containment effects of natural populations and can be cited to refute results obtained in any type of enclosure experiment (see Venrick *et al.*, 1977; Gieskes, 1979; Carpenter and Lively, 1980). Containment effects (reduced photosynthetic capacity, decreases in chlorophyll concentrations, changes in species composition) can occur and should be adequately controlled in experiments. Containment effects are rarely considered to be problems in laboratory experiments with unialgal populations. No attempt will be made to prove that containment effects do not occur, but my approach will be to show that meaningful results can be obtained from enclosure experiments and that some treatment effects can be related to changes that occur in the natural environment.

Conflicting results have been obtained from short-term experiments on containment effects. For example, Venrick *et al.* (1977) found an effect on species composition, whereas Gieskes *et al.* (1979) found decreases in chlorophyll concentration and primary production but only in small bottles and not in 4-l bottles. Initial decreases in chlorophyll concentrations were found in Lake Huron water samples used in laboratory experiments; however, after 9 days changes in species composition resulting from nutrient treatments were much greater than those that could be attributed to effects of containment (Lin and Schelske, 1979).

Studies of species composition in long-term enrichment experiments can be cited to show that population dynamics in enclosures were similar to those that occurred in the water from which the sample was obtained. Data supporting this view have been obtained by Barlow *et al.* (1973) with chemostat experiments on pond water and Cayuga Lake water and by Peterson *et al.* (1974) with chemostat and turbidostat experiments on Cayuga Lake water. No dramatic succession of Cayuga Lake phytoplankton occurred in chemostat experiments, but a change in species composition was obtained in turbidostat experiments. The species change occurred while the phytoplankton density was increasing in the turbidostat, which suggests that either the absence of the initial dilution or the presence of density-dependent factors or both in the turbidostat played a part in the different patterns of succession.

Data from nutrient enrichment experiments on waters from the Great Lakes indicate that levels and types of nutrient enrichment are important in determining species succession. Small differences in the phytoplankton assemblages and no differences in major dominants were found after 15 days of enclosure over a large range of nitrogen and phosphorus ratios and concentrations (Stoermer *et al.*, 1978). At the highest level of enrichment the nitrate nitrogen concentration was 1·1 mg l^{-1}, five times greater than the ambient concentration, and the orthophosphate phosphorus concentration was 60 μg l^{-1}, nearly ten times greater than the ambient total phosphorus concentration. *Fragilaria crotonensis* was dominant in these N and P treatments and was also a major component of the phytoplankton assemblage in the lake. One major difference between assemblages in the laboratory experiments and the field was that blue–green algae comprised a large fraction of the assemblage in the lake from late September to early December while the proportion of blue–greens declined during the laboratory experiment. Diatom growth in the lake was undoubtedly limited during part of this time by lack of adequate supplies of silica (Schelske and Stoermer, 1971), whereas in the laboratory silica concentrations were maintained at 0·74 mg l^{-1} (Schelske *et al.*, 1974) which would not have limited diatom growth.

In this Lake Michigan experiment, adding vitamins, trace metals and EDTA in combination with the highest nitrogen and phosphorus enrichment doubled the chlorophyll production rate to 0·43 doublings per day and drastically altered the species composition in comparison to the parallel treatment with only a nitrogen and phosphorus enrichment (Stoermer *et al.*, 1978). Three diatom species, all species of *Stephanodiscus*, accounted for the increase in chlorophyll production with growth rates that ranged from 0·66 to 0·97 doublings per day and *S. minutus* replaced *Fragilaria crotonensis* as the dominant species in the treatment with vitamins, trace metals, and EDTA. These results indicate that effects of small concentrations of vitamins, trace metals or EDTA were much greater than effects of containment. (The concentration of added EDTA was 0·20 mg l^{-1}, of added iron (10 μg l^{-1}), other trace metals (1 μg l^{-1}) and vitamins ($<0·01$ μg l^{-1}).) In a similar experiment with Lake Huron phytoplankton it was shown that the growth-promoting effect was mainly EDTA and not vitamins or trace metals (Schelske *et al.*, 1978).

In nutrient enrichment experiments with Lake Huron phytoplankton, drastic changes in phytoplankton species composition were obtained (Lin and Schelske, 1979) with phosphorus enrichments of 20 μg l^{-1} and a complete enrichment including trace metals, vitamins and EDTA. However, these were spike experiments, so silica was depleted which caused a shift in species composition in the diatom-dominated assemblages. Other species changes resulted when different components of the complete enrichment were deleted. In the more oligotrophic waters of Lake Superior, enrichment with trace metals, vitamins, and a chelating agent increased phytoplankton growth in comparison to enrichments which only contained phosphorus (Schelske *et al.*, 1972). However, unlike the Lake Michigan experiment discussed above (Stoermer *et al.*, 1978), there were no major effects on species composition. The final standing crops in all treatments including the control were dominated by nanoplanktonic species of diatoms, in the genus *Cyclotella* (*C. glomerata*, *C. stelligera* and *C. comensis*). These are typical oligotrophic diatoms (Hutchinson, 1967) and were the dominant phytoplankton over a considerable part of Lake Superior that was sampled before and after the experiment (Schelske *et al.*, 1972).

Results obtained from nutrient enrichment experiments with Great Lakes phytoplankton substantiate earlier hypotheses about the effects of nutrient enrichment on phytoplankton species composition (Stoermer *et al.*, 1978). First, moderate increases in levels of the major limiting nutrient, phosphorus, will have relatively small effects on species composition. This hypothesis was formulated to explain the lack of species changes over a phosphorus gradient in Grand Traverse Bay (Stoermer *et al.*, 1972) and is discussed generally for the Great Lakes system (Stoermer, 1978). The hypothesis is also supported by results from experiments with Lake Michigan (Stoermer *et al.*, 1978) and Lake Superior phytoplankton (Schelske *et al.*, 1972). Second, synergistic effects of phosphorus and other growth-promoting factors (vitamins, trace metals, and EDTA) cause greater standing crops and species changes than enrichments of phosphorus alone (Schelske, 1979; Stoermer *et al.*, 1978; Schelske *et al.*, 1978). In the Great Lakes system, a wide variety of anthropogenic inputs are associated with the increases in phosphorus loading which undoubtedly act in combination to produce some of the effects that are attributed to eutrophication (Stoermer, 1978). Third, secondary effects of phosphorus enrichment including silica depletion and nitrate depletion can cause shifts in phytoplankton species composition from assemblages dominated by diatoms to those with a smaller proportion of diatoms (Schelske and Stoermer, 1971) and eventually to those with nitrogen-fixing blue–green algae (Vanderhoef *et al.*, 1974).

Investigation of effects at the species level are important and mandate the use of natural phytoplankton assemblages in experiments. Workers who have used natural assemblages in their experiments have pointed out the need and importance of such information (Barlow *et al.*, 1973; Peterson *et al.*, 1974; Jones *et al.*, 1978b; Stoermer *et al.*, 1978). Jordan and Bender (1973a) found that species that responded with increased growth in experiments frequently had increased growth rates in the lake either before or after the experiments were initiated. Work on effects of perturbations on species composition necessarily involves experiments of several days to one or two weeks duration so that growth can be measured from increases in chlorophyll and biomass or changes in cell numbers can be determined from microscopic counting. Knoechel and deNoyelles (1980) have used track autoradiography to study effects of nutrient perturbation at the species level. This technique, although time consuming and little used, offers a means of measuring responses on a shorter time scale than with microscopic counting. In addition obtaining estimates of carbon fixation rates at the species level over short time intervals can perhaps shed some light on the factors that influence species succession and abundance in lakes. Knoechel and deNoyelles state that experiments longer than 24 hours are necessary for initial shock effects to dissipate and for adaptation and selection to occur and that it is the longer-term effects that are important in determining community change in response to environmental perturbation.

Lund (1978) has reviewed results from enclosure experiments. He reports year-to-year variability of results in Lund tubes in Blelham Tarn and offers possible explanations for the variability. Early results showed that the phytoplankton within a tube although qualitatively similar to those in the surrounding water were quantitatively much less (Lund, 1972). Subsequent results have shown that standing crop and species composition can be changed with nutrient additions. Of particular significance is an experiment carried out by C. S. Reynolds in which nitrogen, phosphorus and silica were added at weekly intervals so that the nutrient concentrations were replenished weekly. Under these conditions natural planktonic successions occurred. Gamble et al. (1977) also showed that nutrients must be added to maintain standing crops in tubes.

Some of the negative aspects of containment experiments should be cited. Certain species of phytoplankton apparently cannot be maintained or grown in culture (Carpenter and Lively, 1980) which may only reflect our ignorance or lack of skill in cultivation of species. Certain species apparently do not grow well in enclosures, including *Ceratium hirundinella* and *Mallomonas caudata* in Lund tubes (Lund, 1978), marine flagellates in small bottles (Venrick et al., 1977), *Chroomonas acuta* in 19-l Pyrex bottles (Jordan and Bender, 1973a), and flagellates in 20-l polyethylene bottles (Schelske et al., 1974; Stoermer et al., 1978). Jordan and Bender (1973a) did have one experiment in which *Chroomonas acuta* bloomed, and in experiments with trace metals we have found that some treatments increased the growth of *Rhodomonas minuta* (unpublished results). It is therefore possible that the lack of positive responses in growth by some species is a function of the physical limit on the number of experimental treatments that can be employed.

A great deal of conflicting information on enclosure size can be found in the literature, particularly if one is looking for some standard method. Reducing surface effects is probably the main reason given for the use of large enclosures (Strickland and Terhune, 1961). The need for large enclosures to study periodicity in phytoplankton development was noted by Lund (1978), who also indicated that the 16 000 m^3 tube was not necessarily the ideal size for all purposes or places. At the other extreme is the comment by Verduin (1969) that "big bags" should be used instead of bottles or laboratory aquaria only after experiments have been conducted to show that they are in fact superior to laboratory or bottle tests.

Verduin's objection to the use of large bags was based on the lack of data on eddy diffusivity and on his prediction that eddy diffusivity would be much less in a polyethylene cylinder than in the surrounding water. Bender and Jordan (1970) concluded that reduced eddy diffusion could be used to explain lower rates of primary productivity that were obtained in columns compared to surrounding waters. Boyce (1974) showed mathematically that the apparent vertical transport of

heat in limnocorrals could be explained by horizontal heat transfer. Imboden *et al.*
(1979) calculated eddy diffusion coefficients from profiles of excess radon-222 and
reported that vertical mixing in the upper 10 m of the water column during calm
meteorological conditions was similar inside and outside 12-m diameter limnocorrals used during MELIMEX. Verduin's comments were addressed to small tubes,
but Imboden's results show that eddy diffusivity in limnocorrals was reduced at
least part of the time in comparison to the surrounding water. Steele *et al.* (1977)
have considered these problems in relation to vertical mixing and distribution of
phytoplankton and nutrients in tubes.

VI. Practical Applications

In situ bioassays could be used in aquaculture to determine enrichment
programmes that would be required to stimulate phytoplankton growth. It is
surprising therefore that these procedures have not been applied widely in fisheries
biology. Possibly one of the reasons why the applications have been limited is that
inadequate research has been devoted to the development of techniques that are
appropriate for such practical applications. Boyd (1979) stated that he hoped
"simple procedures for determining fertilizer requirements" for fish ponds "can
eventually be developed, but the prospects are not bright". He also indicated that
the tube technique proposed by Kemmerer (1968) and the algal bioassay procedure
(U.S. Environmental Protection Agency, 1971) were too complicated and laborious
to be of value to field biologists making fertilizer recommendations for most sport
fish ponds. It also has been emphasized in this chapter and by Golterman (1975)
that results obtained with the algal bioassay procedure may be misleading or
difficult to interpret.

Application rates of fertilizers for fish ponds have been determined largely by
empirical methods, with little research on what nutrients and how much of the
limiting nutrients should be added to increase fish production. In general the only
commercial fertilizers that are utilized are those which provide enrichments of
nitrogen, phosphorus and potassium and other substances but which, for example,
may not contain silica required for diatom growth.

Apparently, simple procedures could be developed to determine which nutrients
are limiting phytoplankton and to establish application rates for fish ponds. It
would also seem that *in situ* bioassays with natural phytoplankton assemblages
should be used both for practical and theoretical reasons. Small containers could be
treated with nutrients and be suspended in the water body of interest by field
biologists to determine limiting nutrients. Effects of nutrient enrichment could be
determined with measurements of chlorophyll *a* or cell counts, and either technique
could be carried out by field biologists. It should then be possible to develop
equations or graphs which could be utilized to determine kinds and quantities of
fertilizer or nutrients that should be applied.

Developing techniques and procedures for fertilizing fish ponds would have two advantages over the methods now in use. First, it would undoubtedly prove to be more economical because present rates of fertilizer application are very large relative to the quantities of nutrients that should be required to maintain large algal populations. The second advantage is perhaps more important because it is at least possible based on present theories about nutrient requirements of phytoplankton to control species composition with nutrient concentrations or ratios of nutrients and thus to regulate fish production at the same time.

Another type of practical application has been suggested by Lund (1975). He noted that oligotrophication occurred in waters enclosed in tubes and speculated that this knowledge might be applied in situations where the objective is to control algal problems. A series of tube-like structures or baffles in a reservoir could be expected to reduce phytoplankton growth, and these structures also might be sites for periphytic growth. The combination of increased periphyton growth on surfaces and reduced phytoplankton growth in the water should result in smaller costs for water filtration and treatment.

VII. General Considerations and Conclusions

Experiments with natural phytoplankton assemblages are emphasized in this chapter because algal assays with unialgal cultures to determine limiting nutrients in waters have limited value as a tool in ecological investigations. Such an approach gives no information about what factors may control species succession and abundance, and it is these factors that must be investigated if we are to make advances in understanding algal ecology (see Smayda, 1980). Different types of algal assays have been used comprehensively in a broad range of studies related to algal ecology and eutrophication (Marvan *et al.*, 1979). Algal assay procedures (U.S. Environmental Protection Agency) which use large inocula are not appropriate for experiments related to trophic states of lakes (Golterman, 1975).

Algal assays with unialgal cultures have been used successfully to further our understanding about factors controlling population dynamics and eutrophication in specific situations (Skulberg, 1970). Examples that can be cited are the use of a marine diatom to determine the growth potential of water in Narragansett Bay (Smayda, 1974), a freshwater diatom, *Asterionella formosa* (Lund *et al.*, 1975), to study nutrients that control its seasonal cycle and blue–green algae to investigate factors that control seasonal changes in abundance (Schanz *et al.*, 1979).

Bioassays with algae, either cultures or natural assemblages, can be used to study the effects of toxicants. No attempt has been made to cover this literature comprehensively; however, studies on marine phytoplankton were conducted as part of CEPEX (Hollibaugh *et al.*, 1980; Thomas *et al.*, 1980a) and freshwater phytoplankton were studied in MELIMEX (Gächter and Máres, 1979; Baccini and Sutter, 1979). Other studies of heavy metals on plankton include those by Kerrison *et al.* (1980).

An important question to ask is: are *in situ* enclosures necessary for experiments with phytoplankton? In our laboratory we have concluded that they may not be for some types of questions. Instead we have found that water samples containing natural phytoplankton assemblages could be brought to the laboratory for experiments (Lin and Schelske, 1979, 1981; Stoermer *et al.*, 1978). Results of field experiments with large plastic bags and laboratory experiments with different container sizes have both shown that phosphorus was a limiting nutrient in the upper Great Lakes (Schelske *et al.*, 1978). In addition, it was shown with samples transported to the laboratory that there are seasonal patterns to nutrient limitation in Lake Huron (Lin and Schelske, 1981). In this study a large number of laboratory experiments were conducted, but it would not have been practical to conduct a comparable number of experiments in the field, mainly because the costs would have been much greater. In the future it may be necessary to conduct related experiments in the field, but this hardly appears to be necessary given our present state of knowledge about seasonal responses to phytoplankton to nutrient enrichment. Jones *et al.* (1973b) also concluded that laboratory experiments could be substituted for *in situ* experiments in studies of algal ecology.

Although bioassays have been conducted to determine the growth potential of different waters and to determine limiting nutrients, it should be pointed out that for some purposes the information desired from such bioassays could be obtained in other ways. Chemical analyses of the water obviously can be used to determine whether one of the major nutrients (carbon, nitrogen, phosphorus, or silica) might be limiting. The presence of orthophosphate in readily measurable concentrations probably indicates that phosphorus is not limiting phytoplankton growth. Supplies of silica and nitrate or ammonia in the water can be determined chemically, and the results can be used to assess possible nutrient limitation. On the other hand bioassays for trace metal or vitamin limitation are probably preferable to direct measurements of the substances because direct measurements may not be possible and cannot estimate biological availability. Ionic activity is now being used to estimate the biological availability of trace metals (see Huntsman and Sunda, 1980).

Bioassay results can be interpreted erroneously. Ecologists are generally aware of two types of erroneous interpretations which will be termed either experimental or environmental secondary nutrient limitation. These types of nutrient limitation are mentioned here not because of related theoretical questions but because of possible incorrect application or interpretation of bioassay results.

Experimental secondary nutrient limitation can result from nutrient enrichments if adequate amounts of the primary limiting nutrient are added so that another nutrient or nutrients become limiting (Lin and Schelske, 1981). For example, phosphorus can be added at concentrations that will result in phytoplankton growth limitation by available supplies of nitrogen provided other essential nutrients are added or are present in adequate amounts. Environmental secondary nutrient limitation can result if a body of water has been highly enriched with nutrients. Bioassay results from lakes that are highly enriched with phosphorus, for example, will probably indicate that nutrients other than phosphorus are limiting (Schelske *et al.*, 1978). Results from such experiments can be valid for the water at the time of sampling but should not be interpreted as meaning that phosphorus is unimportant in producing large standing crops of algae. In fact it is quite likely that decreasing phosphorus inputs could reduce the standing crop of algae because the degree of eutrophication of most freshwater lakes is highly correlated with phosphorus concentrations in the water (Schindler, 1977; Powers *et al.*, 1972).

It is perhaps the result of personal bias about the role of silica in phytoplankton dynamics (Schelske and Stoermer, 1971), but it is important to note that diatoms have an obligate requirement for silica. It therefore would appear to be necessary to include treatments with this nutrient in most experiments with natural phytoplankton assemblages. At a minimum silica concentration should be measured over the time course of the experiment to determine whether it is a potential limiting factor. It was shown experimentally that phosphorus added to Lake Michigan waters increases the growth of diatoms which rapidly depletes available or added supplies of silica in the water (Schelske and Stoermer, 1972; Schelske *et al.*, 1974). This "experimental result" of secondary nutrient limitation also occurred on a large scale in waters of Lake Michigan prior to 1970 because increased inputs of phosphorus stimulated diatom production which eventually caused silica depletion in the lake. In stressing the role of silica, it should be noted that the ecological importance of silica depletion caused by increased phosphorus inputs to Lake Michigan could only have been established with nutrient experiments on natural phytoplankton assemblages or with algal assays that used diatom populations.

It also should be noted that phosphorus inputs to water bodies can be controlled but inputs of nitrogen and carbon cannot be controlled because supplies can be obtained from the atmosphere. In North America this has been shown convincingly by researchers working in the Experimental Lakes Area in Canada. Two important results that have originated from these studies are the relative importance of phosphorus and nitrogen in stimulating phytoplankton growth and influencing species composition of phytoplankton (Schindler, 1974) and the lack of importance of internal supplies of carbon in the eutrophication of lakes. Phosphorus added to these unproductive lakes stimulates phytoplankton growth and nitrogen-fixing blue–green algae can at least partially compensate for the lack of added nitrogen (Flett *et al.*, 1980) if adequate supplies of nitrogen are not added. Carbon, which is needed to increase the standing crop of phytoplankton in these low alkalinity and relatively unproductive waters, can be obtained from invasion of atmospheric carbon dioxide and need not be added to the waters (Emerson *et al.*, 1973). The rate of carbon dioxide transfer across the air–water boundary is limited and, as Verduin (1975) has pointed out, may not supply adequate carbon dioxide to support high rates of photosynthesis.

Much can be learned from employing bioassay experiments with natural phytoplankton assemblages in studies of phytoplankton ecology. It would appear from the review in this chapter that the full potential of this approach has not yet been realized. One fruitful line of investigation would be studies of seasonal nutrient limitation and its relationship to phytoplankton dynamics. This is the type of study that cannot be conducted with unialgal assays (bioassays with unialgal cultures) which therefore have limited value in many ecological investigations.

In the introduction to this chapter, several advantages of using the experimental approach instead of descriptive observations in studying phytoplankton dynamics were listed. However, whether these experiments simulate conditions in nature should be a prime consideration. Boyd (1981) stated that there are two tests that can be used to assess "the utility of microcosms in the simulation of selected components of an ecosystem: (i) does the behaviour of the microcosm correspond to the system being simulated, and (ii) do several microcosms, each receiving similar treatments, follow a similar course of events with time?" For experiments with phytoplankton there are related questions. Can specific populations be maintained over the course of the experiment? And what factors must be adjusted in treatments to achieve similarity in phytoplankton populations in the enclosures and in the natural environment?

One of the conclusions from this review is that bigger enclosures are not necessarily better for many types of studies and that it may be advantageous to use laboratory experiments instead of *in situ* experiments. This conclusion supports the view of Verduin (1969) that bottles or aquaria in the laboratory might be better used for experiments than deploying plastic bags in different water bodies. Boyd (1981) also makes a case from reviewing the marine literature that smaller enclosures may be better suited for experimental purposes than large enclosures. Tower tanks and other types of permanent structures that are constructed on land and have been used in experiments were included in his review. Boyd also makes the observation that J. D. H. Strickland, who conceived and pioneered big bag studies, soon dropped them to resume a combination of laboratory and field work.

References

Baccini, P. and U. Suter (1979). *Schweitz. Z. Hydrol.* **41**, 291–314.

Barlow, J. P., W. R. Schaffner, F. deNoyelles, Jr and B. J. Peterson (1973). Continuous flow nutrient bioassays with natural phytoplankton populations. *In* "Bioassay Techniques and Environmental Chemistry" (G. E. Glass, ed.), pp. 299–319. Ann Arbor Science Publishers, Ann Arbor.

Bender, M. E. and R. A. Jordan (1970). *Trans. Amer. Fish. Soc.* **99**, 607–610.

Boyce, F. M. (1974). *J. Fish. Res. Board Can.* **31**, 1400–1405.

Boyd, C. E. (1979). "Water Quality in Warmwater Fish Ponds". Auburn University, Craftmaster Printers, Inc., Opelika, Alabama. 359 pp.

Boyd, C. M. (1981). Microcosms and experimental planktonic food chains. *In* "Analysis of Marine Ecosystems" (A. R. Longhurst, ed.), pp. 627–649. Academic Press, New York and London.

Carpenter, E. J. and J. S. Lively (1980). Review of estimates of algal growth using ^{14}C tracer techniques. *In* "Primary Productivity in the Sea" (P. G. Falkowski, ed.), pp. 161–178. Plenum Press, New York.

Charlton, M. N. (1975). *Verh. Internat. Verein. Limnol.* **19**, 267–272.

Davies, J. M. and J. C. Gamble (1979). *Phil. Trans. R. Soc. Lond. B.* **286**, 523–544.

deNoyelles, F. Jr and W. J. O'Brien (1974). *Limnol. Oceanogr.* **19**, 326–331.

deNoyelles, F., R. Knoechel, D. Reinke, D. Treanor and C. Altenhofen (1980). *Can. J. Fish. Aquat. Sci.* **37**, 424–433.

Dickman, M. and I. E. Efford (1972). *J. Fish. Res. Board Can.* **29**, 1595–1604.

Eide, I., A. Jensen and S. Melsom (1979). *J. exp. mar. Biol. Ecol.* **37**, 271–286.

Emerson, S., W. Broecker and D. W. Schindler (1973). *J. Fish. Res. Board Can.* **30**, 1475–1485.

Fitzgerald, G. P. (1969). *Limnol. Oceanogr.* **14**, 206–212.

Fitzgerald, G. P. and T. C. Nelson (1966). *J. Phycol.* **2**, 32–37.

Flett, R. J., D. W. Schindler, R. D. Hamilton and N. E. R. Campbell (1980). *Can. J. Fish. Aquat. Sci.* **37**, 494–505.

Fournier, R. O. (1966). *Chesapeake Sci.* **7**, 11–19.

Gächter, R. (1979). *Schweiz. Z. Hydrol.* **41**, 165–176.

Gächter, R. and A. Máreš (1979). *Schweiz. Z. Hydrol.* **41**, 228–246.

Gamble, J. C., J. M. Davies and J. H. Steele (1977). *Bull. Mar. Sci.* **27**, 146–175.

Gerhart, D. Z. and G. E. Likens (1975). *Limnol. Oceanogr.* **20**, 649–653.

Gerloff, G. C. (1969). Evaluating nutrient supplies for the growth of aquatic plants in natural waters. *In* "Eutrophication: Causes, Consequences, Correctives", pp. 537–555. Nat. Acad. Sci., Washington, D.C.

Gieskes, W. W. C., G. W. Kraay and M. A. Baars (1979). *Netherlands J. Sea Res.* **13**, 58–78.

Goldman, C. R. (1962). *Limnol. Oceanogr.* **7**, 99–101.

Goldman, C. R. and R. C. Carter (1965). *J. Wat. Poll. Control Fed.* **37**, 1044–1059.

Golterman, H. L. (1975). "Physiological limnology". Elsevier, Amsterdam.

Grice, G. D. and M. R. Reeve (ed.) (1981). "Marine Mesocosms: Biological and Chemical Research in Experimental Ecosystems". Springer, New York.

Hall, D. J., W. E. Cooper and E. E. Werner (1970). *Limnol. Oceanogr.* **15**, 839–928.

Healey, F. P. (1978). *Mitt. Internat. Verein. Limnol.* **21**, 34–41.

Healey, F. P. and L. L. Hendzel (1980). *Can. J. Fish. Aquat. Sci.* **37**, 442–453.

Hollibaugh, J. T., D. L. R. Seibert and W. H. Thomas (1980). *Estuarine and Coastal Marine Sci.* **10**, 93–105.

Huntsman, S. A. and W. G. Sunda (1980). The role of trace metals in regulating phytoplankton growth. *In* "The Physiological Ecology of Phytoplankton" (Studies in Ecology, Vol. 7) (I. Morris, ed.), pp. 285–328. Blackwell, Oxford.

Hutchinson, G. E. (1967). "A Treatise on Limnology. Vol. II Introduction to Lake Biology and the Limnoplankton". John Wiley, New York.

Imboden, D. M., B. S. F. Eid, T. Joller, M. Schurter and J. Wetzel (1979). *Schweiz. Z. Hydrol.* **41**, 177–189.

Jannasch, H. W. (1974). *Limnol. Oceanogr.* **19**, 716–720.

Jensen, A. and B. Rystad (1973). *J. Exp. Mar. Biol. Ecol.* **11**, 275–285.

Jones, K. J., P. Tett, A. C. Wallis and B. J. B. Wood (1978a). *J. Mar. Biol. Assn U.K.* **58**, 923–941.

Jones, K. J., P. Tett, A. C. Wallis and B. J. B. Wood (1978b). *Mitt. Internat. Verein. Limnol.* **21**, 398–412.

Jordan, R. A. and M. E. Bender (1973a). "An *In Situ* Evaluation of Nutrient Effects in Lakes. Project 16010 HIU, Ecological Research Series, EPA-R3-73-018, Office of Research and Monitoring, US Environmental Protection Agency, Washington.

Jordan, R. A. and M. E. Bender (1973b). *Wat. Res.* 7, 189–195.

Kemmerer, A. J. (1968). *Trans. Amer. Fish. Soc.* 97, 425–428.

Kerrison, P. H., A. R. Sprocati, O. Ravera and L. Amantini (1980). *Environ. Tech. Lett.* 1, 169–176.

Knoechel, R. and F. deNoyelles, Jr (1980). *Can. J. Fish. Aquat. Sci.* 37, 434–441.

Kuiper, J. (1977). *Mar. Biol.* 44, 97–107.

Lack, T. J. and J. W. G. Lund (1974). *Freshwat. Biol.* 4, 399–415.

Landers, D. H. (1979). *Limnol. Oceanogr.* 24, 991–994.

Lean, D. R. S., M. N. Charlton, B. K. Burnison, T. P. Murphy, S. E. Millard and K. R. Young (1975). *Verh. Internat. Verein. Limnol.* 19, 249–257.

Lehman, J. T. (1980). Nutrient recycling as an interface between algae and grazers in freshwater communities. *In* "Evolution and ecology of zooplankton communities" (W. C. Kerfoot, ed.), pp. 251–263. Spec. Symp. Vol. 3, Amer. Soc. Limnol. Oceanogr. University Press of New England, Hanover.

Lin, C. K. and C. L. Schelske (1979). "Effects of Nutrient Enrichment, Light Intensity and Temperature on Growth of Phytoplankton from Lake Huron". Univ. Michigan, Great Lakes Res. Div. Spec. Rep. 63 (also is EPA Rep. No. EPA-600/3-79-049, U.S. Environmental Protection Agency, Duluth, MN).

Lin, C. K. and C. L. Schelske (1981). *Can. J. Fish. Aquat. Sci.* 38, 1–9.

Lund, J. W. G. (1972). *Verh. Internat. Verein. Limnol.* 18, 71–77.

Lund, J. W. G. (1975). The uses of large experimental tubes in lakes. *In* "The Effects of Storage on Water Quality", Proc. Symp., pp. 291–312. Reading Univ., March 1975, Bucks, SL7 2HD, England.

Lund, J. W. G. (1978). Experiments with lake phytoplankton in large enclosures. *In* Freshwater Biol. Assoc. Annual Report, pp. 31–39.

Lund, J. W. G., G. H. M. Jaworski and C. Butterwick (1975). *Arch. Hydrobiol. Suppl.* 49, 49–69.

Marshall, J. S. and D. L. Mellinger (1980). *Can. J. Fish. Aquat. Sci.* 37, 403–414.

Marvan, P., S. Přibil and O. Lhotský (eds). (1979). Algal assays and monitoring eutrophication. E. Schweizerbart'sche, Stuttgart.

Menzel, D. W. and J. Case (1977). *Bull. Mar. Sci.* 27, 1–7.

Moss, B. (1981). *Brit. Phycol. J.* 16, 59–76.

Murphy, T. P., D. R. S. Lean and C. Nalewajko (1976). *Science* 192, 900–902.

O'Brien, W. J. and F. deNoyelles, Jr (1976). *Hydrobiologia* 49, 65–76.

O'Connors, H. B., Jr, C. F. Wurster, C. D. Powers, D. C. Biggs and R. G. Rowland (1978). *Science* 201, 737–739.

Owens, O. v. H., P. Dresler, C. C. Crawford, M. A. Tyler and H. H. Seliger (1977). *Chesapeake Sci.* 18, 325–333.

Peterson, B. J., J. P. Barlow and A. E. Savage (1974). *Limnol. Oceanogr.* 19, 396–408.

Powers, C. F., D. W. Schults, K. W. Malueg, R. M. Brice and M. D. Schuldt (1972). Algal responses to nutrient additions in natural water. II. Field experiments. *In* "Nutrients and Eutrophication" (G. E. Likens, ed.), pp. 141–156. Special Symposia Vol. 1, Amer. Soc. Limnol. Oceanogr., Allen Press, Lawrence, Kansas.

Redfield, A. C., B. H. Ketchum and F. A. Richards (1963). The influence of organisms on the composition of sea-water. *In* "The sea. Ideas and Observations on Progress in the Study of the Seas. Vol. 2. The Composition of Sea-water: Comparative and Descriptive Oceanography" (M. N. Hill, ed.), pp. 26–77. Interscience, New York.

Sakshaug, E. and A. Jensen (1978). *Oceanogr. Mar. Biol. Ann. Rev.* **16**, 81–106.

Schanz, F., E. D. Allen and P. R. Gorham (1979). *Can. J. Bot.* **57**, 2443–2451.

Schelske, C. L. (1979). *J. Fish. Res. Board Can.* **36**, 286–288.

Schelske, C. L. and E. F. Stoermer (1971). *Science* **173**, 423–424.

Schelske, C. L. and E. F. Stoermer (1972). Phosphorus, silica and eutrophication of Lake Michigan. *In* "Nutrients and Eutrophication" (G. E. Likens, ed.), pp. 157–171. Special Symposia Vol. 1, Amer. Soc. Limnol. Oceanogr., Allen Press, Lawrence, Kansas.

Schelske, C. L., L. E. Feldt, M. A. Santiago and E. F. Stoermer (1972). *In* "Nutrient Enrichment and Its Effect on Phytoplankton Production and Species Composition in Lake Superior", pp. 149–165. Proc. 15th Conf. Great Lakes Res.

Schelske, C. L., E. C. Rothman, E. F. Stoermer and M. A. Santiago (1974). *Limnol. Oceanogr.* **19**, 409–419.

Schelske, C. L., M. S. Simmons and L. E. Feldt (1975). *Verh. Internat. Verein. Limnol.* **19**, 911–921.

Schelske, C. L., E. D. Rothman and M. S. Simmons (1978). *Mitt. Internat. Verein. Limnol.* **21**, 65–80.

Schindler, D. W. (1974). *Science* **184**, 897–899.

Schindler, D. W. (1977). *Science* **195**, 260–262.

Schindler, D. W. (1980). *Can. J. Fish. Aquat. Sci.* **37**, 313–319.

Sicko-Goad, L., E. F. Stoermer and B. G. Ladewski (1977). *Protoplasma* **93**, 147–163.

Skulberg, O. M. (1970). *Helgoländer wiss. Meeresunters.* **20**, 111–125.

Smayda, T. J. (1974). *Limnol. Oceanogr.* **19**, 889–901.

Smayda, T. J. (1980). Phytoplankton species succession. *In* "The physiological ecology of phytoplankton" (Studies in ecology, Vol. 7), (I. Morris, ed.), pp. 493–570. Blackwell, Oxford.

Soeder, C. J. and J. F. Talling (1969). The enclosure of phytoplankton communities. *In* "A Manual on Methods for Measuring Primary Production in Aquatic Environments" (R. A. Vollenweider, ed.), pp. 62–70. IBP Handbook No. 12. Blackwell, Oxford.

Steele, J. H. (1979). *Phil. Trans. R. Soc. Lond. B.* **286**, 583–595.

Steele, J. H., D. M. Farmer and E. W. Henderson (1977). *J. Fish. Res. Board Can.* **34**, 1095–1104.

Stoermer, E. F. (1978). *Trans. Amer. Micros. Soc.* **97**, 2–16.

Stoermer, E. F., C. L. Schelske and L. E. Feldt (1971). Phytoplankton assemblage differences at inshore versus offshore stations in Lake Michigan, and their effects on nutrient enrichment experiments. Proc. 14th Conf. Great Lakes Res., 114–118.

Stoermer, E. F., C. L. Schelske, M. A. Santiago and L. E. Feldt (1972). Spring phytoplankton abundances and productivity in Grand Traverse Bay, Lake Michigan, 1970. Proc. 15th Conf. Great Lakes Res., 181–191.

Stoermer, E. F., B. G. Ladewski and C. L. Schelske (1978). *Hydrobiologia* **57**, 249–265.

Strathmann, R. R. (1967). *Limnol. Oceanogr.* **18**, 411–418.

Strickland, J. D. H. and L. D. B. Terhune (1961). *Limnol. Oceanogr.* **6**, 93–96.

Strickland, J. D. H. and T. R. Parsons (1972). "A Practical Handbook of Seawater Analysis". Bulletin 167, Fish Res. Board Can., Ottawa.

Takahashi, M., W. H. Thomas, D. L. R. Seibert, J. Beers, P. Koeller and T. R. Parsons (1975). *Arch. Hydrobiol.* **76**, 5–23.

Takahashi, M. and F. A. Whitney (1977). *Bull. Mar. Sci.* 27, 8–16.

Thomas, E. A. (1964). *Verh. Internat. Verein. Limnol.* 15, 342–351.

Thomas, W. H., D. L. R. Seibert and A. N. Dodson (1974). *Estuarine and Coastal Mar. Sci.* 2, 191–206.

Thomas, W. H., J. T. Hollibaugh, D. L. R. Seibert and G. T. Wallace, Jr (1980a). *Mar. Ecol. Prog. Ser.* 2, 213–220.

Thomas, W. H., M. Pollock and D. L. R. Seibert (1980b). *J. Exp. Mar. Biol. Ecol.* 45, 25–36.

Twinch, A. J. and C. M. Breen (1981). *Hydrobiologia* 77, 49–60.

US Environmental Protection Agency (1971). "Algal Assay Procedure Bottle Test". National Eutrophication Research Program.

Vanderhoef, L. N., C.-Y. Huang, R. Musil and J. Williams (1974). *Limnol. Oceanogr.* 19, 119–125.

Venrick, E. L., J. R. Beers and J. F. Heinbokel (1977). *J. Exp. Mar. Biol. Ecol.* 26, 55–76.

Verduin, J. (1969). *Trans. Amer. Fish. Soc.* 98, 335–336.

Verduin, J. (1975). *Limnol. Oceanogr.* 20, 1052–1053.

Wall, D. and F. Briand (1979). *J. Plankton Res.* 1, 103–112.

Wetzel, R. G. (1965). *Mem. Ist. Ital. Idrobiol. Suppl.* 18, 137–157.

3 Qualitative Characteristics of Phytoplankton Assemblages

E. F. Stoermer

Great Lakes Research Division
University of Michigan
Ann Arbor, Michigan, USA

I. Introduction

Any attempt to address the use of qualitative measures in scientific investigation is a task to be approached with a certain amount of trepidation. Our general model of scientific endeavour teaches us to eschew qualitative observation and to always strive for quantitative measurement of phenomena and the elegance of fully extensible mechanistic theory. This model has indeed served us well in the simple sciences and will undoubtedly remain the primary icon of the scientific fraternity. Some have questioned the appropriateness of this model in the context of complex and entrained systems such as organisms (Schrödinger, 1945; Elsasser, 1966) and particularly in the extension to larger, multiphasic ecosystems (Kerr, 1976; Rosen, 1977). Although these fundamental contentions are most interesting, I do not intend to pursue them here, partially from sheer cowardice, but mainly because they are probably untestable in the present state of the art.

Instead, I shall adopt the mainly pragmatic viewpoint that aquatic ecologists are forced, like it or not, to deal with qualitative information. I shall contend that there are logical and productive ways to apply this information to many types of problems. I do not intend to attempt a comprehensive review of the literature, and most of the examples are drawn from one ecosystem, the Laurentian Great Lakes, and one group of algae, the diatoms. This approach contains the obvious bias of topics I am somewhat familiar with, but is also partially defensible from the standpoint of providing some excellent examples. The Great Lakes are a continuous system containing a wide range of natural expressions of the convergence of various forcing functions in particular biological assemblages. They also provide examples of biotic response to gross modifications of input loadings. The diatoms are a large and widely distributed group and are particularly interesting because their remains are often preserved to provide time scale context

ALGAE AS ECOLOGICAL INDICATORS
ISBN 0 12 640620 0

not available for many other groups of algae. They both provide glaring examples of the problems common to the topic and some examples of productive approaches.

As a point of departure, let us reflect momentarily on the current state of the art in aquatic ecology. The science, as we know it today, is a fairly recent invention. It traditionally has been practised by a limited number of the dedicated or foolish and has only recently been infused with the sense of practical urgency which has long fuelled investigations in agriculture and medicine. This history of development has left modern practitioners of the art with a somewhat limited tool kit and a rather curious sense of intellectual indigestion. As an example of the first problem, I doubt that there is a region within the United States for which there is not a reasonably complete and accurate inventory of the terrestrial vegetation. There is no region which has an adequate inventory of the algal flora. Scientists working fields concerned with the practical management of complex biological systems, such as physicians and agronomists, have an established tradition of effective practice based on the interpretation of clinical signs, often preceding quantitative understanding of the basic processes of interest. This tradition results in essentially qualitative information holding a respectable (and publishable) stature in these fields. Aquatic ecologists are also faced with clear signs of system dysfunction which are not strictly quantifiable. Lacking a clearly established empirical tradition, most are rather uncomfortable with the process of verifying, evaluating, and transmitting such information. It is my thesis that this rather vaguely defined intellectual indigestion is a symptom of a fairly major disorder in the practice of our particular art which acts to limit both intellectual growth and practical application of developing knowledge.

II. Related Research

A. Terminology

To avoid any misunderstanding I shall define a few terms used in the following discussion because they tend to be rather loosely applied and often misinterpreted in the literature. I shall speak of a *population* in the sense of Hutchinson (1967) as being all of the organisms of one kind which occur in a single ecosystem. In clearly bounded ecosystems, such as most lakes and streams, a population is, at any point in time, a relatively concrete concept which is correctly and unambiguously perceived by most people if they are provided with appropriate guidance and instrumentation. There are some known exceptions to this generality, and others will probably be discovered as systematic research is refined. For instance, Ruttner (1937) noted morphologically and distributionally distinct sub-populations of several diatom species in his classic study of alpine lakes. Numerous examples are known of algal populations which show gross variations in morphology (e.g. Stoermer *et al.*, 1979) which might be misconstrued as different populations. At a more fundamental level, Gallagher (1980) has demonstrated genetic differences in winter and summer populations of *Skeletonema costatum* (Grev.) Cl. from the same locality. Earlier, Murphy and Guillard (1976) found genetic differences in clones of *Thalassiosira pseudonana* (Hust.) Hasle and Heimdal from different localities, indicating population segregation, but Hasle (Hasle, pers. comm. in Murphy, 1978; Hasle, 1978) determined that these clonal populations would be considered different entities on the basis of morphological criteria by most systematists. The really noteworthy point of these studies is that they are among the first to deal with genetic structure at the population level, and although it may be some time before the implications are sorted out, offer exciting possibilities for refining and improving our perception of population limits (Taylor, 1980).

I shall refer to a *species* as the global set of populations of one kind of organism. This is a considerably less concrete concept. For this reason, it is necessary to define species according to a fairly complicated set of rules (the International Code of Botanical Nomenclature) which sets aside a primary standard (the type), sets priorities in assigning names, and legislates against using different names for the same organism (synonyms) or the same name for different organisms (homonyms). This all sounds rather messy and unscientific, but it actually works rather well for groups of organisms which have been adequately studied. Vascular plant taxonomists, for instance, may carry on protracted debates about the proper circumscription of a species and/or the proper assignment of its name, but they all know what their colleagues are talking about. Unfortunately, this is not always the case with algal species, mainly because there simply have not been enough people working in the field long enough to develop, refine, and catalogue the necessary information. Consider the plight of the poor aquatic ecologist compared to a more fortunate colleague working on a terrestrial system. Suppose that they both need to know about a plant abundant in the particular habitat. At this point, let us recognize that this need to know really consists of two elements. It is both the key to unlocking the data bank accumulated by previous investigators and the address under which the present investigators' information will be filed. Assuming minimal previous knowledge, the terrestrial ecologist can probably obtain his key (n.p.i.) with fairly minimal effort. A stroll to the local university library will undoubtedly reward him with a plethora of local, state, and regional floras. Lacking the confidence to interpret this information satisfactorily, he can almost certainly find a local specialist more than willing to provide consultation. Our aquatic ecologist is unlikely to be equally rewarded. If he is lucky enough to have a major library at his disposal, he may find some incomplete and out-of-date tome, and will be fortunate indeed to find a specialist colleague to guide him. Given this lack of support, it is unsurprising that algal population data are less complete and well labelled than they might be.

An *association* will be referred to as a group of populations occurring in some definable structural and/or functional relationship. The term is often used in a sense that implies more than we really know, but there are enough examples of consistent co-occurrence to justify its use.

An *assemblage* is a group of organisms occurring together; for our purposes let us say in a sample. This co-occurrence may indicate association, but not necessarily. The microscopic algae, as a group, have been finely forged on the evolutionary anvil of adversity and have developed rather remarkable mechanisms to optimize their response to certain conditions, to sequester materials for use under suboptimal conditions (Rhode, 1948), and to survive inimical conditions for extremely extended periods of time (Parker *et al.*, 1969; Stockner and Lund, 1970). Surviving cells may be transported for considerable distances from their primary habitats by currents. Thus the assemblage present in any given sample may consist of populations derived from a number of different habitats and adapted to respond to substantially different suites of conditions.

At this point it might also be well to attempt to define *qualitative information*. Since we are going to be talking about algal populations, I will adopt a rather broad definition. Let us say that qualitative information is knowledge about certain properties or capabilities of integrated systems which are a property of the system. As a crude example, let us suppose that an investigator wishes to study the growth response of two algal populations to the concentration of a particular suite of nutrients. If his populations have no qualitative tags, he is faced with an exceedingly large number of blind trials in order to cover the possible range of conditions under which algae are known to grow. If he knows that one is labelled *Cyclotella comta* (Ehr.) Kütz., and the other *Anabaena flos-aquae* (Lyngb.) Bréb., his range of appropriate choices is automatically limited and he gains some interesting information about the probable responses of these particular populations to the concentration of nitrogen and silicon which might otherwise be difficult to deduce. Our investigator might even become so bold as to deduce certain probable characteristics of ecosystems where these organisms occur. If he assumes too much, he may get into rather serious trouble, but certainly no worse trouble than if he ignored the information entirely.

Our hypothetical investigator is advised to be cautious, because systems which encode qualitative information have certain properties which make the interpretation of this information sometimes less than straightforward. First, the maintenance of system integrity demands that algal populations develop mechanisms to avoid response to certain conditions which might damage the system. Some of these mechanisms are complex and not particularly obvious. Secondly, maintaining the system demands that they have the capability to modify their response to changing input variables over greater or lesser time scales. Thus it is profitable for them to carry, at the population level, more genetic information than is necessary to deal with the input variables of the instant and thus accommodate short-term selective pressures without apparent population modification (Murphy, 1978). This is particularly annoying because it is reasonable to suspect that organisms which have developed this capability to the greatest degree are those most widely distributed and generally recognized in nature. Problems devolving from accommodating to long term selective pressures and resulting in modifications which are sensible at the species level are undoubtedly also with us, but have been little explored in ecologically important species (Lewin, 1976).

B. *Utility of Qualitative Data*

In beginning this discussion, it is probably useful to ask in what situations are qualitative data particularly useful. The most obvious examples are where a particular algal species or population directly causes a particular problem.

Perhaps the most obvious case of this type is direct toxicity. A variety of toxins are produced by members of several algal Divisions (Schwimmer and Schwimmer, 1968). The most notorious examples are the several virulent toxins elaborated by various species of marine dinoflagellates which have been implicated in paralytic shellfish poisoning in humans as well as large-scale mortalities of marine fish and invertebrates. The qualitative aspects of these problems are apparent, as very low quantities of certain toxic dinoflagellates may be sufficiently concentrated by mussels to cause human poisoning (Schantz, 1971). Direct toxic effects produced by freshwater algae are less dramatic and, in so far as we now know, less apt to directly affect human health. Economically important episodes of fish mortality in brackish water fish-culture ponds have been attributed to the flagellate *Prymnesium parvum* Carter (Shilo, 1967; Boney, 1970). In this case it is clear that induction of toxicity is a complex phenomenon, depending on a number of factors which affect cell physiology. Large populations of *P. parvum* may be present without causing toxicity, but it may be triggered by specific variations in ecological conditions. This unpredictability is even more pronounced in the case of blue-green algae responsible for poisoning wildlife and livestock in freshwater localities (Gorham and Carmichael, 1979). A number of toxic factors have been identified in populations of several species of blue-green algae. It is apparent, however, that both toxic and nontoxic strains of the same species may exist in a given habitat and that the production of toxic blooms is highly unpredictable and, at present, poorly understood. It is apparent that toxicity episodes are usually associated with extremely dense blooms. This tends to militate against direct human health effects, since humans would be extremely reluctant to utilize water in such conditions.

This brings us to the next obvious usage of qualitative information, the detection of algal species associated with particular nuisances. Numerous examples are known of species which are associated with taste and odour problems and filter-clogging problems in municipal water supplies (Palmer, 1959). A large number of species belonging to most of the divisions common in freshwater have been noted to produce identifiable obnoxious tastes or odours in drinking water supply sources. These are usually associated with specific metabolic products (Tierney *et al.*, 1976). Filter-clogging problems are usually caused by taxa which have a colonial growth habit and/or excrete large quantities of external sheath material. These occurrences tend to be habitat and treatment system specific. For instance, *Dichotomosiphon tuberosus* (A. Br.) Ernst, which is generally considered to be a rare species because it usually grows at sub-thermocline depths in temperate lakes, occasionally causes nuisances in the Great Lakes when strong fall storms remove masses of it from the bottom (Baylis, 1957). Less specific aesthetic nuisances are caused by high abundance of a number of algal species. Because of their colonial growth habit, gas vacuole formation, and ability to escape nitrogen limitation, blue-green algal species belonging to the genera *Anabaena, Anacystis*, and *Aphanizomenon* are generally associated with undesirable water quality conditions. Like the clinical signs utilized by physicians, the occurrence of such populations conveys some empirically useful information. For instance, I would not particularly care to purchase property on a lake which supports blooms of *Aphanizomenon flos-aquae* L. It is also extensible into a management context. For instance, if *A. flos-aquae* suddenly began to occur in a lake where it was previously absent, management actions might be indicated. The impact of this sort of consideration tends to be underrecognized. Lake Washington is often cited as an example of effective management action in response to incipient eutrophication problems (Edmondson, 1977). In this case, one of the major clues to the developing problem (Edmondson, 1972) was the increasing abundance of *Oscillatoria rubescens* D.C., a species which has often been cited as a classic symptom of eutrophication in oligotrophic lakes (Thomas, 1965, 1969). In the Laurentian Great Lakes there appears to be a characteristic succession of phytoplankton species associated with increasing levels of eutrophication (Stoermer, 1978). Among this suite of species, some of the small, colonial forms of *Stephanodiscus* (*S. binderanus* and *S. tenuis* particularly) have been associated with filter clogging and taste and odour nuisances (Vaughn, 1961). Although these species were noted only after they had become a practical problem (Brunel, 1956), retrospective analysis (Stoermer and Yang, 1969) indicated that they were first introduced to the system approximately a decade before nuisance blooms became common. If this qualitative sign had been correctly sensed and interpreted, as a similar sign was in the better-studied Lake Washington case, the nuisance might have been prevented.

A second major use of qualitative information is in cases where other information is lacking. In the real world, this case is often a subset of the above, since particular problems often occur in a system before sufficient data are developed to allow a mechanistic analysis. In some cases such problems will yield to an appropriate allocation of resources. In other cases systems are so perturbed before studies which could support a mechanistic analysis are initiated that it becomes necessary to ask: (1) what is the natural or quasi-equilibrium state of the system, and, (2) what is the present and near-term historical direction of excursion from this state? As an example, Beeton (1965) compiled data on chemical changes in the Laurentian Great Lakes in an attempt to determine the cause of obviously deleterious qualitative changes in the system. Although these data showed significant changes in concentration of conservative ions, they did not speak to the obvious problem of changes in nutrient ion balance because of deficiencies in analytical methods in the historical chemical data base. This is a common condition. Even more commonly, we are faced with situations where no historic data base exists. In these situations, it is reasonable to attempt to evaluate the system state and changes through retrospective study of biological populations. In some instances, this may be accomplished by recovery and study of preserved collections. Hohn (1969) was able to reconstruct some of the phytoplankton changes which occurred in Lake Erie during the crucial period between 1935 and 1965. Stoermer and Yang (1970) made a less detailed reconstruction of changes in phytoplankton composition in the region of Lake Michigan adjacent to Chicago over a longer time period. Both of these studies treated only with diatoms, the most easily preserved major component of plankton assemblages in these regions. Hickel (1975) made a somewhat more complete reconstruction of historic changes in the phytoplankton assemblages of two well-studied north German lakes, based partly on published records and partly on recovery and analysis of historic preserved material. Unfortunately, freshwater aquatic ecologists have not developed a tradition of documenting their studies with samples deposited in permanent repositories, so this approach is not available in many potentially interesting cases. Because of this, and the fact that available historic accounts are often incomplete and sometimes difficult to interpret due to systematic and nomenclatorial problems, palaeolimnological studies are increasingly being applied to both fundamental, and practically oriented problems. The sediments of many lakes contain recognizable remains of a number of elements of previous biotic assemblages (Berglund, 1979). Although these include representatives of many algal groups (Van Geel, 1972, 1976, 1978) and their characteristic biochemical constituents (Vallentyne, 1956, 1960; Zullig, 1961; Gorham and Sanger, 1975; Griffiths *et al.*, 1969; Sanger and Crowl, 1979), most research has centred on species with mineralized cell walls. Of the major groups common in fresh water, diatoms have received the most attention, partially because of their abundance in many types of lake sediments and partially because of the technical difficulty in identifying the scales of Chrysophytes (Piernaar, 1980) and the difficulty of connecting cysts or statospores to living forms (Nygaard, 1956). Studies of diatom remains have been applied to a large number of problems (for a recent review, see Battarbee, 1979) including effects of natural environmental modifications such as climate change (Haworth, 1976) and resultant sea level (Cleve-Euler, 1951–1953) and lake level (Stoermer, 1977) changes. These studies are of particular value as they allow us to contrast the natural development of lakes toward eutrophic (Pennington, 1943) or dystrophic (Nygaard, 1956) condition with the modifications brought about by sewage pollution (Stockner and Benson, 1967), acid precipitation (Almer *et al.*, 1974), and catastrophic events, such as forest fires (Andresen, 1976). Andresen's study is especially notable because of its objective analysis of system response to varying degrees of perturbation and partial relaxation of these stresses. In some cases it has proven possible to validate diatom assemblages in lake sediments against long-term records of phytoplankton abundance in a lake (Nipkow, 1927; Haworth, 1980) which provides a connection between the two approaches.

Another use of qualitative information is in providing identification of major physiological compartments or capabilities of biotic assemblages. This, of course, is the general case of the problem-associated set discussed earlier. It is particularly important when dealing with algal assemblages because of the great differences in physiological requirements and capabilities encompassed. It is perhaps generally realized, but seldom explicitly stated, that physiological requirements and successional pressures operate at two distinct levels in algal assemblages. Not all physiological groups of algae have the same essential requirements (Stewart, 1974). The most obvious cases are the silicon requirement of diatoms (Darley, 1974; Werner, 1977) and certain Chrysophyta and the nitrogen-fixation capability of blue-green algae (Burris, 1969; Stewart, 1973; Fogg, 1974). These limitations may operate as a sort of gross shunt in both natural and artificial systems in addition to competitive optimization around certain ranges of relevant environmental parameters. Typically, an aquatic primary producer community will react to increasing stress in three distinct phases. If phosphorus loading to a lake system is increased, the first reaction is accommodation through increased growth of indigenous populations (Parsons *et al.*, 1972). If this stress is prolonged or severe, it will eventually result in replacement of the indigenous populations by entities more fit to operate under the modified condition, classical succession at the population or species level. Further forcing will eventually result in reaching the support threshold for some other essential major nutrient which will result in a gross modification of the assemblage at the Divisional level (Schelske and Stoermer, 1971). This final act has no precise analogue in terrestrial producer communities, as it may result in significant biochemical differences in the food base produced. For these reasons, it is particularly important to have an accurate assessment of the operators present in an aquatic system if accurate predictions are to be made about its response to a given perturbation.

The final, and I think most important, argument for accurate and relatively complete qualitative information is the fact that a stable population structure is the most conservative measure of ecosystem stability. When dealing with any complex and structured system, it is always possible that the set of variables addressable contains unmeasured, poorly measured, or inappropriately measured elements. Further, the array of potentially important variables is seldom satisfactorily filled. Although this is generally recognized by workers active in the field, it is a huge temptation to extend mechanistic extrapolations (models) beyond their reasonable utility. Unfortunately, such simplifications may convey a false sense of quantitative certainty to the naive, so the temptation looms even larger. Fee (1971) presented an interesting commentary on specific applications, and more recently Ehrenfeld (1978) explored the general philosophical context of this problem at some length. The practical expression of these concerns is found in the National Environmental Protection Act mandate for protection of indigenous populations. Although this sound principle is often rather shamefully trivialized in application, it is the "bottom line" for most applied studies and should not be lost from sight.

C. Data Synthesis and Interpretation

If we accept the assumption that qualitative information is of some potential utility, we must explore the means by which such information may be effectively transmitted and interpreted. The earliest approach, as in other sciences, was observation and narrative synthesis, but this approach has some obvious deficiencies. It is difficult to imagine a human observer who is entirely unbiased, either by inability to sample any significant portion of the total range of possible conditions or by the human tendency to weigh observations unevenly. The results of such work are also subject to difficulties in translation, as it is difficult to imagine two independent observers who share the same range of experiences, cultural and educational biases, and observational and deductive abilities. Because of these deficiencies, this approach has been widely and righteously decried. In fairness, however, it should be pointed out that it also has some attractive features. The human brain is a remarkable instrument which has processing capabilities that have not been duplicated and indeed are not fully understood. Human language, despite of or because of its imprecisions, is a very rich medium of expression, completely unmatched by simplified logical formalisms. Curiously, this very subtlety and power of expression tends to operate against utilization of this particular method, simply because almost any scientist is able to decode the simplified formalisms, but relatively few are able to interpret adequately the terminology that has developed around the study of algal populations and associations. For those who care to take the time to become familiar with this terminology, the observational literature which grew from the discovery and early synthesis phases of phycological study contains a wealth of information and detail somehow sadly lacking in many modern accounts. Because of the history of development of phycological studies, very few individuals have been trained to interpret and appreciate this literature. Unlike study of terrestrial plants, which extends into human prehistory, effective study of microscopic algae was effectively organized only after development of relatively modern and high quality light microscopes. Indeed, for certain groups, effective recognition and study did not arrive until development and general availability of the electron microscope (Hibberd, 1980a,b). With the development of adequate light microscopes, study of microorganisms in general, and the microscopic algae in particular, became the cutting edge of science of the day and attracted the attention of many individuals of truly prodigious energy and intellectual capacity. Relatively few of these individuals were associated with universities, and very few left behind them students to extend and synthesize their work and discoveries. Following the grand period of growth in phycology, from about 1830 to 1910, the subject, lacking clear utility or association with the practical arts, sank into a period of relative decline. Each generation had a few notable figures, a surprising number of whom, particularly those of ecological bent, were self-trained and followed their particular science as a secondary avocation. For instance, G. Huber-Pestalozzi (Jaag, 1966), compiler of one of the most comprehensive and still commonly used works on phytoplankton systematics and ecology, was trained in medicine as well as biology. His collaborator, and the author of several classic studies in palaeolimnology, F. Nipkow, was trained as an apothecary (Huber-Pestalozzi, 1963). Another collaborator, and in his own right perhaps the most notable authority on diatom systematics and ecology of this tradition, F. Hustedt (Behre, 1970), spent a large portion of his long and distinguished career as a high school teacher. Hustedt's occasional adversary, B. Cholnoky (Archibald, 1973), although trained in botany, followed various occupations owing to the vicissitudes of World War II before settling in South Africa to produce his numerous contributions to the diatom flora of the region and its reflection of water quality and his compilation of observations (Cholnoky, 1968), which is sort of a monument to this type of approach. None of these individuals, like many of their contemporaries, formally trained students, although they all corresponded widely and entertained many laboratory visitors. The resultant of this history is an extremely diffuse, idiosyncratic, and largely unassimilated literature which is a source of frustration and hence is studiously avoided by many modern students.

One of the signposts in the search for systematization of this information was the construction of systems of classification of ecological tendencies of algal taxa. The basic rationale of such efforts is to reduce the complexities of observational data to a standardized nomenclature which might be more readily interpreted by individuals who do not share the observational experience or mode of expression of a particular investigator. Numerous classification "spectra" of observed responses to common physical and chemical conditions have been composed, refined, and debated over the years. It is interesting to note that the earliest and perhaps best known of these, the so-called saprobien spectrum or system (Kolkwitz and Marsson, 1908), was developed from practical concern about the effects of organic pollution on stream communities and has been widely utilized (Margaleff, 1955; Caspers and Schulz 1960; Fjerdingstad, 1965; Palmer, 1969). Relationships of algal taxa to commonly measured chemical parameters such as pH (Hustedt, 1937; Meriläinen, 1967; Cholnoky, 1968), oxygen (Hustedt, 1937), salinity (Kolbe, 1927; Carpelan, 1978), mineral nutrient concentration (Rawson, 1956; Sparling and Nalewajko, 1970), calcium (Niessen, 1956), and physical variables such as temperature and water movement (Cholnoky, 1968) form the basis for similar systems. In addition, the occurrence of organisms has been classified according to general habitat type, specific habitat type, seasonality, and a number of other less easily specified variables (Hutchinson, 1967). This literature has been summarized for particular groups of algae, particularly diatoms (Cholnoky, 1968; Lowe, 1974) and blue-greens (Van Landingham, 1977). Although this approach has some practical utility, it suffers from a number of deficiencies. Classification categories are rather broad and, at best, furnish a minimal representation of the information content present. It is also clear that the classification schemes developed do not adequately cover the range of causal factors. This is a very damaging shortcoming, especially in a world where many biological systems face exotic chemical insults whose effects may be subtle and operate over long time periods.

Another approach to representing algal data in an easily assimilable and comparable form is the construction of ratio indices. This approach has been applied by Thunmark (1945) and Nygaard (1949, 1955) to phytoplankton assemblages in Scandinavian lakes. In their approach, the trophic status of lakes is rated according to the abundance ratio of major physiological groups, or sub-groups such as centric and pennate diatoms. This approach is interesting in that it speaks to succession at the major physiological group level discussed earlier. Attempts to extend this concept to within Division shifts of major classes or genera have been, as might be expected on first principle, less successful. For instance, Stockner and Benson (1967) proposed the ratio between araphid pennate and centric diatoms (A/C ratios) as an index of eutrophication, based on their analysis of sedimentary assemblages in Lake Washington. Stockner (1971, 1972) later extended and generalized this argument. Brugam (1979) has pointed out that a contrary trend exists within post-settlement diatom assemblages in many localities. This is well illustrated by the studies of Haworth (1972) and Andresen (1976). Andresen worked in a classically eutrophic lake and studied the effects of deforestation and fire in the surrounding landscape on the lake ecosystem. In this particular situation, centric diatoms were completely dominant in both presettlement and postsettlement horizons and became significantly *more* abundant during the period of increased nutrient loading to the lake. These difficulties illuminate the problems associated with the ratio index approach. Ratios of major physiological Divisions have some information content, because they are reflective of major nutrient ion availability, but this is also obviously strongly determined by the regional geological context in which a given water body is found. Thus lakes receiving, for instance, increased phosphorus loading may become either nitrogen or silica limited, depending on their natural loadings of these materials. Hence, the observed response of major Divisions to this perturbation may be different, and ratios different, or indeed opposite. This approach may be instructive if restricted to regions with similar and uniform physiography, but will produce misleading results if broadly extrapolated. The potential of schemes using categories intermediate between the Divisional and species level is even less promising. Most groups of microscopic algae are very broadly classified at the generic level and many common genera have species which occur over most of the range of habitats available to photosynthetic organisms. Thus epithets like *Navicula* and *Oscillatoria* convey very little unique information, as they can apply to organisms which occur in desert soils, arctic ice, and a wide variety of intermediate habitats. In the Laurentian Great Lakes one suite of species of the diatom genus *Cyclotella (C. comta, C. ocellata, C. operculata)* is abundant in the least perturbed portions of the system (Stoermer, 1978), which is in agreement with observations from other oligotrophic lakes (Hutchinson, 1967); however, another suite of species of the same genus (*C. atomus*,

C. cryptica, C. meneghiniana fo. *plana, C. pseudostelligera*) dominates extremely perturbed regions of the system. In this case, a simple ratio index, such as the A/C ratio discussed above, is apt to be grossly misleading.

Assemblage diversity estimates have been widely employed in practically oriented phycological studies. They are particularly attractive because they provide a concise statement of floristic effects resulting from extreme system insults such as gross toxicity or extreme and prolonged nutrient imbalance. Diversity measures based on specific population abundance model assumptions (e.g. Patrick *et al.*, 1954) or more generalized measures of population frequency distributions (Margalef, 1956; Pielou, 1977) have been widely employed as an index of disturbance. In general, such indices are most easily interpreted in cases which approach some limit of tolerance and the forcing functions are clear and unambiguous. Examples include the depauperization of communities existing under extreme light limitation, high temperature, or severe toxic stress. In cases where more subtle effects are present and competitive interactions predominate (Grenny *et al.*, 1973), simple diversity measures are much more difficult to interpret. In some respects, algal assemblages probably constitute a worst case for this type of summarization. Due to their short generation times and capacity for luxury consumption, algal populations are particularly subject to hysteresis effects. Populations are also subject to physical displacement from their primary growth habitat. Not surprisingly, these difficulties render interpretation of single assemblage diversity estimates of freshwater phytoplankton particularly tenuous. Community diversity responses are more appropriately addressed through integration over reasonable temporal or spatial scales, either by averaging (Stoermer, 1975) or physical integration as occurs in fossil assemblages.

Hysteresis and population displacement problems also demand caution in interpreting the results of traditionally applied univariate statistical techniques. In most cases, however, the capacity of current, generally available tools to manipulate and analyse algal assemblage data greatly exceeds our current capacity to produce appropriate data. During the past decade, a variety of multivariate statistical techniques have become generally available (Allen, 1971; Levandowsky, 1972; Allen and Skagen, 1973; McIntire, 1973; Holland and Claflin, 1975; Schelske *et al.*, 1976; Stoermer and Kreis, 1980) which allow objective analysis of algal assemblage data. The potential power of these methods has not yet been fully explored, but they offer the possibility of providing an objective summarization of population data in the commonly understood language of science. If this potential is to be satisfactorily exploited, improvements will have to be made in the way population information is developed and processed.

D. Nomenclature

It is clear that a certain degree of ambiguity is associated with the names presently associated with algal populations. There is an obvious need for refinement of the taxonomic standard of phycology at almost all levels. There are numerous examples of ecologically important and commonly occurring taxa which are so poorly circumscribed that the names applied to them are almost meaningless. Taking a notorious example from the group I am most familiar with, "*Stephanodiscus astraea* (Ehr.) Grun.*" is a very commonly reported freshwater diatom. According to the rules of nomenclature, no such organism exists! Its nomenclatural type is a rare and obscure member of the genus *Cyclotella*, as generally conceived. Further, the apparent wide distribution of this organism is based largely on an extremely ambiguous secondary description which is broad enough to encompass nearly any large member of the genus *Stephanodiscus*. This confounded pattern of mis-identification has led to some extended ecological analyses, based on the occurrence and distribution of *S. astraea*, which are substantially without foundation. This, and many other similar lamentable cases, provide a substantial argument that we must find a way to improve the general level of algal taxonomy at all levels. There is considerable work to be done in exploratory floristics, formal nomenclature, basic systematic study, and especially comprehensive and revisionary treatments before the general standard of algal taxonomy will approach the level of better-studied groups of organisms. This is perhaps the most serious barrier to useful employment of qualitative information at the present time. In this area, we have a clear case of potential application seriously outrunning its fundamental research base. This should provide some food for sober reflection by academic phycologists.

An associated problem is the difficulty in addressing the information which is available in this area. Because of the scattered and fragmentary form of the literature in this field, it is exceptionally difficult to develop the literature resources necessary to achieve an acceptable standard of taxonomic practice. The general lack of comprehensive identification manuals and systematic treatises places a disproportionate burden upon persons entering the field or upon workers at institutions which do not have extensive primary literature holdings.

E. Data Acquisition and Storage

At the present time the potential utility of qualitative information is limited by the fact that data production is almost entirely dependent on a highly skilled human observer. Thus the rate of data acquisition in this area has not significantly increased in either quantity or quality over a very considerable period of time. This stands in stark contrast to the refinements and automation of methods for acquiring chemical and physical information which have taken place in the past thirty or so years. If anything, the phycological analyst's task has been made more difficult by the growing realization that specialized and divergent techniques must be applied to achieve accurate identification of all populations which might occur in a given assemblage. This tends to reinforce the human tendency to turn from the known, and therefore mundane, to novel and therefore "exciting" aspects of the work. If one disregards a few truly exceptional personality types, it is a virtual truism that an analyst's productivity in terms of pure raw data tends to be inversely proportional to his skill level. This tendency to be distracted by novelty is, of course, also encouraged by normal academic institutional conventions.

Clearly, development of automated methods of data production, or even automated methods which would aid skilled analysts, would have an extremely salutory effect on progress in this field. Automation would both improve the speed and accuracy of routine identification, which humans tend to do slowly and with appreciable error rates, and free skilled personnel to apply their talents to some of the neglected topics discussed earlier. A properly designed, automated system would also have the capability to acquire and store the requisite information to refine and systematize taxonomic work. This need has long been recognized. It is interesting that the National Science Board (1976) "in order to point out research areas of particular concern where a less than adequate effort is currently being made" noted that: "There will be an associated need for development of technologies for the rapid, automatic counting and sorting of the smaller and more numerous living elements of the system, namely algae and zooplankton. Technologies (such as image analysis) are available for transfer to this task ... ." Unfortunately, this need has not been adequately filled. Some interesting approaches have been explored (e.g. Price et al., 1978; Cairns et al., 1979), but we still appear to be some distance from devices which will be routinely useful in the average laboratory. Although the National Science Board seems to have substantially overestimated the ease of technological transfer, there is a family of reasonably mature techniques which can be transferred to specific components of taxonomic problems and this area deserves intensive research.

In contrast to the problems with data production, the capacity to effectively store, display, and objectively analyse complex biological data sets has increased enormously during the past few decades. Historically, the sheer mass and complexity of biological data sets of meaningful size has operated as a deterrent to their useful employment. Investigators working with algal assemblages are usually faced with arrays of several hundred taxa which display subtly different responses. Because of the amount of labour involved even to reduce such data to a form where first-order comparisons can be made by manual methods, it is a great temptation to restrict interest to some limited subset of the original data. This both limits the ultimate power of any subsequent objective analysis and totally submerges additional information which might eventually prove useful. One must suspect that a great many, if not most, of the biological population data actually collected lie in some forgotten file and never become available to the general community of investigators interested in such problems. In the present state of technology this is inexcusable. Fully mature technologies exist which are capable of storing, displaying, and making available to a wide array of objective analysis tools almost unlimited amounts of biological population data. Utilization of such a system also conveys the additional advantages of making it extremely easy to transfer data between laboratories and encouraging systematic collection of long term compatible data sets. Despite the relative maturity of this technology, it is my observation that they have not been adequately utilized by most workers in the field. Development of a successful system demands close cooperation between the biologist user and an expert programmer. Indeed, much of the current resistance among biologists to utilizing such systems comes from unfortunate experiences with primitive systems which were unduly difficult to use or inappropriate for their particular needs. The potential advantages, however, are so great that it behooves anyone seriously interested in population level studies to avail themselves of this tool.

F. Quantitative Methods

Finally, further effort must be devoted to interfacing population data with quantitative data derived from other sources. The inherent difficulties in relating population estimates to quantitative measures of constituents have long been recognized (Lohmann, 1908). Despite these problems, it is most tempting to convert population estimates to estimates of volume or mass even though the associated errors must be very large. In many cases, it may be argued that acceptance of the unavoidable associated errors are adequately compensated for by the additional insights gained through retaining the physiological compartmentalization which exists in natural assemblages (Bierman *et al.*, 1980). It appears that this dilemma can be resolved by resorting to available methods of quantitative cytology. It is possible within the present state of the art to determine the actual volumes of cytoplasmic compartments in algal cells with a high degree of precision (Sicko-Goad *et al.*, 1977) and to determine the presence and sites of accumulation of particular chemicals within algal cells (Sicko-Goad *et al.*, 1977; Sicko-Goad and Stoermer, 1979; Silverberg, 1975, 1976). Application of these techniques will undoubtedly provide many significant insights into the function of biological communities and cannot help but serve to bridge the gap between purely qualitative and quantitative approaches.

III. Future Research and Conclusions

Although speculation regarding the future paths of science always contains an element of danger, it is perhaps appropriate to summarize this discussion with some thoughts about how the topics discussed may relate to future directions of research in the field. It is my conjecture that, as the limits of short term process studies and models of ecosystem function based on gross assemblage parameters are explored, there will be an increasing need to adequately sense and interpret the information contained in the occurrence and distribution patterns of biological populations. Many of the technical tools to accomplish this are already at hand, if properly and thoughtfully employed. Most of the real barriers to making significant advances in this area are at a very fundamental level, which probably says something about the way phycologists are trained in the US. Academic phycologists in the US have traditionally been disproportionately concerned with the relationship of their speciality with the rest of botany, particularly evolutionary trends on the grand scale. It somehow seems that the real and urgent ecological problems of the present day have passed over this discipline without leaving any fundamental change in approach to the science or the way in which phycologists are trained. Parker (1977) has commented upon the rather incongruous trend towards the decline of field-oriented phycology during the past thirty years despite the increase in general interest in the area. I conclude that reversal of this trend will signal the beginning of real progress in this area.

References

Allen, T. F. H. (1971). *J. Ecol.* **59**, 803–828.

Allen, T. F. H. and S. Skagen (1973). *Brit. Phycol. J.* **8**, 267–287.

Almer, B., W. Dickson, C. Ekström, and U. Miller (1974). *Ambio* **3**, 30–36.

Andresen, N. A. (1976). Recent diatoms from Douglas Lake, Cheboygan County, Michigan. Doctoral Dissertation, University of Michigan, Ann Arbor. 419 pp.

Archibald, R. E. M. (1973). Obituary: Dr B. J. Cholnoky. *Rev. Algol.*, n.s. **11**, 8–14.

Battarbee, R. W. (1979). Diatoms in lake sediments. *In* Berglund (1979).

Baylis, J. R. (1957). *Pure Water* **9**, 47–74.

Beeton, A. M. (1965). *Limnol. Oceanogr.* **10**, 240–254.

Behre, K. (1970). *Beih. Nova Hedwigia* **31**, XI–XXII.

Berglund, B. E. (1979). "International Geological Correlation Programme Project 158, Paleohydrological Changes in the Temperate Zone in the Last 15,000 Years. Subproject B. Lake and Mire Environments. Vol. II. Specific Methods." Dept. Quaternary Geology, Lund, Sweden. 340 pp.

Bierman, V. J., Jr, D. M. Dolan, E. F. Stoermer, J. E. Gannon and V. E. Smith (1980). The development and calibration of a spatially simplified multi-class phytoplankton model for Saginaw Bay, Lake Huron. Great Lakes Planning Study Contribution No. 33. Great Lakes Basin Commission, Ann Arbor, Michigan. 126 pp.

Boney, A. D. (1970). *Oceanogr. Mar. Biol. Ann. Rev.* **8**, 251–305.

Brugam, R. B. (1979). *Freshwat. Biol.* **9**, 451–460.

Brunel, J. (1956). *Le Naturaliste Canadien* **83**, 89–95.

Burris, R. H. (1969). *Proc. R. Soc.*, B. **172**, 339–354.

Cairns, J., Jr, K. L. Dickson, P. Pryfogle, S. P. Almedia, S. K. Case, J. M. Fournier and H. Fuji (1979). *Wat. Res. Bull.* **15**, 1770–1775.

Carpelan, L. H. (1978). *Oikos* **31**, 112–122.

Caspers, H. and H. Schulz. (1960). *Internat. Rev. Hydrobiol.* **45**, 535–565.

Cholnoky, B. J. (1968). "Die Ökologie der Diatomeen in Binnengewässern." J. Cramer, Lehre, W. Germany. 699 pp.

Cleve-Euler, A. (1951–1953). Die Diatomeen von Schweden und Finnland. K. Sven Vet.-Akad. Handl., fjärde ser., **2**, 1–163 (1951); **3**, 1–153 (1952); **4**, 1–158 (1953).

Darley, W. M. (1974). Silicification and calcification. pp. 655–675. *In* "Algal Physiology and Biochemistry" (W. D. P. Stewart, ed.). Univ. Calif. Press, Berkeley, California. 989 pp.

Edmondson, W. T. (1972). Nutrients and phytoplankton in Lake Washington, pp. 172–193. *In* "Nutrients and Eutrophication" (G. Likens, ed.). American Society of Limnology and Oceanography Special Symposium No. 1.

Edmondson, W. T. (1977). The recovery of Lake Washington from eutrophication. pp. 102–109. *In* "Recovery and Restoration of Damaged Ecosystems" (J. Cairns, Jr, R. L. Dickson and E. E. Herricks, eds). University Press of Virginia.

Ehrenfeld, D. (1978). "The Arrogance of Humanism." Oxford University Press, New York.

Elsasser, W. M. (1966). "Atom and Organism; a New Approach to Theoretical Biology." Princeton University Press, Princeton, New Jersey.

Fee, E. J. (1971). A numerical model for the estimation of integral primary production and its application to Lake Michigan. Doctoral Dissertation, University of Wisconsin. 169 pp.

Fjerdingstad, E. (1965). *Internat. Rev. Hydrobiol.* **50**, 475–604.

Fogg, G. E. (1974). Nitrogen fixation. *In* "Algal Physiology and Biochemistry" (W. D. P. Stewart, ed.), pp. 56–682. Univ. Calif. Press, Berkeley, California.

Gallagher, J. C. (1980). *J. Phycol.* **16**, 464–474.

Gorham, E. and J. E. Sanger (1975). *Verh. Internat. Verein. Limnol.* **19**, 2267–2273.

Gorham, P. R. and W. W. Carmichael (1979). *Pure Appl. Chem.* **52**, 165–174.

Grenny, W. J., D. A. Bella, and H. C. Curl (1973). *Amer. Nat.* **107**, 405–425.

Griffiths, M., P. S. Perrott and W. T. Edmondson (1969). *Limnol. Oceanogr.* **14**, 317–326.

Hasle, G. (1978). *Phycologia* **17**, 263–292.

Haworth, E. Y. (1972). *Freshwat. Biol.* **2**, 131–141.

Haworth, E. Y. (1976). *New Phytol.* **77**, 227–256.

Haworth, E. Y. (1980). *Limnol. Oceanogr.* **25**, 1093–1103.

Hibberd, D. J. (1980a). Prymnesiophytes (= Haptophytes). *In* "Phytoflagellates" (E. Cox, ed.), pp. 272–317. Elsevier/North-Holland, New York.

Hibberd, D. J. (1980b). Eustigmatophytes. *In* "Phytoflagellates" (E. Cox, ed.), pp. 319–334. Elsevier/North-Holland, New York.

Hickel, B. (1975). *Verh. Internat. Verein. Limnol.* **19**, 1229–1240.

Hohn, M. H. (1969). *Bull. Ohio Biol. Surv.*, NS 3 (1), 1–211.

Holland, R. E. and L. W. Claflin. (1975). *Limnol. Oceanogr.* **20**, 365–378.

Huber-Pestalozzi, G. (1963). *Naturf. Ges. Zürich.* **108**, 470–473.

Hustedt, F. (1937). *Archiv. Hydrobiol.*, *Suppl. Bd.* **15**, 131–177.

Hutchinson, G. E. (1967). "A Treatise on Limnology. Vol. II. An Introduction to Lake Biology and the Limnoplankton." J. Wiley, New York. 1, 115 pp.

Jaag, O. (1966). *Schweiz. Zeitsch. Hydrol.* **28**, 97–103.

Kerr, S. R. (1976). *J. Fish. Res. Board Can.* **33**, 329–332.

Kolbe, R. W. (1927). *Pflanzenforsch.* **7**, 1–146.

Kolkwitz, R. and M. Marsson. (1908). *Ber. Deutsch. Bot. Ges.* **25A**, 505–519.

Levandowsky, M. (1972). *Ecology* **53**, 398–407.

Lewin, R. A. (ed.) (1976). "The Genetics of Algae." Blackwell, London. 360 pp.

Lohmann, H. (1908). *Wiss. meeresuntersuch. Abt. Kiel, N.F.* **10**, 131–370.

Lowe, R. L. (1974). Environmental requirements and pollution tolerance of freshwater diatoms. US E.P.A., Environmental Monitoring Series Publ. No. EPA-670/4-74-005. 333 pp.

Margaleff, R. (1955). "Los organismos indicadores en la limnologica. Biologia de las Aguas Continentales. XII." Min. Agr. Forestal Invest. Exp., Madrid. 300 pp.

Margaleff, R. (1956). *Invest. Presq.* **3**, 99–106.

McIntire, C. D. (1973). *J. Phycol.* **9**, 193–215.

Meriläinen, J. (1967). *Ann. Bot. Fenn.* **4**, 51–58.

Murphy, L. S. (1978). *J. Phycol.* **14**, 247–250.

Murphy, L. S. and R. R. L. Guillard (1976). *J. Phycol.* **21**, 9–13.

National Science Board (1976). "Strengthening Environmental Programs." US Government Printing Office, Washington, D.C. 27 pp.

Niessen, H. (1956). *Archiv. Hydrobiol.* **51**, 281–375.

Nipkow, F. (1927). *Schweiz. Z. Hydrol.* **4**, 11–120.

Nygaard, G. (1949). *Kondl. Dansk Vid. Selsk., Biol. Skr.* **7**, 1–293.

Nygaard, G. (1955). *Verh. Internat. Verein. Limnol.* **12**, 123–133.

Nygaard, G. (1956). *Fol. Limnol. Scand.* **8**, 32–94.

Palmer, C. M. (1959). Algae in water supplies. An illustrated manual on the identification, significance, and control of algae in water supplies. US Dept. HEW, Public Health Service, Taft Sanitary Engineering Center, Cincinnati, Ohio. 88 pp.

Palmer, C. M. (1969). *J. Phycol.* **5**, 78–82.

Parker, B. C. (1977). Preliminary analysis of the trends in algal systematics and taxonomy in North America. Special Symposium on the History of Phycology. IX Internat. Seaweed Symposium, Santa Barbara, California.

Parker, B. C., N. Schanen and R. Renner (1969). *Ann. Missouri Bot. Gard.* **56**, 113–119.

Parsons, R. T., K. Stephens and M. Takahashi (1972). *Fish. Bull.* **70**, 13–23.

Patrick, R., M. H. Hohn and J. H. Wallace (1954). *Notul. Nat., Acad. Nat. Sci. Phila.* **259**, 1–12.

Pennington, W. (1943). *New Phytol.* **43**, 1–27.

Pielou, E. C. (1977). "Mathematical Ecology." J. Wiley, New York.

Pienarr, R. N. (1980). Chrysophytes. *In* "Phytoflagellates" (E. Cox, ed.), pp. 213–242. Elsevier/North-Holland, New York.

Price, B. J., V. H. Kollman and G. C. Salsman (1978). *Biophys. J.* **22**, 29–36.

Rawson, D. S. (1956). *Limnol. Oceanogr.* **1**, 18–25.

Rodhe, E. (1948). *Symb. Bot. Upsaliens.* **10**, 1–149.

Rosen, R. (1977). *Bull. Math. Biol.* **39**, 663–678.

Ruttner, F. (1937). *Int. Rev. Hydrobiol.* **35**, 7–34.

Sanger, J. E. and G. H. Crowl (1979). *Quat. Res.* **11**, 342–352.

Schantz, E. J. (1971). The dinoflagellate toxins. *In* "Microbial Toxins Vol. VII. Algal and Fungal Toxins" (S. Kadis, A. Ciegler and S. J. Ajl, eds). Academic Press, New York and London.

Schelske, C. L. and E. F. Stoermer (1971). *Science* **173**, 423–424.

Schelske, C. L., E. F. Stoermer, J. E. Gannon and M. S. Simmons (1976). Biological, chemical and physical relationships in the Straits of Mackinac. US EPA, Ecological Research Series EPA 600/3-76-095. Environmental Research Laboratory, Duluth, Minnesota.

Schrödinger, E. (1945). "What is Life? The Physical Aspects of the Living Cell." The Macmillan Co., New York. 92 pp.

Schwimmer, M. and D. Schwimmer (1968). Medical aspects of phycology. *In* "Algae, Man and The Environment" (D. Jackson, ed.), pp. 279–358. Plenum Press, New York.

Shilo, M. (1967). *Bact. Rev.* **31**, 180–193.

Sicko-Goad, L. and E. F. Stoermer (1979). *J. Phycol.* **15**, 316–321.

Sicko-Goad, L., E. F. Stoermer and B. G. Ladewski (1977). *Protoplasma* **93**, 147–163.

Silverberg, B. A. (1975). *Phycologia* 14, 265–274.

Silverberg, B. A. (1976). *Phycologia* 15, 155–159.

Sparling, J. H. and C. Nalewajko (1970). *J. Fish. Res. Board Can.* 27, 1405–1428.

Stewart, W. D. P. (1973). Nitrogen fixation. *In* "The Biology of Blue-green Algae" (N. Carr and B. A. Whitton, eds), pp. 260–278. Blackwell, Oxford.

Stewart, W. P. D. (ed.) (1974). "Algal Physiology and Biochemistry." Uni. Calif. Press, Berkeley, California. 989 pp.

Stockner, J. G. (1971). *J. Fish. Res. Board Can.* 28, 265–275.

Stockner, J. G. (1972). *Verh. Internat. Verein. Limnol.* 18, 1018–1030.

Stockner, J. G. and W. W. Benson (1967). *Limnol. Oceanogr.* 12, 513–532.

Stockner, J. G. and J. W. G. Lund (1970). *Limnol. Oceanogr.* 15, 41–58.

Stoermer, E. F. (1975). *Verh. Internat. Verein. Limnol.* 19, 932–938.

Stoermer, E. F. (1977). *J. Phycol.* 13, 73–80.

Stoermer, E. F. (1978). *Trans. Amer. Microsc. Soc.* 97, 2–16.

Stoermer, E. F. and R. G. Kreis, Jr (1980). Phytoplankton composition and abundance in southern Lake Huron. US Environmental Research Reporting Series EPA 600/3-80-061. Environmental Research Laboratory-Duluth, Duluth, Minnesota. 384 pp.

Stoermer, E. F. and J. J. Yang (1969). Plankton diatom assemblages in Lake Michigan. Univ. Michigan, Great Lakes Res. Div. Spec. Rep. No. 47. 168 pp.

Stoermer, E. F. and J. J. Yang (1970). Distribution and relative abundance of dominant plankton diatoms in Lake Michigan. Univ. Michigan, Great Lakes Res. Div. Publ. No. 16. 64 pp.

Stoermer, E. F., J. C. Kingston and L. Sicko-Goad (1979). *Beih. Nova Hedwigia* 64, 65–78.

Taylor, F. J. R. (1980). Basic biological features of phytoplankton cells. *In* "The Physiological Ecology of Phytoplankton" (I. Morris ed.), pp. 3–55. Studies in Ecology Vol. 7. Univ. Calif. Press, Berkeley. 625 pp.

Thomas, E. A. (1965). *Mitteil. Österr. Sanitätsverwaltung* 66, 1–11.

Thomas, E. A. (1969). The process of eutrophication in central European lakes. *In* "Eutrophication: Causes, Consequences, Correctives." pp. 29–49. National Academy of Sciences, Washington, D.C.

Thunmark, S. (1945). *Folia Limnol. Scand.* 3, 1–66.

Tierney, D. P., R. Powers, T. Williams and S. C. Hsu (1976). Actinomycete distribution in northern Green Bay and the Great Lakes. Taste and odor relationships in eutrophication of nearshore waters and embayments. US EPA, Great Lakes Initiative Contract Program. EPA-905/9-74-007. US EPA, Region V, Chicago, Illinois. 168 pp.

Vallentyne, J. R. (1956). *Limnol. Oceanogr.* 1, 252–262.

Vallentyne, J. R. (1960). Fossil Pigments. *In* "Comparative Biochemistry of Photoreactive Systems" (M. B. Allen, ed.), pp. 83–105. Academic Press, New York and London.

Van Geel, B. (1972). *Acta Bot. Neerl.* 21, 261–284.

Van Geel, B. (1976). *Rev. Paleobot. Palynol.* 222, 337–343.

Van Geel, B. (1978). *Rev. Paleobot. Palynol.* 25, 1–120.

Van Landingham, S. (1977). Guide to identification and environments of blue-green algae (Cyanophyta) significant in water quality evaluation. US EPA, Cincinnati, Ohio. 279 pp.

Vaughn, J. C. (1961). *Pure Water* 13, 45–49.

Werner, D. (1977). Silicate metabolism. *In* "The Biology of Diatoms" (D. Werner, ed.), pp. 110–149. Blackwell, Oxford.

Züllig, H. (1961). *Verh. Internat. Verein. Limnol.* 14, 263–270.

Part Two

Marine Ecosystems

4 Phytoplankton as Indicators of Sea Water Quality: Bioassay Approaches and Protocols

Serge Y. Maestrini

Centre de Recherche en Ecologie marine et Aquaculture de l'Houmeau Case 5, 17137 Nieul-sur-Mer, France

Daniel J. Bonin

Station Marine d'Endoume, Chemin de la Batterie des Lions, 13007 Marseille, France

M. R. Droop

Dunstaffnage Marine Research Laboratory, Oban, Argyll, Scotland

I. Introduction

In the sea, as in inland waters, research into the relation between phytoplankton and its nutritional environment encounters the problem of determining precisely which factor or group of factors may be controlling primary production and population successions at any time.

The classical approach and indeed that most commonly followed in this field is an indirect, descriptive one, drawing information from three categories of measurements: (i) concentration of the nutrients, (ii) phytoplankton biomass and (iii) phytoplankton biochemical activity. This approach, however, gives only tentative results, because the very fine correlations observed *in situ* between chemical parameters and biomass yield or growth rate allow one neither to define the true relation existing between them nor to explain, for instance, why some waters are fertile and others are not. This last is not a recent problem: Russel (1939) observed that apparently chemically identical waters encountered at different times in the English Channel differed in their productivity and that the chaetognath *Sagitta setosa* served as an indicator of the less and *S. elegans* of the more productive water. De Valera (1940), Kylin (1941) and Wilson (1951, 1966) also noted that there were "good" and "poor" waters for rearing algal plantlets. More recently there have been reports both of unproductive waters rich in nutrients and of productive waters completely exhausted of mineral salts (Menzel and Ryther, 1960; Menzel *et al.*, 1963; Strickland *et al.*, 1969; Berland *et al.*, 1973a; Thomas, 1979).

These discrepancies could be explained by any of a number of possibilities, but their very existence serves to underline the inadequacy of the descriptive approach and totally refutes Redfield's (1958) statement challenging the experimental approaches in marine biology.

The first possibility is that not all the essential nutrients have been taken into account (Wetzel, 1965; Bonin and Maestrini, 1981). For example, for a long time the potential of other forms of nitrogen than NO_3 or NH_4 was neglected or ignored. Now it is known that urea (McCarthy, 1972a,b) and various amino acids and purines (Wheeler *et al.*, 1974, Antia *et al.*, 1975) can equally well be used by various phytoplankton species, at least under certain conditions. The same is true of phosphate esters, which some algae can use in place of phosphates (Kuenzler and Perras, 1965). Furthermore, it is now recognized that numerous trace elements and organic growth factors are necessary for the proper development of an algal assemblage (e.g. Provasoli, 1963), but these are seldom included in the analyses.

Moreover, as Provasoli (1963) wrote, "the deficiency of trace metals depends quantitatively far more on the physical status governing their availability to the cells than on the total amount". If it is possible to determine a theoretical relationship between light and major nutrients and the upper limit of the organic yield of a marine ecosystem (Steele and Menzel, 1962), frequently the predicted yield is not reached in nature because other "conditioning factors", such as organic ligands, are needed by the natural populations in order to utilize all the supply of light, nutrient anions and oligo–elements (Barber and Ryther, 1969; Barber *et al.*, 1971; Lin and Schelske, 1981; Smith *et al.*, 1982).

The second possibility is that different populations in an identical environment and having the same biomass and species composition can show very different potential for growth. Growth potential depends on the physiological state of the population, which in turn is a function of previous history (Droop, 1974; Harrison *et al.*, 1976). Thus up-to-date knowledge of a population at any instant is insufficient to forecast its future development, even given a perfect knowledge of the nutritional environment. This is because it is the intracellular rather than ambient concentrations of the nutrient elements that determine which element is limiting (Caperon, 1968; Droop, 1968). This explains why cells can be limited apparently even in the presence of plenty (for a limiting element may yet to be taken up, or may be in a form not easily assimilated (Goldberg, 1952; Guillard, 1968)) or, on the other hand, grow in the complete external absence of a nutrient (Kuhl, 1974).

Consequently, in most cases a simple measure of the external concentrations of all the mineral and organic nutrients known to be required for growth of phytoplankton does not permit one to say unequivocally which is limiting growth. This applies to unispecific algal populations and even more to complex natural assemblages whose components may have different requirements and growth characteristics.

The biochemical approach to the problem of limiting factors relies on measurements of chemical composition and of any metabolic activity known to be associated with specific deficiencies (Healey, 1973a, 1975). For example, it is known that phosphorus deficiency is accompanied by a lowering of the concentration of cell phosphorus, by an increase in alkaline phosphatase activity and an accelerated phosphorus uptake from a phosphorus-rich medium (Healey, 1973b, 1975). Likewise severe nitrogen deficiency is accompanied by lowered cell nitrogen and an increase of NH_4 uptake from an NH_4-rich medium (Syrett, 1953, 1962; Eppley and Renger, 1974). Lowering of the chlorophyll to carbon (Steele and Baird, 1961; Holmes, 1966) and protein to carbohydrate or carbon ratios (Myklestad and Haug, 1972; Haug *et al.*, 1973; Myklestad, 1974) are general indicators of nutrient deficiency or very high light excess.

The decrease of the ratio of the enzymes associated with carbon fixation, namely RuBPCase (ribulose-1,5-diphosphate carboxylase) and PEPCase (phospho(enol)pyruvate carboxylase) (Beardall *et al.*, 1976; Morris *et al.*, 1978), the decrease of the ATP content (Sakshaug and Holm-Hansen, 1977) and the decrease of the adenylate energy charge (Falkowski, 1977) are interpreted in the same way. A shift in the mean cell volume could also indicate a variation in the nutritional conditions (Berland *et al.*, 1970; Davies, 1976). However, although observations such as these can give a general indication of the nutritional state of an alga when maintained in pure culture in constant illumination and temperature, there is so much variation from species to species when physical conditions are varied that, in practice, it is very difficult to use these methods for studying natural populations, the physiological state of which will as a rule result from the interaction of a multitude of factors.

These are the reasons why ultimately there is advantage to be gained in adopting an experimental approach that makes use of the response of the organisms themselves, namely bioassay.

The principle of a bioassay is very simple: if one cultivates algae under conditions in which only one factor (or group of factors) is varied, and the response is proportional, then that factor (or group) is considered to be limiting. Growth thus obtained is the summation of all variables, measurable or not by other methods, that have any direct or indirect influence on growth. This approach is in principle very close to that of the conventional biological assay, which uses the response of an organism to measure the concentration or activity of a particular metabolite that cannot be measured by the chemical means; for example, the vitamin assay of body fluids (Welsch, 1947; Finney, 1952) or the standardization of antibiotics. There is a distinction, for although experimental procedures are often similar, the objectives differ. The bioassays discussed in this chapter are concerned less with measuring actual concentrations of compounds than estimating comparative nutritional capacity.

These bioassays can be used in different ways. (i) To measure the total yield potential (AGP = Algal Growth Potential) of a given body of water by means of a well-defined protocol. It is possible thus to characterize and compare different waters in terms of the growth of a chosen organism. Such an approach is commonly adopted in freshwater studies with *Selenastrum capricornutum*, as test organism. (ii) To establish which nutritional factor is primarily limiting growth, by using either a test alga or the natural population of water under study. It is possible in this way to establish which nutrient would increase productivity of the ecosystem most effectively and at least cost. In contrast to freshwaters, where it has been adopted both with large enclosures (Lund, 1972) and full-size lakes (Parsons *et al.*, 1972), this approach has been rarely attempted in the marine environment (Berland *et al.*, 1975; Parsons *et al.*, 1977; Granéli, 1981). When the natural population is used it is possible also to observe qualitative (species) changes following particular nutrient enrichments (see e.g. Dunstan and Tenore, 1974; Frey and Small, 1980). (iii) To estimate the effect of inhibitory substances, originating from domestic or industrial waste waters, on population successions and the productive capacity of a natural ecosystem (see Chapters 8, 9, 11 and 12).

Bioassays have been applied less extensively in the marine than in the freshwater field, probably because for a long time nitrogen has been thought of as the primary, if not only, limiting factor in the sea and also because of the acknowledged greater capacity for self-purification of marine waters.

In this chapter, reference will be made to numerous seawater assays by biological methods, but the review should not be considered exhaustive. Rather it is intended to be an account of the evolution of ideas and successive improvements in the methods used by oceanographers and physiologists working with marine unicellular algae. A brief summary of all published protocols, outlines of suggested procedures and prospects appears in Chapter 5.

II. Bioassays Used in Marine Algal Research

A. *Definitions*

The term *bioassay* does not refer to a unique well-defined method but rather to a large assemblage of techniques varying in practical detail with every author (see Appendix Table, in Chapter 5). The variations encountered in marine bioassays fall into five main categories: (i) the sea water may be used without enrichment or may be enriched with a mixture of nutrients whose number, selection and concentration depends on the objectives and the character of the water under study; (ii) the test algae may be the natural phytoplankton population or cells from a unispecific culture; (iii) the volume of the experimental cultures may be anything from small (10 to 20 ml) to very large (several dozen m³); (iv) incubation of the test cultures may be *in vitro* or *in situ* and may be of long duration (about a week) or quite short (several hours); and (v) the criterion of growth may be a biomass measurement or its equivalent or a measurement of some biochemical activity. The number of combinations of methods employed is very great, and it is impossible to describe any protocol without reference to its objectives and limitations. On the contrary, it is possible to evaluate a method only in relation to the objectives and questions posed, since the method is all-embracing and some information or measurements will always be precluded. Therefore, *before undertaking a bioassay it is absolutely necessary to establish exactly what one wants to know.*

In the first place, we believe that it is equally important to distinguish between two aspects all too frequently confused under the headings "(algal) growth" and/or "primary production". (i) The expression "Algal growth potential" (AGP) was introduced by Oswald and Golueke (1966) for "the dry weight of algae which grew in a given water sample in the laboratory when no factor other than dissolved nutrients is limiting growth". More simply, AGP is the bioassay maximum biomass potential of a given body of water in the absence of grazing or toxic substances. It is expressed as biomass per unit volume. AGP is more precise a term than fertility. (ii) Primary production is by definition "the creation by photosynthetic plants of organic matter incorporating solar energy" (see the review by Whittaker *et al.*, 1975). It is measured as a rate (of CO_2 fixation, of O_2 liberation, or of increase in algal biomass) and as such is expressed as mass per unit volume per unit time.

This distinction is essential in the choice of approach to two of the foremost questions confronting oceanographers: to know (i) which is the nutritional factor in shortest supply that limits algal "growth" and (ii) which factors present in the medium would eventually halt growth.

It is possible, in fact, for the concentration of material to be sufficient not to limit the growth rate of an alga and yet be low enough in concentration to limit the final yield (Morris *et al.*, 1971; Maestrini and Bonin, 1981). Furthermore, AGP, specific growth rate and species successions could be controlled by different nutrients (Officer and Ryther, 1980).

Finally, according to the kind of information that follows from their use one finds that biological tests fall into two main groups: those carried out with unenriched water on cultured test algae and those depending on differential enrichments carried out either with cultured algae or the natural population.

B. Bioassays with Unenriched Water

Since the water being assayed is used without modification, the experimental techniques are aimed at protecting the alga from grazing and creating conditions of illumination and temperature that will allow its growth (= AGP and/or growth rate) to be a reflection of the "quality" of the water, inasmuch as that quality depends on the quantity of nutrient elements or of toxic substances inhibiting utilization of these nutrients. There are two general methods, batch cultures and dialysis cultures. The latter is a recent development and has not yet given rise to any published results in the marine field. Its merits are discussed in Chapter 5; here we limit ourselves to batch culture bioassays.

The practical realization of this assay follows from the actual definition of AGP. It consists in cultivating the test algae under suitable conditions of light and temperature in a filtered sample of the sea water and measuring the maximum biomass produced.

This approach is particularly useful for making comparisons between different waters. For this it suffices simply to set up all the samples at the same time with the same test alga in identical conditions. Besides providing comparisons between the samples the responses also allow one to compare AGPs obtained with other sets, provided of course that experimental conditions and test alga are identical. This method was used by Smayda (1971) to study the euphotic zone of a large region of the North East Atlantic, by Berland et al. (1973a) to differentiate between water masses in the North West Mediterranean, and by Charpy-Roubaud et al. (1982) to assess the potential of the waters of the North Patagonian bays (Argentina) for shellfish culture. One can also demonstrate by this method temporal changes in the fertility of a location as it might concern commonly widespread algae, for example the investigations of Robert et al. (1979) on the oyster cleansing ponds on the French Atlantic coast (Vendée).

Used in this way this approach has limited scope and is just a quicker way of characterizing water masses according to the sum of the factors contributing to their quality (see, however, Appendix Table in Chapter 5). However, it is always possible to take a larger sample and analyse for nutrient concentrations before and after growth and calculate the amount of each major nutrient taken up by the test algae and thus the percentage of that originally present. This approach is not without pitfalls: on the one hand, a lack of uptake could be due to inhibitors and/or the absence of other indispensable elements not included in the analysis, and on the other, luxury consumption could lead to a distorted picture. But it is true that for each element and each test alga the biomass formed per molecule assimilated can give an indication of the one in shortest supply. Thus Droop (1973) has suggested that "the potential final crop respecting each nutrient would be given by the quotient of the sum of the cell and dissolved components and the cell subsistence quota . . . and the nutrient that yields the smallest biomass potential is the limiting one".

Maestrini and Robert (1979) have employed a "species yield index" as being more representative than raw AGP of the unknown properties of a water mass. Based on the fact that it takes a nearly constant quantity of each nutrient (the subsistence quota, sensu Droop) to produce a given algal biomass, the "species yield index" is defined as the ratio of biomass produced by a test-alga versus the nutrient apparently taken up (i.e. analysed-nutrient disappearance). Hence, when all chemical forms of a nutrient are analysed, the yield index meets a species constant, whereas when all forms which were taken up by the cells were not analysed, the ratio increases. Therefore, while AGP only allows one to say generally whether the water is poor or rich respecting the mineral nutrients, by contrast, the species yield index can indicate reduced or increased availability of any one of those nutrients and/or the uptake of other unsuspected chemical species. Bages et al. (1978), for example, have shown that, in a water in which nitrogen was the factor limiting AGP, the species yield index varied very little between one test alga and another but nevertheless distinguished clearly between two types of water (Table 1). By contrast, with phosphorus and silicon, both of which were in excess, the volume of the species yield index depended more on the test algae than the water quality, presumably because luxury uptake varied from one algal species to another. Furthermore, while investigating the circumstance of two raised values of species yield index in oyster ponds, Maestrini and Robert (1981) were able to show indirectly that a significant proportion of the algal biomass had been produced by forms of nitrogen not accounted for by chemical analyses, yet nevertheless assimilated (i.e. dissolved organic nitrogen). These results were later confirmed by direct analysis of the stored samples (Maestrini and Robert, unpublished). They are also consistent with those obtained by Shiroyama *et al.* (1976) and Forsberg *et al.* (1978) for lake water.

Bioassays undertaken with unenriched water are simple but they are very limited and imprecise. In the first place the large number of simultaneous samples required presents the experimenter with the difficult choice between adequate sample volume and adequate frequency of measurement. Too small a sample is likely to be unrepresentative and moreover does not permit the most sensitive methods of biomass estimation, while at least daily measurements are needed to ensure that the biomass peak is not missed. However, a more serious criticism of the method concerns the usual methods of storage and preservation of the seawater samples. Certainly the problems posed by differing treatment of samples are not confined to the fertility (AGP) test, although with that test it is virtually impossible to achieve uniformity. In fact, if one wants to compare waters of different origin, except in the rare case when samples are collected rapidly by air, they must be preserved until all have been collected. It has been known for a long time (Redfield and Keys, 1938) that simple preservation in the dark or refrigeration will not prevent changes in the concentration of nutrient salts. Such changes are mainly due to decomposition of mineral complexes and microbial mineralization of organic compounds, to the lysis of algal cells and the absorption of ions on suspended matter and on the vessel walls. To avoid or greatly diminish these effects Heron (1962) recommends storage in polythene vessels lightly preimpregnated with iodine, while Gilmartin (1967) suggests an addition of chloroform and Jenkins (1967) of a mercury salt to the samples. The last obviously cannot be used, because the algae are inhibited or killed by the mercury, which cannot be removed from the samples. Strickland and Parsons (1960) and Thayer (1970) later suggested rapid freezing and keeping at less than $-10°C$. Low temperature storage at $-20°C$ is the method currently most frequently used, but even this is not entirely free of problems: many workers have observed the appearance of a precipitate on unfreezing. As yet little has been done to establish exactly what is happening, but Brindle (personal communication) reports that the fluctuations are of the order of $\pm 20\%$ in NO_3, PO_4 and SiO_4 analyses when the samples are filtered (0·45 µm) prior to freezing. Also, MacDonald and McLaughlin (1982) have reported that quick freezing is only required for turbid

Table 1 Species yield index (µg chlorophyll a produced per µg-atom taken up) for 3 test algae *in vitro* cultured in neritic unenriched sea water and oyster-pond water. Δ∞: value variable, more often very high. Simplified after Bages *et al.* (1978)

Water origin	Nutrient taken up	Biomass produced (µg chlorophyll a)		
		Navicula ostrearia	*Phaeodactylum tricornutum*	*Skeletonema costatum*
Neritic sea water	N ($NO_3 + NO_2 + NH_4$)	1·9	1·1	1·4
	P (PO_4)	49·3	Δ∞	21·0
	Si (SiO_3)	3·2	Δ∞	3·1
Oyster-pond water	N ($NO_3 + NO_2 + NH_4$)	3·9	2·4	3·0
	P (PO_4)	38·0	Δ∞	16·2
	Si (SiO_3)	2·3	Δ∞	2·3

samples and phosphorus analysis. For nitrate, quick freezing does not improve precision and even turbid samples store acceptably without filtering; however, thawing techniques become important after two months storage. We also have shown (see Berland *et al.*, 1974) that precipitate causes a $\frac{2}{3}$ drop in the carbonate alkalinity, a drop which makes absolutely essential a direct measurement of ΣCO_2 (method of Van Slyke and Neill, 1924) before every estimation of the rate of carbon assimilation. Furthermore Fitzgerald and Faust (1967) have already suggested that samples should be filtered before preservation or freezing. But filtration itself causes some modification of the water, notably in respect of ammonia, which is easily adsorbed to the filter (Eaton and Grant, 1979) and of phosphorus, which in contrast, is either liberated from certain types of filter (Jones and Spencer, 1963) or adsorbed by others (Marvin *et al.*, 1972). This is why in order to determine the effects of these sources of error we recommend, whenever possible, that three series of analyses should be made: one at the time of taking the samples, one immediately after unfreezing and before inoculation with the test algae, and finally one at the end of the algal growth. Even this can provide no more than an indication, because it is unlikely that the analyses will include all the nutrient factors that were controlling growth at the time the samples were taken. Taking all this into account, one concludes that it is preferable also to include differential enrichment with these bioassays as if the objective were to establish the factor limiting AGP.

C. *Bioassays with Experimentally Enriched Water*

There are many methods aimed at evaluating the role of particular nutrients in the sea on the basis of the algal biomass changes resultant upon experimental enrichment. It is therefore very important here to differentiate between bioassays that purport to examine a natural situation and those that predict, by means of a model, the effects of variations in concentration of the nutrients on assimilation or growth characteristics. The relation most frequently used is represented by a hyperbola similar to that proposed by Monod (1942) for bacteria (Fig. 1). Moreover essential nutrients that become toxic at higher concentrations (e.g. essential heavy metals) are best represented by an arched curve (Fig. 2) or possibly a plateau but the algal responses are sometimes more complex, as showed by Gavis *et al.* (1981), who described a sudden increase in the slope of growth rate versus copper–concentration increase. In either case a necessary condition is that all other essential nutrients are present in sufficient amounts, so that their concentrations do not influence the observed rates. It is therefore of primary importance to separate the toxicity tests and the measurements of kinetic constants from the enrichment bioassay which puts into play several nutrients at a time and takes no account of growth rate but only of yield (AGP).

The enrichment bioassays conducted in batch conditions with the aim of elucidating a natural situation have been employed over a long period; their use in toxicity surveying has also been attempted (see, for instance, the reviews of

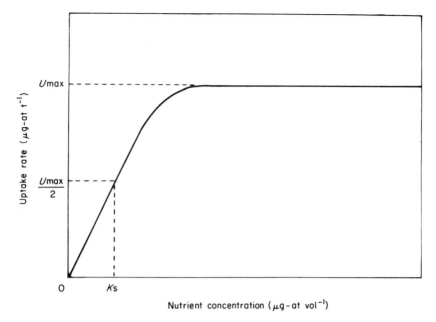

Fig. 1 Variation in uptake rate (in µg-at per unit volume per unit time) as a function of the concentration of the limiting nutrient (in µg-at per unit volume). U = uptake; K_s = half saturation constant. Redrawn from Dugdale (1967).

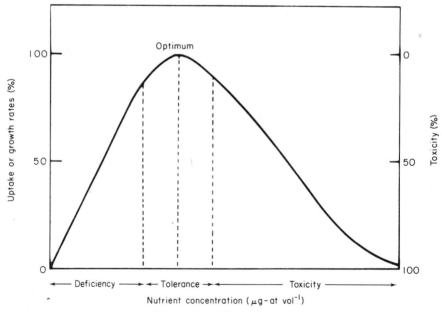

Fig. 2 Variation in growth rate (as a percentage of the maximum) as a function of the concentration of an essential nutrient toxic in high concentrations. Redrawn from Perkins (1976).

Miller *et al.*, 1978 and Dixon *et al.*, 1979). Because our purpose is to focus attention on bioassays applied to naturally occurring processes and our opinion is that toxicity tests would be more conveniently run with continuous cultures (see Chapter 5), our present comments will only concern the differential enrichments with the aim of establishing the nature of the nutrient limiting the maximum standing crop (= AGP).

The term *differential enrichment* is given to bioassays performed with aliquots of the same sample differentially enriched with nutrients in order to determine which was in short and which adequate supply in the sample.

There are two complementary procedures:

(*i*) Adding the nutrients singly, when each positive response indicates that the nutrient in question was in short supply while a response no greater than the unenriched control indicates that one of the other nutrients or some other factor was limiting.

(*ii*) Adding all the nutrients less one, when, provided no inhibitors are present, the response is maximal except when the omitted nutrient coincides with that in shortest supply in the sample. In this case the response is proportional to the titre of that nutrient and is thus a bioassay of the identified limiting nutrient.

Thus, by multiplying the combinations and working with two assays corresponding respectively to non enriched (AGP bioassay) and enriched optimally with all nutrients, one can accumulate sufficient information to establish the order in which the nutrients would come to limit AGP *in situ* at the time the sample was taken. Figure 3 shows an example of this.

It sometimes happens, especially with neritic waters, that the cultures with one or more nutrient elements omitted from the enrichment produce larger maxima than those receiving the complete complement (Fig. 4). This is nearly always associated with the omission of the metal mix and implies that the samples contained sufficient or excessive quantities one or more of these elements and that the enrichment resulted in toxic concentrations.

The enrichment method is by no means recent: Allen and Nelson (1910) enriched sea water with various mixtures to find the most favourable combination for algal growth. But it was Schreiber (1927) who first employed the technique with the express purpose of identifying the elements limiting the growth of phytoplankton *in situ*. His "*physiologische Meerwasseranalyse*" performed with 5 diatoms of the genera *Biddulphia, Chaetoceros* and *Melosira* and one chlorophyte, *Carteria*, enabled him to show that sometimes nitrogen and sometimes phosphorus was limiting in the vicinity of Heligoland. Later Gran (1931, 1933), Riley (1938) and Matudaira (1939), then Ryther (1954) and Conover (1956) follow the same line but the very limited number of nutrients and combinations included prevented their work from significantly advancing the use of this approach. Finally, it is during the last two decades and through the work of Ryther and Guillard (1959), Thomas (1959) and

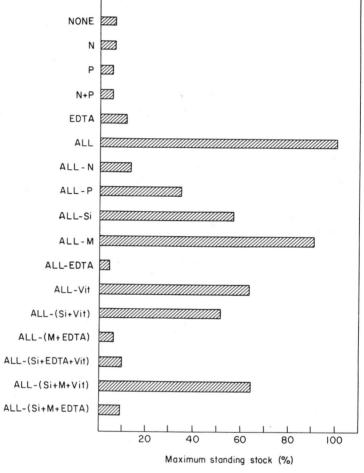

Fig. 3 Results from a series of differential enrichments of a sample of Mediterranean water. In this water the factor limiting AGP of the test algae, in this case natural phytoplankton, was chelation (EDTA) followed, in order of importance, by N, P, Si, vitamins and metals. All = all nutrients added (N, P, Si, M, Vit, EDTA). M = metal mix (B, Co, Cu, Fe, Mo, Mn, Zn); Vit = vitamin mix (B_1, B_{12}, biotin). Redrawn from Berland *et al.* (1973a).

especially Smayda (1964), that differential enrichments have been used in a less sporadic and more rational manner and with the adoption of experimental procedures more appropriate to the aims of the research.

The procedures adopted by the different authors vary in complexity, but one can make the primary distinction between those that employ natural phytoplankton and those that employ cultures.

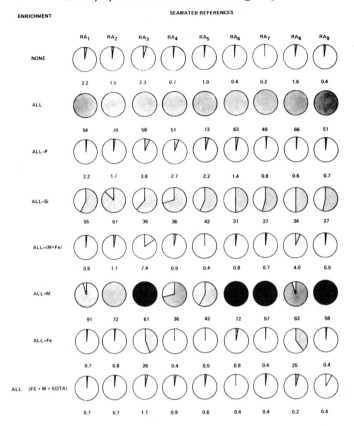

Fig. 4 Algal growth potential (AGP) for the diatom *Skeletonema costatum* cultivated in differentially enriched samples of water from off the Rhône Delta, Mediterranean Sea. The angle of the stippled area is proportional to the AGP relative to that of the complete enrichment (All). A black area indicates a response superior to the latter. Numbers below the circles refer to the amount by which the inoculum increased. Redrawn and simplified from Berland *et al.* (1973d).

1. Enrichment of sea water containing its natural population By differentially enriching samples of sea water filtered free of micro zooplankton (300 μm per pore) and treating them immediately, one can identify the nutritional factors limiting the AGP of the natural population at the time the sample was taken. This is the method most frequently used. However, here again, one can obtain more information, for instance, by analysing the species composition before and after growth and expressing the growth of each species in each culture as number of divisions and then expressing this number as a percentage of the total obtained for the species over all the treatments. Then for each species one can calculate the mean percentage for all the cultures having had a particular enrichment. The highest mean percentages correspond to the presence of the element(s) most favourable to each species in interspecific competition. An example of this is summarized in Table 2.

When differential enrichments are undertaken in this way one must always keep in mind that two essentially different kinds of information have been obtained. The first concerns the element limiting AGP *in situ* at the time of taking the sample; this is of relevance to the study of natural processes. The second is an experimental indication of which species would dominate upon a particular artificial enrichment. This is of especial value in aquaculture (Dixon *et al.*, 1979), since it allows one to plan enrichment of the feed water as to favour the most nutritional species for the higher links in the food chain, that is, in the absence of interference by factors other than nutrients (Lane and Levins, 1977).

Changes in specific composition of enriched sea water need not necessarily occur. Thus Jacques *et al.* (1973, 1976) and Fiala *et al.* (1976) reported no change in cultures after up to a month's incubation. Nevertheless, theoretical models (see review by Maestrini and Bonin, 1981) and experimental cultures in large volumes (Ryther *et al.*, 1975; Stoermer *et al.*, 1978) show that some change is the rule. For this reason we consider that this extended use of differential enrichments with natural populations has potential for future development.

So far there has been no concern on the part of authors to standardize methods for these assays. Nevertheless, methodological limitations have occasionally led to the adoption of identical solutions. Estimation of growth, for example, is most frequently based on the rate of carbon assimilation simply because the ^{14}C method is simple, very sensitive, and quick. Again, many workers practise either actual or simulated *in situ* incubation in an attempt to maintain natural conditions as far as possible. When the sample volumes have been small (33–300 ml), most authors have used the ^{14}C method as the sole criterion of growth (Glooschenko and Curl, 1971; Perry, 1972; Berland *et al.*, 1978; Charpy-Roubaud *et al.*, 1978b, 1982; Corredor, 1979). However, small samples nearly always imply no replication and therefore no control over measurement errors. Also (as we shall see in Section V) to measure the rate of carbon uptake is not to measure algal biomass but merely some activity over a short period, and the experimenter has no certainty of choosing the best time for AGP measurement, namely the unique instant when algal biomass is maximal with limiting nutrient exhausted and yet the population still healthy. Precise monitoring of algal growth is more easily achieved when the culture volume is large (0·5–100 l), but few workers have troubled to do this. They are: Fiala *et al.* (1976) (cell counts on alternate days); Vince and Valiela (1973) and Jacques *et al.* (1973, 1976) (daily counts) and Barber and Ryther (1969) (12-hourly ^{14}C measurements over the whole of the incubation period). Recent improvement in certain analytical techniques have enabled several analyses to be made on the same sample. For example, Graneli (1978) was able to make a cell count, a ^{14}C measurement, a chlorophyll *a* measurement, and ambient nutrient analysis all on the same 300-ml culture.

The advantage of incubating natural populations *in situ* is obvious. Certainly the

Table 2 Relative numbers of division (percentage) by species, of a natural population after enrichment (14 different treatments plus an unenriched control) and incubation to biomass stabilization, as a function of the added elements (see text). Cell counts after incubation by Utermöhl's method

Nutrients added		*Nitzschia longissima*	*Nitzschia* species	*Rhizosolenia fragilissima*	*Thalassionema nitzschioides*	*Thalassiosira hyalina*
		(11)	(7)	(5)	(5)	(14)
No nutrient added (control)		20·3	9·3	0	0	8·9
All nutrient added		0	12·1	13·1	15·0	10·9
Nutrient(s) most favouring[a]	Nature	Fe and P	Chel	Vit	Chel	Vit
	$\bar{x}$	21·2 18·7	16·4	21·4	17·0	16·2
	s	3·2 0·4	3·3	3·3	2·4	3·0
Mean value for non-favouring enrichments	$\bar{x}$	14·4	9·2	0	11·7	12·4
	s	3·1	0·9			2·8

[a] Chel: chelating agent (EDTA); Vit: vitamin mix (B_1, B_{12}, biotin). $\bar{x}$: mean percentage; s: standard error. Respective numbers of cultures in which growth occurred (over 15 cultures made) are mentioned within parentheses. Simplified after Charpy-Roubaud et al. (1983).

cells are placed immediately in natural conditions of light and temperature, but the frequent raising of the cultures for readings is a nuisance. Few workers have used the *in situ* technique, and those who have done so with a small number of differential enrichments (Gran, 1931; Riley, 1938; Conover, 1956; Glooschenko and Curl, 1971; Lännergren, 1978), which is of course a disadvantage when the aims are to determine which of many possible factors is limiting AGP. Indeed, one is frequently limited by the need to find conditions that are not only as near natural as possible but also conducive to clear-cut responses; furthermore, one is usually forced to compromise between the ideal and what can be achieved with the means at hand. This doubtless explains the great variety of procedures employed (see Chapter 5).

The use of the natural population as test organisms has the undeniable advantage that the result relates directly to organisms properly representative of the water mass under study (Goldman and Mason, 1962; Wetzel, 1965). If the enrichments respect certain conditions, there is no lag phase such as one frequently gets in cultures maintained in a very rich medium. For this reason this type of bioassay is very well suited to identifying the factors limiting the AGP of the total population at the time of sampling. On the other hand it does not allow one to predict changes in the limiting factor because the species composition of the phytoplankton changes in response to changing conditions according to the differing nutritional needs of the various species. Comparisons both in time and space are more easily made with the same test organisms throughout; hence the interest in differential enrichments using *cultured algae*.

2. *Differential enrichments with cultured algae*   In the marine field this approach is historically the oldest, dating back to Schreiber (1927). However, workers since then have preferred to use natural populations, in the conscious or unconscious belief that they alone were the legitimate tools for studying natural processes. Saving a short paper by Ryther (1954) and one, of ill-defined objectives, by Johnston (1963), the use of unialgal cultures for the assays in the manner described below was mainly due to Smayda (1964). The improvement of culture nutrients "effected" during the previous decade (e.g. Provasoli *et al.*, 1957) and the considerable increase in knowledge of the physiology of some of the most easily cultivated species have no doubt largely been responsible for spreading the use of pure cultures in marine bioassay.

Two quite distinct objectives call for the choice of this type of bioassay: (i) to understand the distribution of a given species in the successions in terms of water quality, when of course that species has to be the test organism, and (ii) to assess water quality in respect of the whole phytoplankton population, in which case the test species has to be one that is typical of the whole assemblage.

The first approach above is autecological. Wilson (1966), for example, used an axenic culture of the dinoflagellate *Gymnodinium breve* to determine the nutrient characteristics of waters supporting toxic blooms of this alga. He showed that this species multiplies in tropical waters rich in complexed iron, and contrary to previous ideas (Smith, 1948) it is this element rather than phosphorus that limits its development in these regions. Later research undertaken with a greater variety of enrichments (Collier *et al.*, 1969) demonstrated the presence in the bloom waters ("red tides") of significant quantities of sulphides, complexing compounds, and salts of nitrogen and phosphorus carried down by rivers and suggested that they are the cause of the dangerous red tide. Similarly Ignatiades and Smayda (1970) attempted to understand the distribution of the diatom *Rhizosolenia fragilissima* in terms of open and inshore waters. They were able to show that the neritic character of this species is a reflection of certain nutritional requirements only satisfied by coastal waters and that in Narragansett Bay its absence in late autumn and winter and in late spring was correlated with the poor availability of certain trace metals, cobalt and molybdenum in particular.

The second approach serves quite different aims: the species is now chosen according to its ability to respond to a particular property of the waters under study without its necessarily being a member of the natural plankton of the regions. Although, as we have seen previously, differential enrichments always serve to identify the factor limiting AGP at the time of taking the sample, but when the tests use the natural population only one answer is possible. When cultured algae are used there are in principle as many possible AGPs as algae, since the physiological characteristics and nutritional needs may differ from one species to another. Thus, for example, the AGP of a strict autotroph depends solely on the mineral nutrients and availability of trace elements. In that case the response obtained in cultures enriched with a mixture lacking only vitamins would be the same as that with the complete mixture. If one ignored the fact that the alga was an autotroph one might conclude that the vitamin concentration *in situ* was adequate for all algae. It is only with auxotrophs whose requirements for this particular vitamin are known can one establish whether there is a deficiency of any of them relative to the major nutrients (N and P). Similarly, a diatom could show up an eventual lack of silicon, while an unsilicified alga in the same conditions would not.

Furthermore, nutrients may be in a form available to one species and not to another. Nitrogen is a good example of this. There are no species in culture able only to use inorganic nitrogen (NO_3, NO_2, NH_4), but the inventory of available organic compounds of nitrogen varies greatly from one species to another. For the chlorophyte *Chlamydomonas palla* for example, it is very reduced, whereas for *Tetraselmis striata* or the diatom *Phaeodactylum tricornutum* it is much larger (Antia *et al.*, 1975). One must be careful to choose algae not only for their ease of handling in this kind of work (see Section IV) but also for their ability to respond to all the nutrients found in sea water or at least to those that on previous experience are likely always to be in excess. In practice in most instances it requires more than one species to achieve the desired coverage. Various workers have long recognized this: Sakshaug and Myklestad (1973) used 4 algae, Berland *et al.* (1973d) 3, Specht and Miller (1974) 3, Berland *et al.* (1978) 10, Maestrini and Kossut (1981) 6, and

Charpy-Roubaud et al. (1982) 5. Also one can make use of particular nutritional idiosyncrasies; for instance Hemiselmis virescens utilizes neither NO_3 nor NO_2 for growth (Droop, 1957) and could be used with profit alongside P. tricornutum, which has a wide spectrum, to study a water's potential respecting forms of nitrogen other than NO_3 and NO_2.

The method used to "sterilize" the samples is an important consideration when the test organisms are cultured algae. First the natural population must be eliminated. Elimination could in principle be effected either by killing or removal. Destruction of algal cells could be effected either by freezing, by heat sterilization or by chemical means, but all such methods are objectionable unless the killed cells are also removed, since killed cells lyse or are lysed and release their contents while their remains tend to absorb ions differentially. Indeed, for this reason whenever it has been necessary to preserve samples after lifting the practice has always been to filter them prior to freezing (Wilson, 1966; Collier et al., 1969; Ignatiades and Smayda, 1970; Ryther and Dunstan, 1971; Berland et al., 1973a, 1974; Melin and Lindhal, 1973; Sakshaug and Myklestad, 1973; Tarkiainen et al., 1974; Rinne and Tarkiainen, 1975; Hitchcock and Smayda, 1977; Yoneshigue-Braga et al., 1979). Sterilization by autoclaving prevents any bacterial action but is otherwise objectionable since it is accompanied by precipitation of iron and calcium phosphates and the consequent removal by absorption of much of the larger organic component (Droop, 1961; Jones, 1967; Lewin and Chen, 1971). Nevertheless, autoclaving should be used when one desires to determine the amount of algal biomass that can be produced from all additional sources of nutrient, including organisms, even if only a part of re-solubilized nutrients appeared in analysable forms (Saldick and Jadlocki, 1978). A greater algal yield is reached when the water is filtered than when it is autoclaved, while on the other hand, in nutrient-poor waters, algal yields are greater after autoclaving than after filtration (Schanz, 1977); yet some data are conflicting (Schanz, 1982). In short, the best procedure is to filter samples immediately on lifting and if possible use them at once.

Even filtration is not without problems. Sterilizing membranes (0.22 μm) for example retain some of the colloids and adsorb larger molecules (Holm-Hansen, 1969; Filip and Middlebrooks, 1975; Fitzgerald, 1975; Robarts and Southall, 1977). For this reason Wilson (1966) and Smayda (1970) preferred the less complete sterilization of 0.45 μm membranes. This practice was followed by all the authors cited above, but for our part and on the evidence of some of our work (see Fig. 5) we prefer an even larger pore size (i.e. 0.8 μm), notwithstanding that many bacteria will pass through (Anderson and Heffernan, 1965) and the study of growth factors is made more difficult.

III. Nature and Concentrations of Nutrient Enrichment

All known natural chemical elements occur in sea water. It is therefore not

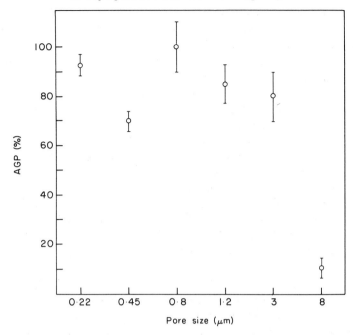

Fig. 5 Algal growth potential (AGP) of *Phaeodactylum tricornutum* cultivated in pond water, filtered but not enriched as a function of the filter pore size. AGP as a percentage of the maximum obtainable. After unpublished data of E. Gonzalez-Rodriguez, S. Maestrini and J.-M. Robert.

surprising, from a nutritional standpoint, that planktonic algae are found in all marine waters. Nevertheless a look at any of the numerous exhaustive lists of seawater composition (e.g. Sverdrup *et al.*, 1942; Goldberg, 1963, 1965; Culkin, 1965; Hood and Pytkowicz, 1974; Wilson, 1975) cannot help but impress one of the extreme diversity of their concentrations. Some elements are never present in more than trace amounts. According to Liebig (1840), the element that limits growth of a plant is that which is in shortest supply relative to the needs of the plant. It is therefore obvious that if algae expressly require these trace elements in large amounts, algal biomass will be small. It is therefore imperative to know all their specific requirements.

However, lists of algal requirements comparable to those of seawater composition do not exist, although various mineral elements are known to be essential for algae (see the reviews by Arnon (1960), Wiessner (1962), O'Kelley (1974), Spencer (1975) and Bumbu (1976). Martin and Knauer's (1973) table (see Brewer, 1975) furnishes one of the few lists of algal elemental composition. This area of phycology is generally neglected and it is significant that the review of Smith and Kalber (1974) only gives one table, attributed to Ryther (1956), who himself had only collected together the data from three previous articles and which dealt with a very limited number of essential elements (C, H, O, N and P). This neglect can be attributed to the fact that many elements occur in the sea in such high concentrations that it is inconceivable that they should ever be limiting. Table 5 of Sverdrup *et al.* (1942, p. 234) illustrates this well. (Note, however, the ratio $Si/C = 0.4\%$ is incorrect; it should be 7.0%.) The table shows, for instance, that there is 4.3 times as much calcium as carbon (on an atomic basis), but in the cells of diatoms and dinoflagellates there is, respectively, 27 and 125 as much carbon as calcium, which represents a concentration of carbon against calcium of, respectively, 100-fold and 500-fold. Silicon, on the other hand, is relatively sparse in sea water (at·C:at·Si $\simeq 21.8$) but in algae, and especially diatoms with their silicified frustules, it is very abundant (at·C:at·Si in diatoms $= 2.5$). This represents a 13-fold concentration of silicon against carbon, and the same is true of dinoflagellates, though their silicon requirement is very small. By similar reasoning and by analogy one can eliminate Cl, Na, S, Mg, K, Br and B from the list of possible limiting elements, because the concentration of the least abundant of these, i.e. boron, is of the order of $370 \, \mu g\text{-at·l}^{-1}$ which is 1000 to 10 000 times that of the remaining constituents of sea water.

Finally, at the end of a long evolution of ideas, which cannot be discussed here but a sketch of which will be found in Riley's article on the history of marine chemistry (Riley, 1965), one comes to the conclusion that phytoplankton biomass in the sea is controlled by four groups of nutritional factors. (i) Nitrogen, phosphorus and silicon, the major nutrients which are required in relatively large amounts relative to the carbon requirement (106 C:15 N:1 P, the so-called Redfield ratio) and which occurs in rather low concentrations in the sea. (ii) Trace metals, so called because they both occur and are required in trace amounts. (iii) Vitamins, namely cyanocobalamin (B_{12}), thiamine (B_1) and biotin. (iv) Finally, chelating compounds, which are not nutrients *sensu stricto* but which control the availability of the heavy metals. All those elements are included in the mixture used in differential enrichment assays.

A. Composition of the Enrichment Mixtures

Having defined the four groups of nutrients, it is now necessary to decide which of these components must be treated individually and which can be grouped together for the sake of economy; the criterion for the choice being previous experience. The literature in this field is rich in bald assertions, e.g. "In the sea nitrogen limits primary production" (Morris, 1974). Generalizations such as this are unwarranted, for although it is undeniable that nitrogen is often mentioned as the element limiting AGP, the evidence has often been work of limited scope which has acquired the respect of accumulated proof through repeated citation. The reality is otherwise, and a detailed examination of *all the published articles* concerning nutritional factors limiting AGP, primary production or phytoplankton growth leads to the following conclusions:

(*i*) The majority of investigations have concerned a few favoured sites (see Appendix Table, Chapter 5). Huge areas of ocean, such as the Tropical Atlantic, South Atlantic, Indian Ocean, the whole South Pacific, the Atlantic seas and the Arctic Ocean, all the African and Asian coasts, and all the coasts of Australia and South America, to mention only the large regions, have either not been examined or have been the subject of unrepeated spot checks.

(*ii*) The primary factor limiting AGP mentioned by these isolated researches is by no means always nitrogen. Phosphorus regularly figures in the Mediterranean and also in the Central Pacific Gyre. Now, silicon, organic growth factors and inhibitors have just as often been mentioned. (For key references, see Berland *et al.*, 1980.)

(*iii*) Other elements would become limiting should the concentration of the limiting element increase. Elements classed as second or third are frequently nitrogen, phosphorus or complexing substances; silicon, iron or vitamins likewise but less frequently. Trace metals are rarely mentioned (probably because of methodological drawbacks).

(*iv*) In a complex population even when the overall AGP is controlled by one element, it is possible that the partial AGP for the various components may be controlled by others. Vitamins and silicon are frequently "recessive" in this way.

(*v*) The primary limiting nutrient at a particular site can vary over the course of an annual cycle.

Thus, in the light of past experience, it would seem logical to single out nitrogen, phosphorus, silicon and complexing compounds (EDTA) for individual treatment. Vitamins were nearly always present in sufficient quantity in the sea waters studied thus far (Bonin *et al.*, 1981), and it would be unnecessary to involve them in the enriching mixtures were it not that most of the previous studies were limited to coastal areas, so that the possibility of a greater role in the open oceans cannot be ruled out (see Menzel and Spaeth, 1962). It is, however, unnecessary to treat them individually.

The vitamins required by auxotrophic planktonic algae are generally accepted to be limited to cyanocobalamin, thiamine and biotin (see Provasoli and Carlucci, 1974). These have been included by all authors since Ryther and Guillard (1959), and we do not suggest any change in the procedure.

By contrast it is quite otherwise with trace metals. Riley (1938) only added iron in addition to phosphorus and nitrogen, while Ryther and Guillard (1959) were the first to introduce cobalt and copper, manganese, molybdenum and zinc. This was a judicious initiative, since these elements, together with iron, are required by algae and yet occur in very low concentrations in sea water. The tables in Hood and Pytkowicz (1974) give the following ranges: iron, $0\cdot5$ ng-at$\cdot$l^{-1} to $1\cdot1$ μg-at$\cdot$l^{-1}; cobalt, 85 pg-at$\cdot$l^{-1} to 70 ng-at$\cdot$l^{-1}; copper, 3 ng-at$\cdot$l^{-1} to 425 ng-at$\cdot$l^{-1}; manganese, 4 ng-at·l^{-1} to 156 ng-at·l^{-1}; molybdenum, 3 ng-at·l^{-1} to 127 ng-at·l^{-1}; zinc, 15 ng-at·l^{-1} to 764 ng-at·l^{-1}. Furthermore, some of these elements have occasionally been mentioned as being limiting for natural phytoplankton, e.g. iron (Menzel and Ryther, 1961; Menzel *et al.*, 1963; Tranter and Newell, 1963; Skulberg, 1966, 1970; Glooschenko and Curl, 1971; Glover, 1978), cobalt (Ignatiades and Smayda, 1970), molybdenum (Ignatiades and Smayda, 1970; Sournia and Citeau, 1972), manganese (Sanders, 1978; Sunda *et al.*, 1981) and zinc (Anderson *et al.*, 1978). These elements should be included in the metal mix while the others previously used should be omitted. This applies especially to magnesium, bromine and sulphur, which have been occasionally included, and also to boron, used by many authors. The presence of these elements in enrichment mixtures is no doubt due to the fact that they are included in most culture media based on synthetic sea water (e.g. Provasoli *et al.*, 1957; Turner and Droop, 1978; Harrison *et al.*, 1980) without taking into account the fact that their concentration in natural sea water far exceeds the needs of algae. Carbon, a very essential element, occurs in sea water at a mean concentration of 2·07 mg-at·l^{-1}. This equally constitutes an atomic reserve far in excess of other essential nutrients. For example for a POC:PON ratio of 7·7 to 10 (Healey, 1975) there would need to be a reserve of nitrogen of the order of 200 to 268 μg-at·l^{-1}. Such concentrations of nitrogen would only be associated with pollution, for even the richest coastal waters and areas of upwelling never show more than 60 μg-at N·l^{-1} (Armstrong, 1965a,b; Broenkow, 1965; Vaccaro, 1965; Strickland *et al.*, 1969). Furthermore, the marine carbonate system, being in partial equilibrium with atmospheric CO_2, is buffered against biological depletion (for review, see Skirrow, 1975). Mineral carbon therefore should be omitted from enrichments. Similarly, except in the study of disequilibrated waters and/or algal species with particularly high requirements for trace elements the following should also be omitted even though some of them are known to be essential (e.g. Arnon, 1960); barium, titanium, zirconium and vanadium (Wilson, 1966; Collier *et al.*, 1969); also aluminium, cadmium, chromium, nickel and tungsten (Granéli, 1978). *We emphasize that the objective is not to know whether an element plays an important part in algal metabolism, but to establish whether the concentration in marine waters is such as to diminish or arrest metabolic activity.*

As we have seen previously, sea water can be enriched in two different ways. The first is to add each nutrient separately and measure the response of a test alga. Then, if the result for a particular enrichment shows better growth than the unenriched control, one concludes that that nutrient was limiting the AGP of the test alga. The biomass obtained does not relate to that nutrient but to the one next in order of limitation. Growth no better than the control indicates that either that element is not limiting or that others are equally or more so; and one cannot be more precise than this. The second way to make enrichments is to add all the nutrients less one. In this case the response is proportional to the concentration (in the sample) of the omitted nutrient and the nutrient yielding the lowest biomass when omitted is the one limiting in the sample. Because of the great adaptability of algal metabolism and consequent variability of overall biomass response, one should not base one's conclusions on the theoretical minimum of treatments. Confirmation is always necessary, which is the reason for employing the two enrichment techniques simultaneously. For the same reason the numerous papers that used less than ten combinations (see Appendix Table, Chapter 5) should not in our opinion be used as models. It is very evident, in extreme cases, for example when only nitrogen or phosphorus have been included in the enrichments and the numbers of combinations have been limited to two or three (Ryther, 1954; Ryther and Dunstan, 1971; Vince and Valiela, 1973; Specht and Miller, 1974; Specht, 1975; Lännergren, 1978; Teixeira and Tundisi, 1981) that the range of possible responses is equally limited. Frankly, such bioassays are without interest, at least unless previous work had shown that all other nutrients were without influence.

Thus it is necessary, in our opinion, on the basis of the six nutrient elements or groups described above, to have as many as 13 different enrichments in order to establish with confidence the factors limiting AGP. These are:

None	(0)
+N	(1)
+P	(2)
+Si	(3)
+M	(4)
+Vit	(5)
+Chel	(6)
All	(7)
All−N	(8)
All−P	(9)
All−Si	(10)
All−M	(11)
All−Vit	(12)
All−Chel	(13)

Where None = unenriched sea water sample; All = all 6 nutrient units present (N, P, Si, M, Vit, Chel); M = metal mix (Co, Cu, Fe, Mn, Mo, Zn); Vit = vitamin mix (B_{12}, B_1, biotin); Chel = chelator (EDTA).

This minimal set of enrichments has sometimes been exceeded (Menzel and Ryther, 1961; Smayda, 1974; Thomas, 1969; Thomas *et al.*, 1974; Berland *et al.*, 1978; Maestrini and Kossut, 1981). In fact if one suspects that two nutrients play a major role it is of advantage to combine them in two extra enrichments, e.g.

N+P	(14)
All−(N+P)	(15)

Furthermore, the heavy metal complexing action of many organic compounds results in their controlling the availability of the metal (see the reviews by Jackson and Morgan, 1978, and Huntsman and Sunda, 1980). This is particularly true of iron. Thus where previous results indicated a possible role for this metal it was not possible to distinguish between the metal's influence and that of EDTA in one's interpretation unless one also added the two factors together:

$$+ Fe \qquad (16)$$
$$+ (Fe + EDTA) \qquad (17)$$
$$All - Fe \qquad (18)$$
$$All - (Fe + EDTA) \qquad (19)$$

Other combinations are also possible in response to particular needs. For instance molybdenum is necessary for nitrate reductase activity (Wiessner, 1962), so that a lack of that metal would prevent nitrates from being assimilated and one would be led to the erroneous conclusion that the AGP was being limited by nitrogen (Sournia and Citeau, 1972). Similarly, cobalt is one of the constituents of cyanocobalamin, so that one would need to distinguish its role from that of the vitamins in those cases where both appear to be limiting AGP. There is also the inhibiting effect of zinc on silicon assimilation (Rueter and Morel, 1981) and of copper and manganese on photosynthetic activity (Wiessner, 1962; O'Kelley, 1974). One might be led to conclude that advantage would be had from employing metals and vitamins singly rather than as mixtures, but this is not so in practice, for experience has shown that metals and vitamins are seldom the primary limiting factor. The grouping into six as described is normally adequate, and it suffices merely to modify the range when necessary, that is, when the results suggest that one of the pooled factors might be influencing the outcome. This, however, could occur more frequently in the future, since metal–growth interaction is a complex field requiring an accuracy not always achieved in the past (Morel *et al.*, 1979; Bewers and Windom, 1982). Recent work of Anderson *et al.* (1978) and Sunda *et al.* (1981) has reported previously unnoticed growth-limiting effects of such trace metals as zinc and manganese.

In practice a distinction has to be made between neritic and ocean waters. For instance coastal waters are nearly always so rich in heavy metals (Bryan, 1976) that metallic additions are not only pointless but often harmful (e.g. Charpy-Roubaud *et al.*, 1983). Then one can omit the metal mix from the combinations, only keeping one (21) with the mix as a control (equivalent to the previous enrichment "All")

$$All \; (= N + P + Si + Vit + Chel) \qquad (20)$$
$$All + M \qquad (21)$$

The new complete enrichment (20) serving as reference now contains no metal additions. This range of enrichments is only useful if in effect the biomass obtained with No. 21 is clearly less than with No. 20. If it is not, it is to be concluded that trace metals were not in excess and one should proceed normally and refer all biomasses to enrichment No. 21.

Taking all into consideration, there is no strict rule governing the number of treatments one should give: the experimenter can always rely on his experience to eliminate useless combinations, bearing in mind however that at least two different enrichments per element are needed and that in this field an excess of precautions does not harm.

B. Concentration of Nutrients in Enrichments

The choice of final concentrations depends on several more or less conflicting needs.

(*i*) The nutrients supplied as enrichments have to be in greater concentration than in sea water in order to be able to differentiate clearly between the effect of the added and native nutrient.

(*ii*) Also they must be large enough to allow sufficient enhancement for the observed differences to be significant and attributable to real phenomena.

(*iii*) On the other hand, concentrations should be as low as possible so as not to modify assimilation mechanisms nor introduce toxic effects associated with excessive concentrations (of trace metals particularly).

(*iv*) The ionic equilibrium between the various elements should not be altered.

Addition of massive amounts of nutrients to sea water cannot but gravely perturb the species equilibrium by bringing into play absorption mechanisms that only operate under artificial conditions. The results of such assays are always of doubtful relevance to the natural situation (Holm-Hansen, 1969). Also, the higher the nutrient concentration, the higher the algal and bacterial biomass. A high bacterial population can modify the initial composition of the medium and consequently bias the results of the tests. It is also important to prevent the build-up of large amounts of excretory products, which may act as toxins or chelators, and to keep the pH from becoming excessively high, which can affect both the physiology of the cells and the chemistry of the medium.

Berland *et al.* (1973c) have shown that the biomass produced is no longer proportional to nutrient concentration when the latter is very high. Thus with the diatom *Skeletonema costatum* non-linearity is observed with concentrations above the following: $N = 40$ µg-at·l^{-1}; $P = 15$ µg-at·l^{-1}; $Si = 250$ µg-at·l^{-1}; $Fe = 175$ ng-at·l^{-1}; Vitamin $B_{12} = 87$ pм. Also it is always necessary to avoid running into carbon limitation caused by having so large a biomass that CO_2 diffusion into the medium cannot keep pace with the algal assimilation. The titre of nitrogen should be kept below 250 µg-at·l^{-1}, and the other elements correspondingly low. For this reason the practice of these authors (Fournier, 1966; Collier *et al.*, 1969; Berland *et al.*, 1973d, 1974; Lindahl and Melin, 1973; Melin and Lindahl, 1973; Malewicz, 1975; Rinne and Tarkiainen, 1975; Graneli, 1978 and Dufour *et al.*, 1981), who used initial concentrations of nitrogen of 500 µg-at·l^{-1} or more, should not be followed, even if certain of them were studying very rich waters.

Excessive enrichments must be avoided at all cost. The low biomass resulting from very low enrichment is not really very difficult to estimate, for it is always possible to count all the cells in a sample of a few dozen ml by Utermöhl's sedimentation method (Utermöhl, 1958) or more quickly by that of Steemann Nielsen (1978a). Moreover, the recent increase in sensitivity of biochemical techniques will henceforth permit analyses to be performed on a reduced number of cells (see Section VB), so that enrichments can be made at "ecological" concentrations.

There is no precise definition of the limit of a "natural" nutrient concentration, unless it is the highest value previously encountered at the site under study. If the integrity of the water under study is to be preserved, nutrient additions should not exceed the natural optimal concentrations. However, with very oligotrophic waters, such as the Mediterranean or Sargasso Seas, the natural reserve of nutrients is so low that if this rule were strictly adhered to the enrichments would be too low to be able to discriminate. With such waters a little more than the "natural" amount must be added.

In addition to the overall concentration there is the question of the relative concentration of the nutrients. In this connection a distinction has to be made between ratios thought to be optimal for algal metabolism (e.g. Healey, 1975, 1978) and those typical of natural sea water. Obviously, the latter should be our model, but unfortunately there is no general agreement as to what ratios are typical, so great is the variation to be found in the sea. In the circumstances one can but adopt the ratios most frequently reported.

Thus, the mean N:P ratio of 15, originally reported by Redfield (1934) and Fleming (1940), should perhaps be retained, even though the scatter was large. Unfortunately there are no similar observations for the N:Si and P:Si ratios, although it is known that there is generally more silicon than nitrogen in sea water. Similarly, we know of no publication giving a clear indication of what the "normal" proportions of the metals used in bioassay might be. The only information is the very broad range given in various reviews; that of Hood and Pytkowicz (1974) (Table 3) has already been referred to. We have calculated the concentrations of trace metals from these values relative to a nitrogen concentration of 50 μg-at$\cdot$l^{-1} and compared them with those proposed by authors working with culture media based on enriched sea water (e.g. Provasoli *et al.*, 1957; Guillard and Ryther, 1962; Provasoli, 1968; Antia and Cheng, 1970). These concentrations ratios we suggest are sufficient representative of sea water for our purpose. The concentration of the chelating agent (EDTA) is adjusted so that the ratio [chelator]:[metal] = 2 (at:at), although according to Droop (1961), 50 would be more logical, because of the massive amounts of magnesium and calcium in sea water.

In summary, taking account of all the constraints and alternatives justified above we propose that three different ranges of concentrations should be used (Table 4):

(*i*) The concentrations range corresponding to double the highest found in oligotrophic waters (e.g. Mediterranean water). We consider that this is the correct one to be used for assays with the natural populations of this type of water, since it is the range most likely to respect the physiological habits of

Table 3 Minimum, medium and maximum concentrations of Co, Cu, Fe, Mo, Mn and Zn occurring in natural sea waters, on the basis of data given by Hood and Pytkowicz (1974), and possible concentrations for bioassay use

Element	Natural sea waters			Bioassays		
	Minimum	Maximum	Mean	Calculated	Culture	Proposed
N	0	40 000	20 000	50 000	50 000	50 000
Co	ε	70	5	1	5	5
Cu	3	63	19	47	10	10
Fe	0	1110	46	115	270	200
Mo	2	127	104	260	100	100
			10	25		
Mn	4	156	27	67	500	100
Zn	15	740	99	247	200	200

[a] Calculated = mean *in situ* values multiplied by 2·5; Culture = media values of Antia and Cheng (1970) divided by 10; proposed = our suggested values. All concentrations as ng-at·l^{-1}.

Table 4 Suggested concentrations in the low (i), middle (ii) and high (iii) ranges of enrichment mixtures for batch bioassays (differential enrichments).

Element	Ranges of enrichment		
	Low (i)	Middle (ii)	High (iii)
N	15	25	50
P (μg-at l^{-1})	1·0	1·7	3·3
Si	18	30	60
Co	1·5	2·5	5
Cu	3	5	10
Fe (ng-at·l^{-1})	60	100	200
Mn	30	50	100
Mo	30	50	100
Zn	60	100	200
EDTA (μM)	0·4	0·6	1·2
Thiamine (nM)	44	74	148
Biotin (pM)	123	205	410
B_{12} (pM)	40	67	134

these algae. It is probable, however, that the concentrations will be too low when cultured algae are used for the assays, for, unless carryover has been reduced to an absolute minimum, its effect will swamp that of the added nutrients and the assay will be useless.

(*ii*) The concentrations of the middle range are 25% above the means shown in
       Table 3. This formulation is suitable for mesotrophic waters with all types
       of test algae and also for oligotrophic waters treated with cultured algae.
(*iii*) The concentrations of the highest range are 25% above those found in the
       richest waters and these should be used to assay such.

IV. Test Algae, Inoculum

A. Criteria in the Choice of Species

A great variety of species is available for bioassay, and the information actually
obtained may depend on the individual nutritional needs. Moreover the physiological state of the alga at the time of the assay may influence the results. This should
always be taken into account in the interpretation, as also should the ecological
status of the alga.

   If the problem is to understand the conditions that allow particular species to
dominate, that species or similar autochthonous species should be used (it is
unlikely that any of these will be readily available in culture). If it is to measure the
available nutrient content or toxicity of one water and to compare it with others,
then it is necessary to use species whose nutritional and cultural characteristics are
well known and appropriate; the choice will depend *inter alia* on which nutrients
one is interested in and it may be necessary to use more than one species for a
complete picture. Taken to the extreme this would involve the use of species
sensitive to particular factors employed in the manner of conventional biological
assays with calibrated linear dose–response relations.

   The choice of culture has rarely been the subject of preliminary investigation.
Usually the criteria have been availability and ease of handling. The question has
been discussed in limnology, however, (Komarek and Lhotsky, 1979; Komarek and
Marvan, 1979; Rosen, 1981), and it appears that a degree of standardization has
been achieved in this area. Species most often used belong to the genera
Microcystis, *Oscillatoria*, *Euglena*, *Chlamydomonas*, *Scenedesmus*, and above all,
Selenastrum, figuring in more than 200 reports from the pioneering work of
Skulberg (1964) to Leischman *et al.* (1979). Most of the cultures have been available
from culture collections. They are well known nutritionally so that optimal
conditions can be set up for their use in bioassay (Anonymous, 1971a,b, *inter alios*).

   By contrast, in the marine field, in spite of tentative attempts at standardization
(Anonymous, 1974; Gargas and Pedersen, 1974; Specht and Miller, 1974;
Chiaudani and Vighi, 1978) very many cultures of diverse origin, embracing some
30 species (see Appendix Table, Chapter 5), have been used, generally without any
clear justification for the choice. Many have been maintained for a very short time
and are not available from culture collections. Some studies on marine waters have
even been carried out with freshwater species.

The choice of test alga should not depend solely on convenience; other criteria are also important, and a culture should:

(*i*) Be of sufficiently representative ecologically and geographically to permit a realistic assessment of productive capacity of the water under study.

(*ii*) Possess characteristics in keeping with the nutritional status of the water, which may vary from the rich and often unbalanced conditions of neritic regions to extreme impoverishment of some parts of the open oceans.

(*iii*) Well-known nutritionally. Also, algae with well-defined contrasting requirements for certain factors (e.g. vitamins and elements such as N, P, and Si) can be used simultaneously to provide a more complete picture.

(*iv*) Not be genetically variable, or if so, at least stable for the duration of the tests. If variability manifests under certain culture conditions (Murphy, 1978; Necas, 1979; Brand *et al.*, 1981), one must maintain the cultures in conditions to which they have been accustomed right up to the time of the tests (see Komarek and Marvan, 1979).

(*v*) Further to (*iv*), depositing cultures with a culture collection makes them available with absolute security as to origin and maintenance, to different workers over an extended period. Furthermore, the designation of cultures used in bioassays should be reported as precisely as the results of the tests themselves.

(*vi*) As far as possible, be easily maintained in defined media to ease standardization by different laboratories.

(*vii*) Be fast-growing to cut the time required for the assays.

(*viii*) Lend themselves easily to standard methods of assessment (see Section V). Chain-forming species are inconvenient, especially for electronic counting, as also are species whose cell volume or indeed chemical composition (e.g. cell chlorophyll *a*) is very sensitive to culture conditions.

(*ix*) Not show too great a tendency to settle on the walls of the culture vessels. This is especially important if small culture volumes are to be used.

Obviously no single species could satisfy every criterion above, and the choice will be a compromise dependent on one's objective and means. Physiological and cultural characteristics (see Bonin *et al.*, in preparation) suggest a group of six species as most favourable for AGP assays, each for a different reason. They are the haptophytes *Emiliania* (= *Coccolithus*) *huxleyi* and *Pavlova* (= *Monochrysis*) *lutheri*, the chlorophyte *Dunaliella tertiolecta* and the three diatoms *Skeletonema costatum*, *Phaeodactylum tricornutum* and *Thalassiosira pseudonana*. Other species with particular nutritional needs may be included as, but only if, occasion demands. For example, the examination of AGP associated with nitrogen sources other than nitrate requires the use of a strain lacking the ability to reduce nitrate such as the cryptophyte *Hemiselmis virescens* (Droop, 1955a, 1957; Antia and Chorney, 1968; Antia *et al.*, 1975).

B. *Inoculum*

Bioassays of all kinds require, by definition, that the response should depend solely on the factor under investigation, which is why procedures are aimed at optimizing all factors other than this. But a reliable estimation demands that the test cells should not modify the medium and that their response should not depend on their previous physiological condition or at least that the dependence should be known and reliably reproducible.

It is possible to satisfy these requirements in one of two ways. The first is to use an extremely small inoculum, down to the theoretical limit of a single cell, so that the medium is not modified even after several divisions. The final response should then be independent of the previous history and it should only be necessary to ensure that the inoculum is viable. The second approach is to get the inoculum cells into a standardized reproducible condition, preferably as near that of the natural population as possible. This can be done either by depleting cell reserves in batch cultures or by using dialysis or chemostat cultures.

1. Use of extremely low initial cell densities   Growth of a small inoculum (from a mature culture) in an inorganic nutrient medium generally follows several recognizable phases (see for example Fogg, 1965), the most important of which from our point of view is the first, namely the lag phase. During this phase there is no increase in cell numbers, even when the initial count is as high as 1 to 5×10^6 cell·l^{-1} for cells of medium size (i.e. 7–15 μm diameter, equivalent to *c.* 60–500 μm^3). One reason for this is that the cells require some time to reassemble their enzymes and reserves depleted in the final phases of the source culture. Romano (1976a,b) has shown that after as little as 24 h after inoculation cells of *Dunaliella marina*, *Phaeodactylum tricornutum* and *Gonyaulax tamarensis* have increased in size (double in the case of the first two, which do not have a rigid case). The enlargement is accompanied by an increase in adenylate and to a lesser extent chlorophyll *a* and other components accelerating the formation of high energy compounds. The lag phase is in fact a period of feverish biochemical activity preceding the onset of cell division. An apparent lag phase might be due to a large proportion of the inoculum being permanently damaged, though apparently alive, so that the component growing exponentially takes a while before it has any influence on the cell count. Thus, the initial history of a culture is not independent of the state of the parent culture. Spencer (1954), in a now classic paper, showed how important the phase of development of the parent culture can be to the course of the lag in a subculture.

The use of very small inocula has its own statistical danger, for variability between individuals can cause spurious response differences and delays due to inoculum failure. A propos of this, Steemann Nielsen (1978a,b) proposed a method which, when adapted to the marine field, could constitute a decisive advance. This author stated that the high cell densities associated with the end of the logarithmic phase and stationary phase of test cultures are suitable only for determining "maximum standing stock". It is quite wrong to use them for true growth rate estimations, i.e. those appertaining *in situ*. For the latter, experiments should employ cell densities no greater than the natural ones so that the population rates of nutrient uptake are not large enough to change nutrient concentrations perceptibly during the life of the experiment. Such conditions are difficult to achieve by conventional methods and Steemann Nielsen proposed the use of successive dilutions of the algal material with the experimental medium prior to its use. He showed that it was possible with this method to estimate growth of *Nitzschia palea* or of the chlorophyte *Selenastrum capricornutum* down to 10^4 cells·l^{-1} and to assay down to 0.02 μg-at·l^{-1} nitrogen and 0.06 μg-at·l^{-1} phosphorus.

Specific growth rate measurement is not, however, in our opinion, the best criterion for the type of bioassay under discussion, especially as the counting technique proposed by Steemann Nielsen (Steemann Nielsen and Brand, 1934, in Steemann Nielsen, 1978a) is very time-consuming and is impractical for daily monitoring of a large number of test cultures. Nevertheless, this work does show clearly that with due care bioassays can be performed with very sparse initial populations.

The advantages to be gained from the use of very small inocula are outweighed to a large extent by the increased time needed for the growth of the cultures. This is particularly true for bioassays conducted at sea. Practical constraints cannot be neglected, because they largely govern the conduct of research. One should not, for example, by pushing perfectionism to the extreme, reduce the number of samples to such an extent that they no longer adequately reflect the properties of the water under study. Most workers carrying out bioassays have therefore adopted procedures that allowed them to attain the maximal crop in their cultures fairly quickly. Mostly, initial cell densities have been two orders of magnitude larger than those recommended by Steemann Nielsen, namely at least 10^6 cells·l^{-1}.

Several workers, however, have attempted to start their cultures with lower densities than this. Thus, Wilson (1966) used initial counts of 10^3 to 2×10^4·l^{-1}, but the organism, *Gymnodinium breve*, was large. Smayda (1973) has used the smallest initial densities, 2×10^4 to 7×10^4 cells·l^{-1}, of in this case *Skeletonema costatum*. Sakshaug and Myklestad (1973) and Hitchcock and Smayda (1977) with the same species used 20×10^4 and 50×10^4 cells·l^{-1}, respectively, and Graneli (1978) used 50×10^4 cells·l^{-1} of *Phaeodactylum tricornutum*, a species of comparable size. Ignatiades and Smayda (1970) with *Rhizosolenia fragilissima*, Specht and Miller (1974) with *Dunaliella tertiolecta*, and Berland *et al.* (1978) with *Chaetoceros lauderi*, *Thalassiosira rotula* and *Prorocentrum micans*, did not go above 40×10^4 cells·l^{-1}, but all these algae were large celled, 500 μm^3 and above in volume. For all these authors and especially for all those who have employed initial densities more than ten times as high, the biomass at the start would no longer be a negligible fraction of that at the finish, so that carryover of reserves would have interfered with the assays had not precautions been taken to eliminate it. This applies especially when the water assayed was low in nutrients.

2. Need to predeplete test cells of internal reserves　The physiological state of an algal cell depends largely on its previous history; the cell composition in particular is a result of post-nutritional conditions (see for example Eppley and Strickland, 1968; Fuhs, 1969 and Healey, 1975, 1978). Cells placed in a rich medium take up mineral phosphorus, for instance, at a rate far exceeding the needs of synthesis (Ketchum, 1939; Scott, 1945; Mackereth, 1953); this is the so-called luxury consumption by which when opportunity presents cells can quickly accumulate reserves for use by their progeny (see Rhee (1972) and Kuhl's (1974) review). Kuenzler and Ketchum (1962), for example, showed a 33-fold range in cell phosphorus in *Phaeodactylum tricornutum*, according to whether the cells were grown in a poor (2·2 µg-at P·l^{-1}) or rich (32–80 µg-at P·l^{-1}) medium.

According to these authors, a cell grown in a medium rich in nutrients could divide four or five times solely on the strength of its reserves. Droop (1974, 1975) reported a similar range in *Monochrysis* (*Pavlova*) *lutheri*. Similarly in this organism internal iron varied over a 40-fold range according to conditions in chemostat cultures (Droop, 1973), and vitamin B$_{12}$ (Droop, 1968, 1974, 1975) over a 200-fold range. The latter implies that on occasions seven to eight divisions would be possible on reserves alone. One infers, from experience with the serial transfer technique for establishing growth factor requirements, that the situation is even more extreme with thiamine, but no quantitative data exist.

Since Ketchum's work with phosphorus a fair amount of evidence has accumulated to the effect that depleted cells can take up at least some nutrients at rates far in excess (> × 100) of the highest possible rates of utilization (e.g. Droop, 1968, 1973). Recently McCarthy and Goldman (1979) report that this is true also of NH$_4$ in the diatom *Thalassiosira pseudonana*, the luxury specific rate of uptake being some 30 times the maximum specific growth rate. Thus, a cell nurtured in very oligotrophic conditions, suddenly finding itself in a eutrophic environment, could take up all the nutrients required for one generation in a fraction of its generation time. For these reasons the depletion of test cells prior to their use in bioassays is essential to the success of procedures and should be carried out rigorously and without fail.

3. Depletion of test cells　The most common method of depleting cells prior to this use is to grow them in batch culture in a minimal medium. Thus Smayda (1964) assayed the very oligotrophic Sargasso Sea water with *Thalassiosira pseudonana* cells that had grown up in an unenriched sample of the same water, but he did not stipulate how long it took. Ignatiades and Smayda (1970) preconditioned their test cells for four days in only one of the waters they were studying, even though the latter differed greatly one from another. The water used, that from Narragansett Bay, was manifestly richer than that from the Sargasso Sea included in the study.

Hitchcock and Smayda (1977), likewise, kept their parent culture for six days in unenriched water. Sakshaug and Myklestad (1973) are the only workers to have employed prolonged (10–14 days) preconditioning in differentially enriched samples, a procedure which is only possible if the numbers of enrichment mixtures is greatly reduced (to 4). But these authors did achieve depletion of their inocula with very poor water. Ryther and Dunstan (1971) also used this method, washing their test cultures for two days in very poor water prior to using them to assay eutrophic waters. Following them, Berland *et al.* (1973d, 1974) and Charpy-Roubaud *et al.* (1978a,b) did likewise for seven days. Only one of these authors (Berland *et al.*) justified the choice of procedure by a preliminary study.

Berland *et al.* (1973b) had in fact followed the time course of changes in cell contents and viability of seven planktonic algal species kept in a sample of Mediterranean water very low in nutrients. They found that the amounts of POC, PON, chlorophyll *a*, protein and carbohydrates all reached a minimum within a period of two to seven days of the start; an example of this is shown in Fig. 6. Furthermore, by taking these cells at various times and in various amounts to start subcultures in a normal medium, it was possible to set limits to the duration of this depleting treatment and inoculum sizes for adequate depletion of reserves coupled with maintenance of viability (Fig. 7). The results obtained (Table 5) showed that with the exception of *Phaeodactylum tricornutum*, which is very resistant, the duration of the depleting process should be short, corresponding to a minimum of two and a maximum of four divisions.

Preliminary experiments such as these are, in our opinion, essential to a rigorous bioassay, especially as the losses suffered during the depleting treatment varies both with species and parameter. As a general rule chlorophyll *a* content diminishes the most, but the composition of the water being used for the treatment has a considerable influence on the outcome. In this connection, in a later study with *Skeletonema costatum*, Berland *et al.* (1973c) showed that a water relatively poorer in phosphorus has the effect of reducing the number of cell divisions, while at the same time the cells become larger and richer in chlorophyll *a* and PON, whereas one poorer in nitrogen produces small pale-coloured cells. It is therefore necessary to choose the water for the depleting procedure with care.

Accordingly, waters such as coastal water rich in nutrients and waters from upwelling regions should be avoided. Poor summer waters or open ocean water offer the best medium, but they should contain at least traces of all the essential mineral nutrients. Organic compounds can always be absorbed on active charcoal; thus Berland *et al.* (1973b) recommend agitation for 20 to 30 minutes with $10 \, \text{g} \cdot \text{l}^{-1}$ charcoal (pre-washed carefully to avoid introducing mineral impurities with the charcoal). Even treated thus, water of the required specification is not easy to find. Therefore, when one had been found a sufficient quantity should be held to allow treatment of all the test cells to be used for a batch of samples however large it may be. In our opinion this solution is preferable to the use of a synthetic seawater medium because, unless spectroscopically pure and very expensive chemicals are

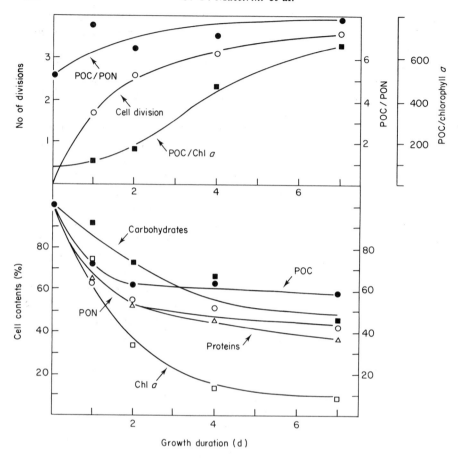

Fig. 6 Time course, during batch culture of *Phaeodactylum tricornutum* in poor Mediterranean water, of POC, PON, cell protein, carbohydrate, chlorophyll *a*, and POC: PON and POC: chlorophyll *a* ratios, as percentages of the initial values. The cells were from the exponential phase of batch culture in a rich medium. Redrawn from Berland *et al.* (1973b).

used, more mineral impurities will be introduced with the major salts than are found in poor natural sea water.

4. Preparation of test cells by dialysis culture For certain bioassays one should endeavour to have available a large enough number of test cells of a single species whose physiological characteristics are at once known and as close as possible to those of the wild population of the water being assayed. Except in those rare cases of a unispecific population these conditions can only be realized with appropriately treated cultured cells. Batch (Section B2) or continuous culture (Section B5) are not

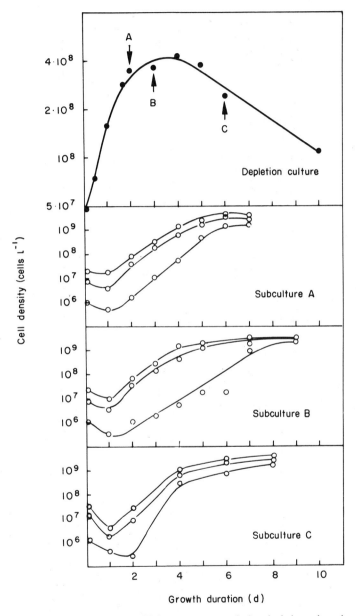

Fig. 7 Time course of cell counts of *Skeletonema costatum* during depletion culture in very poor Mediterranean water (see Fig. 6) and of the triplicate subcultures in a rich medium taken from it at times A, B and C. After Berland *et al.* (1973b).

Table 5 Depletion cultures of six algal species in very nutrient poor water, showing (i) time necessary for attainment of minimal cell constituents, (ii) limit of viability, (iii) estimated optimum for depletion and (iv) initial cell density for the bioassay (10^6 cell·l^{-1})

Species	Minimum cell constituents (day)	Limit of viability (day)	Optimum-depletion duration (day)	Initial cell density for bioassay (10^6 cell·l^{-1})
Asterionella japonica	4	4	4	6-8
Phaeodactylum tricornutum	4-7	4-7	4-7	10-50
Skeletonema costatum	2	2-4	2	1·2-50[a]
Exuviella mariae-lebouriae	2	2	2	1·5
Chlamydomonas palla	2	2	2	6-40
Monallantus salina	2-7	4	2-4	6-90

[a] Mean of 4 different cultures. Simplified from Berland *et al.*, 1973b.

necessarily the best way of depleting such cells, since it is never possible for the experimenter to mimic exactly natural conditions, whatever the care taken. To avoid this difficulty some workers have attempted direct cultivation of the algae *in situ*. The idea is that by leaving the cells sufficiently long in the sea they would integrate all the affects produced and thereby come into perfect equilibrium with natural conditions.

The normal way of doing this is by dialysis culture, sometimes termed cage culture. The principles involved in this technique are summarized in Schultz and Gerhardt (1969). Dialysis sacks whose walls allow passage in both directions of small molecules (MW < 12 000) are the simplest means of maintaining large numbers of algal cells *in situ* without their being subject to grazing, sinking or dispersal. The use of dialysis membranes in algal culture dates from Pratt (1944). Trainor (1965) first applied the method *in situ*, but the work of Jensen *et al.* (1972) has been responsible for its greatest development (see the review by Sakshaug and Jensen, 1978), mainly as a means of monitoring toxic compounds or limiting nutrients in sea water, on the basis of the maximum growth rates obtained (Sakshaug, 1977). Dialysis as a means of pre-conditioning inocula as discussed above has been attempted only by Berland *et al.* (1976, 1978), Kossut and Maestrini (1977) and Maestrini and Kossut (1980, 1981).

Berland and colleagues, in fact, kept ten different algal species in coastal waters of the Gulf of Lions for two weeks and then assuming the period to be long enough for their purpose, used the cells for differentially enriched batch culture bioassays. This procedure was not based on any preliminary study; in particular they had not followed the changes in cell contents of the immersed algae and therefore were not absolutely sure that at the end their cells corresponded to the wild state. Such a preliminary study was finally carried out by Maestrini and Kossut in the Ligurian Sea.

These authors followed the history of five cultured species during immersion. Two of them, *Skeletonema costatum* and *Hemiselmis virescens*, did not survive immersion in summer water, which, however, was rich enough in nutrients by Mediterranean standards (e.g. PO_4 content 0.3–0.48 µg-at·l^{-1}), but was rather warm (21–25°C). The three other species survived the *in situ* culture very well and had their reserves depleted by cell divisions (phosphorus by 67–88%; chlorophyll *a* by 59–89%, but nitrogen only by 24–53%) (Table 6). At the end of the immersion period all the cells had an elementary composition typical of the natural population. The period of immersion was more than 21 days, which was longer than the time these authors showed normally to be necessary, namely that sufficient for five divisions. This corresponds to the four divisions of Berland *et al.* (1973b) needed to deplete the reserves plus one to account for any nutrients taken up from the depleting medium. With similar material Maestrini and Kossut (1981) have recently shown by three different methods, namely (1) differential enrichments, (2) comparing the chemical composition of experimental and wild cells and (3)

Table 6 Decreases in PON, POP, chlorophyll *a* and ATP contents per cell during immersion in Ligurian sea water, on June and July 1976, of the chlorophyte *Platymonas suecica* and the diatoms *Phaeodactylum tricornutum* and *Thalassiosira pseudonana* cultured in dialysis sacks. Losses are expressed in per cent of initial values (inocula). From Maestrini and Kossut (1981)

Species	Period	PON	POP	Chlorophyll *a*	ATP
Platymonas suecica	June	24	74	80	54
	July	27	81	89	92
Phaeodactylum tricornutum	June	38	88	59	86
	July	47	68	75	86
Thalassiosira pseudonana	June	53	87	72	87
	July	52	80	85	85

applying the Cell Quota model (Droop, 1974), led to the same conclusion regarding the identity of the nutrient limiting the AGP of each of the species *in situ*.

This method, therefore, looks very promising and seems to be the one to use when the objective is more to characterize thoroughly a single sea water than to effect a simultaneous comparison of many waters. There are, however, physical limitations; for example the need to immerse the test cells at a fixed station for the required time can be administratively difficult and expensive when working from ocean-going research vessels whose task is usually to cover wide areas.

5. Use of the chemostat to produce cells of known composition   The steady-state chemostat is a continuous culture apparatus in which the specific growth rate and dilution rate are equal (Monod, 1950; Novick and Szilard, 1950) and constant. Matters are normally arranged so that one factor (usually a nutrient) clearly limits the growth rate, all others being in excess of requirement. The relation between nutrient concentration in the inflow, dilution rate biomass and cell composition respecting both the limiting and the other nutrients is complex (see Droop, 1974). To summarize: the cell content (cell quota) of the limiting nutrient is uniquely related to the dilution rate ($=$ specific growth rate) and is *independent of that nutrient's concentration in the inflow*, whereas that of non-limiting nutrients depends both on the dilution rate and their input concentration. Population biomass in the reactor is related both to the dilution rate and to the concentration of the limiting nutrient in the inflow. This can be expressed as follows

$$\frac{\mu}{\mu'_m} = 1 - \frac{\rho^k Q}{Q}$$

and for all practical chemostat settings,

$$s_R \sim Qx$$

($\mu = D =$ specific growth rate $=$ dilution rate; $\mu'_m = $ a "maximum" specific growth rate; Q = cell quota; ^kQ = subsistence quota = minimum possible cell quota; ρ = the luxury coefficient, the standardized non-limiting:limiting nutrient cell quota (i.e. standardized by division by the respective subsistence quotas); s_R = nutrient concentration in inflow; x = biomass in reactor). In general the lower the dilution rate the lower the cell nutrient quotas, while the ratios between the cell quotas of the various nutrients approximates to the ratio between the concentrations of those nutrients in the inflow. The correct use of the chemostat (for our purpose) depends on a proper understanding of these relations.

Since the objective is to produce cells equally poor in all cell constituents, the obvious procedure would seem to be to use the chemostat at a very low dilution rate with an input of all the nutrients as low as possible and in amounts proportional to the respective subsistence quotas. The difficulty here is that the latter are in general not known. However, one might hazard (from the work of Droop (1973) and Rhee (1978) that the ratios of subsistence quotas respecting N, P, Fe, vitamin B_{12} are of the order of $4 \cdot 5 \times 10^6 : 1 \cdot 5 \times 10^5 : 8 \times 10^3 : 1$. Lists of probable subsistence quotas of N, P and Si, gleaned from the literature, are to be found in Lehman *et al.* (1975) and Shuter (1978). The Redfield N:P ratio (Redfield, 1934), namely that reported to be most frequently encountered in natural populations, is 16:1, rather less than that above. However, the ratios in natural populations would in general have been influenced by luxury uptake to an unknown extent and are therefore not as good a guide as that between the true subsistence quotas. Furthermore, in the bioassays under discussion the subsistence ratios of the cultured test alga are of more relevance than those of natural populations, since it is the former's response that is being used.

In spite of the present general lack of subsistence quota data, continuous culture holds great promise as a means of predepleting algae for use in bioassay. Indeed, it is surprising that more attention has not been given to the method in view of the fact that it is the only one capable of producing a continuous supply of absolutely identical cells of a desired composition and physiological state.

6. *Advantages and uses of the different methods of preparing test cells* It is apparent that the different methods of preparing test cells do not lead to identical results. Care has therefore to be exercised in the choice. Summarizing the various constraints, one can say in short that, for all assays designed to compare many different waters, perhaps from widely separated regions, it is essential for the test algae used throughout the series to be absolutely identical. Bioassays with unenriched water to estimate AGP are in this category. Also some other methods, which are not bioassays in the narrow sense used by us, have the same requirement; for instance biological assays (e.g. vitamin assays, Ford *et al.*, 1955; Droop, 1955b; Carlucci, 1973a,b) and studies of the role of particular nutrients such as nitrogen (Grant *et al.*, 1967; Eppley and Thomas, 1969), phosphorus (Borchardt and Azad, 1968; Thomas and Dodson, 1968; Stewart *et al.*, 1970) or iron (Hayward, 1968, 1969). On the other hand, for bioassays aimed at studying a single station with the

maximum precision possible, for example to follow the changes over a period in the factors limiting AGP, "wild-like" test cells are the most appropriate. Of course the two types of test cells are not mutually exclusive, and indeed advantage is often to be gained from their simultaneous use.

Taking cells from the exponential phase of batch cultures is by far the simplest method of obtaining normalized algae, but that condition is short-lived and in practice can only provide cell of very approximate constancy. Furthermore, unless the inoculum is extremely small it must be predepleted of nutrient reserves. Use of larger inocula is usual because they do not need so long an incubation time. The work of Berland *et al.* (1973b) with *Chaetoceros lauderi* and *Skeletonema costatum* underlines the necessity of eliminating the carryover of reserves and the subsequent need for predepletion. But this should not be carried too far, for over-depleted or delicate cells have difficulty in adapting, particularly under oligotrophic or otherwise adverse conditions, and can give very erratic and misleading results. With eutrophic waters depleted cells, being more sensitive, show up differences better than non-depleted cells. Thus a compromise, a correct degree of predepletion is necessary.

Depletion by batch culture demands constant observation, and one is never sure of the viability of the cells produced unless one tests for this before the assay proper, which is not practicable, and furthermore, even when the right moment is chosen there is no guarantee that all relevant nutrients are equally depleted, so that the outcome of an assay, particularly of nutrient poor water, may depend to a large extent on the composition of the depleting medium.

By contrast, in a chemostat the production of viable cells is assured and because of the nice control of nutrient input and growth rate obtainable, with proper use of a chemostat, it should be possible to produce viable cells evenly depleted to any desired extent. Because of this and because production is continuous and indefinite and the product unvarying in composition, the chemostat ought, in our opinion, to become the sole supplier of preconditioned cells for bioassay.

The chemostat itself presents no problems, for quite a simple and cheap glass and silicone apparatus more than suffices (Droop, 1966), especially as it can be built compactly for shipboard use.

As we have seen, test cells having the properties of the wild population can only be obtained by immersion *in situ*, i.e. by dialysis culture, yet the long-term adaptation which allows algae to grow in previously toxic media (Stockner and Antia, 1976) or to take up new nutrient sources (Antia *et al.*, 1977) or to shift from a set of "slow-adapted" to "fast-adapted" kinetic constants (Droop, 1974; Conway *et al.*, 1976; Harrison *et al.*, 1976) is still largely ignored. Direct immersion by means of a simple attached cellulose bag is unsatisfactory, for, as Maestrini and Kossut (1980) found, this produces insufficient agitation for efficient exchange to take place, especially if the diameter exceeds 2·5 cm; also there is a fouling problem. In fact, continuous mechanical agitation as described by Jensen *et al.* (1972) or Eide *et al.* (1979) (Fig. 8) is absolutely necessary. This is certainly an inconvenience when

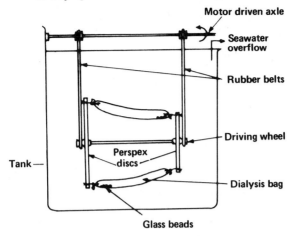

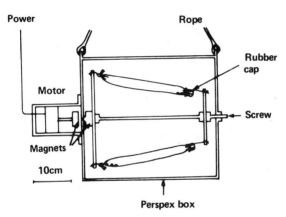

Fig. 8 Apparatus for mechanical agitation of dialysis bags. (*Upper*) for use on land (Jensen *et al.*, 1972); (*lower*) for use immersed (Eide *et al.*, 1979).

operating from a ship, but, provided the ship is large enough to accommodate several m³ tanks of sea water on deck, the dialysis could be carried out on board. The laboratory equipment described by Skipnes *et al.* (1980) and Marsot *et al.* (1981) is not suitable for the present purpose because it does not allow immersion *in situ*.

To summarize inoculum preparation

(*i*) Use of large inocula of undepleted cells is inadmissible.

(*ii*) If long incubation times can be tolerated very small inocula ($10^3 \cdot l^{-1}$) of untreated cells can be used.

(*iii*) For assays requiring an inoculum of constant physiological state the cells must be provided by a chemostat.

(*iv*) For assays requiring "wild-like" cells, i.e. cells in equilibrium with the water to be assayed, one should use dialysis culture with *in situ* immersion.

V. Culture Incubation: Estimation of Growth of Test Cultures

Having inoculated the test cultures they must be incubated in conditions of light and temperature to allow biomass increase and cell division. However, there remains the question of whether one should attempt to reproduce as natural conditions as possible and if so should one go as far as *in situ* cultures. There are two possible answers to this: yes, if the assays are based on growth rate; and no, if the criterion is the maximum standing crop. The reason is that under suboptimal conditions the rates of many of the biochemical reactions leading to growth are sensitive to variations in light and temperature. Thus only *in situ* or, failing that, simulated *in situ* cultures can provide answers that can be correctly applied to the natural situation. Dialysis culture, as discussed by Maestrini *et al.* in Chapter 5, is of particular relevance here.

In contrast, maximum biomass potential of any given water is generally not dependent on rates (that is provided such metabolic processes as grazing and autolysis are unimportant). For this measurement it would be pointless, indeed inefficient, to adopt the constraints of *in situ* culture. *Batch bioassays must be measured at the time of maximum biomass* since the maximum is the basis of this approach. It is not as easy an operation as would appear, as not all measured variables maximize simultaneously nor do they subsequently decay at the same rate, so in the following sections we discuss timing and the choice of variables open to one. Other aspects—the choice of vessels, volumes, light and temperature—have been summarized by Maestrini *et al.* (Chapter 5); see also Schelske (Chapter 2).

A. Timing of Maximum Biomass Measurement

Methodological progress during the last two decades has greatly simplified the use of a number of biomass estimations, for instance POC and PON, and has introduced completely new methods such as ATP. Nevertheless, accounts of the evolution of such variables during the course of a unialgal or even mixed culture are still rather rare.

McAllister *et al.* (1961) and Antia *et al.* (1963) have reported results of an *in situ* incubation of a phytoplankton population from the coast of British Columbia filtered through a 300 μm mesh into a 121 m³ plastic sphere (Strickland and Terhune, 1961). Dry weight, rate of carbon assimilation, chlorophyll *a*, POC, PON and protein were all measured over a three-month period to the point when all the products of photosynthesis were remineralized. We have taken note of the periods immediately preceding and immediately following the algal bloom and have recalculated the data as percentages of the maximum reached by each variable. Treated thus and regrouped in a single figure (Fig. 9) the variables exhibit a marked succession of maxima. In particular chlorophyll *a* shows a very sharp peak showing that the period of maximum is very short lived. Carbon assimilation increases rapidly up to the chlorophyll peak and remains high and even increases slowly during the period when chlorophyll *a* and PON fell by 50% and 30% respectively.

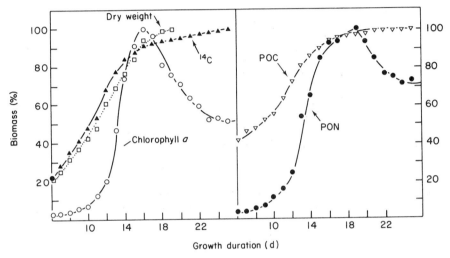

Fig. 9   Time course of dry weight, POC, PON, chlorophyll *a* and ^{14}C uptake rate in batch cultures of natural populations from the coast of British Columbia. All values are expressed as a percentage of the maxima attained. Redrawn and simplified from Antia *et al.* (1963).

In contrast, the experiments of Charpy-Roubaud *et al.* (1983), undertaken with natural populations from the Patagonian coast, show clearly that the rate of carbon assimilation peaked sharply earlier than chlorophyll *a* (Fig. 10). The rapid decrease after a stationary period of three or four days of carbon uptake and to a lesser extent of POC is in marked contrast to what Antia *et al.* (1963) found. Other experiments of this type have not included a sufficient range of measurements to permit useful comparisons. On the other hand, the analyses made by Berland *et al.* (1970) during the life of a pure culture of the Xanthophyte, *Monallantus salina* covered a large range of variables and are very relevant to our present discussion.

After treatment of these authors' data as described above to bring the peaks to the same height, one observes (Figs 11 and 12) that the curves for chlorophyll *a*, proteins, ATP, population volume and ^{14}C uptake all have a very sharp maximum, while the cell counts and PON, by contrast, show a plateau with a continuing slow

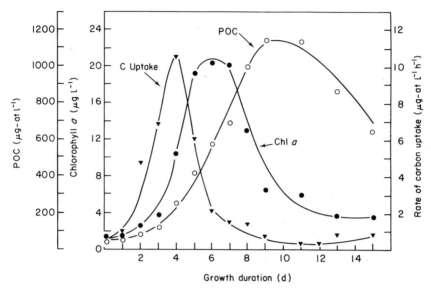

Fig. 10 Time course of chlorophyll a, POC and ^{14}C uptake rate in batch cultures of natural populations from the coast of Patagonia. Redrawn after Charpy-Roubaud *et al.* (1983).

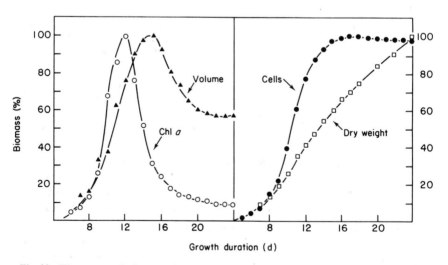

Fig. 11 Time course of cell count (Cells), dry weight population, volume, and chlorophyll a of batch culture of *Monallantus salina*. All values as percentages of the maxima attained. Redrawn and simplified from Berland *et al.* (1970).

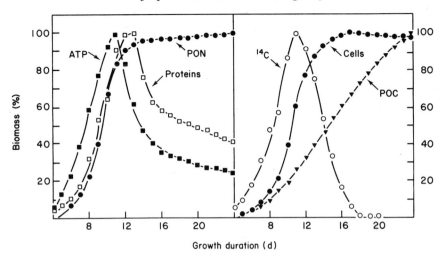

Growth duration (d)

Fig. 12   Time course of cell count (Cells), POC, PON, proteins, ATP and ¹⁴C uptake rate of *Monallantus salina* in batch culture. All values as percentages of the maxima attained. Redrawn and simplified from Berland *et al.* (1970).

increase to the end of the incubation period. ¹⁴C uptake and ATP peaked first, followed a day later by chlorophyll *a*, two days later by protein, and after a further three days by population volume. The maxima of population volume and protein could be looked as on plateaux of very short duration, since their values hold to within 5% on two successive days. Those of the other three variables (chlorophyll *a*, ATP and ¹⁴C) on the other hand, were very short lived, almost certainly less than 24 h, so much so that the values quoted were perhaps not the true maxima. In this connection, there is probably a better chance of obtaining the true value with species having a relatively low rate of division, i.e. ~ 0.6 div. d^{-1}, than with faster growing species. *Skeletonema costatum* and *Pavlova lutheri* with rates according to Sakshaug and Holm-Hansen (1977) of 2.9–3.1 div. d^{-1} and 1.2–1.9 div. d^{-1} respectively, are a case in point. In the cultures with N concentration comparable to those used in bioassays (i.e. 50 µg at·l^{-1}) but with an excess of P (Fig. 1, in loc. cit) one observes that the ATP maximum was obtained 21 h after the start of incubation, the maxima of *in vivo* fluorescence and chlorophyll *a* after 22 h and the PON maximum after 24 h. The cell count, on the other hand, took four days to reach a maximum and POC, five days. Here also a rapid drop after the peak was observed with *in vivo* fluorescence and chlorophyll *a* and with ATP, which however increased again later. Only the cell count POC and PON held up for at least two consecutive days. With *P. lutheri* the sequence was similar but spread over a longer period.

   Thus one can summarize the usual batch culture evolution of the different variables used for biomass estimation as follows:

   (*i*) ^{14}C assimilation and ATP reach their maxima first but decline quickly afterwards with no plateau.

  (*ii*) Chlorophyll *a* and *in vivo* fluorescence follow a similar course but slightly more slowly.

 (*iii*) For all the above the period over which the maximum may be estimated is less than 24 h.

  (*iv*) The protein maximum occurs immediately after that of chlorophyll *a* and persists for more than 24 h with a slower decline thereafter.

   (*v*) The cell count, population volume and PON maxima occur after a delay of at least three or four generation times of at the very least 24 h. The maxima are followed by plateaux of duration at least as long, so that one has a minimum of two consecutive photoperiods in which to measure the maxima.

  (*vi*) Dry weight and POC are the last variables to maximize, and their subsequent decline is very slow by comparison with the other variables.

If these properties are taken into account, it is of no consequence which variable is used to measure biomass. It is also equally clear that *biomass estimation in a series of test cultures cannot be effected at a set time and without intermediate monitoring without seriously affecting the validity of the results.* Continued monitoring is absolutely essential and one should use a measurement whose methodology is at once reliable, sensitive and rapid. It is equally true that consideration of material constraints must influence the final choice of measurement.

B. Relative Merit of the Different Biomass Measures

It is not necessary to examine the general validity of each method in detail, as this would go beyond the scope of the chapter and has been already frequently reviewed (see e.g. the last comprehensive papers by Sakshaug (1980) and Leftley *et al.* (1983), but it will be useful here to highlight the advantages and drawbacks of each in the context of bioassay.

The three "state" measurements, dry weight, cell count and population volume, that involve the whole cell are theoretically the most representative of biomass, though in practice they are not exempt from bias, which significantly reduces their reliability.

Dry weight, in the first place, is dependent on the degree of dehydration of the material. Dehydration may be carried out after filtration either by drying oven (Krey, 1950, 1964; Banse *et al.*, 1963) or by lyophilization. In either case the material is subsequently weighed. This operation can be carried out with great precision, but the dried filtrates absorb water vapour so quickly from the atmosphere that the weight can be observed increasing during the course of the weighing. Precise standardization of techniques and timing is therefore necessary, and the weighings must be carried out in a desiccating atmosphere. With lyophilization the problem is not so acute, but much more material is needed because of the difficulty of recovering the cells from the drying ampoules. All in all, this method requires too much material to allow more than one or two readings on any test culture, at least with the culture volumes normally employed. As desiccation and lyophilization usually take 24 h the dry weight method is inconvenient for daily monitoring of test cultures; it has the status of a back-up rather than an essential measurement.

Cell count is the most immediately accessible measure, especially of unispecific populations: all that is required for an estimation to be made in a few minutes is a microscope and counting cells of appropriate dimension. Direct cell counts on natural populations are not as easy because of the very large spread of cell size, but it is the low numbers ($< 100 \times 10^6$ cells·l^{-1} of average sized species) that is the real difficulty. Concentration by sedimentation (Utermöhl, 1958) or centrifugation (Steemann Nielsen and Brand, 1934, in Steemann Nielsen, 1978a) is necessary, or even the use of an electronic particule counter such as the Coulter counter (Sheldon and Parsons, 1967; Parsons, 1973). Each has its drawbacks: sedimentation involves a 24 h delay; centrifugation with a precipitating agent involves a delicate and time consuming operation; Coulter counting requires an expensive apparatus and much larger samples than optical counting methods and moreover is only reliable for unispecific populations of isolated (not colonial) unicells. The extreme variability in cell volume in a single species (see Bonin *et al.*; in preparation, for example) poses problems in a method that assumes numbers to be equivalent to biomass. Chances of error are greater when cell size is linked with nutrient limitation; it is well known, for instance, that phosphorus limitation produces large, highly coloured cells, while with nitrogen limitation the cells become small and pale. Since most bioassays involve limitation by diverse nutrients, one cannot equate numbers to biomass without biasing the results (Conover, 1956).

Population volume, as given by the Coulter counter or by sedimentation in a hematocrit or by centrifugation in calibrated capillaries (Sorokin, 1973), is independent of the volume of the individual cells, so this is a better measure than the cell counts although it is not without difficulties. Haematocrits are not sensitive and require a large amount of material and are therefore contra-indicated here. The Coulter counter is more sensitive, but it does not distinguish between living cells and detritus. This indeed is the main failing of total cell volume as a biomass measurement. One must also remember that the Coulter counter functions on electrical impedance difference and will give true volume only if it has been calibrated with material of the same impedance as the particles to be measured. Setting this aside, and the question of detritus, one has still to bear in mind that population biomass and population volume are proportional only for cells of constant density, and constant density is the exception rather than the rule, especially under conditions of nutrient deficiency.

In summary, therefore, methods based on whole cells do not necessarily afford a better measure of the true living algal biomass than methods based on certain individual constituents.

Any one of the three major cell constituents, namely carbon nitrogen and phosphorus, should logically be suitable for biomass estimation. However, carbon is a component of all organic compounds and is therefore also contained in the lipid and carbohydrates of the cell reserves and so cannot be regarded as solely representative of the metabolically active part of the biomass. Nitrogen is a better representative of the active compound, since it is not stored nearly to the same extent as carbon. Nevertheless PON and POC have nearly always been measured together since the introduction of the CHN Analyser (Simon et al., 1962). Accurate and sensitive, this apparatus is well adapted to measuring maximum algal biomass, but because of samples for the Analyser need to be dessicated, which takes time, it is not suited to daily monitoring large numbers of cultures. Also the CHN Analyser is expensive to use and should therefore be reserved for the more important analyses. Protein nitrogen can most easily be measured colormetrically (Lowry et al., 1951; Kochert, 1978; Rausch, 1981), but the method is not very sensitive and requires too much time for our purpose. On the other hand, more sensitive techniques such as fluorescence (Udenfriend et al., 1972; Böhlen et al., 1973; North, 1975; Castell et al., 1979) or high performance liquid chromatography (HPLC) (Hearn and Hancock, 1979; Hollaway et al., 1979) are difficult to carry out and require expensive specialized equipment. The estimation of protein, like PON, should be reserved for the maximum standing crop once other methods have shown that it has been attained. Phosphorus, even more than carbon and nitrogen, is exclusively associated with active molecules, and since it is quickly remineralized on cell death, it should be the most valuable measure of biomass. Unfortunately this is not so, for cells can accumulate phosphorus in great quantity above their immediate needs (Ketchum, 1939; Mackereth, 1953; Kuenzler and Ketchum, 1962). Particulate organic phosphorus (POP) is in fact measured as orthophosphate after lengthy preliminary mineralization (Hansen and Robinson, 1953; Strickland and Parsons, 1972), so that the POP and PO_4 accumulated within the cells as luxury uptake are not differentiated. Mackereth (1953) illustrated this very clearly (Fig. 13) by showing that a phosphate-limited alga in the stationary phase, when exposed to an addition of PO_4, showed an instantaneous increase in POP, while cell division recommenced much later.

Chlorophyll a being exclusively a plant constituent, has naturally been used extensively to express phytoplankton biomass. Great improvements in methodology, such as fluorometric analyses (Yentsch and Menzel, 1963; Marker et al., 1980) or HPLC analyses (Brown et al., 1981), permit measurements of the very low concentrations in cultures of unenriched sea water. There are two serious drawbacks to chlorophyll: (1) chlorophyll a content drops rapidly once synthesis ceases or, more important, (2) it varies from species to species and with physical conditions, especially light (Halldal, 1970; Falkowski, 1980; Perry et al., 1981; Prezelin, 1981). Banse (1977), de Jonge (1980) and above all Gieskes and Kraay (1975) provide further evidence of this and have concluded that chlorophyll a is a most unreliable measure of algal biomass.

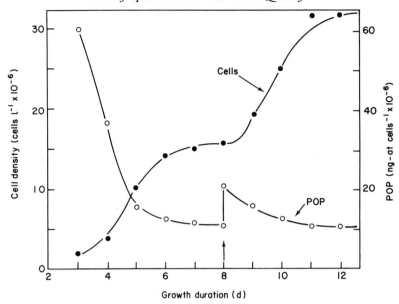

Fig. 13 Time course of cell count (closed circles) and POP (open circles), before and after the initial concentration of PO_4 (on the 8th day indicated by the arrow), in a culture of *Asterionella formosa*. Redrawn from Mackereth (1953).

ATP is not so much an indicator of biomass as of metabolic activity. Nevertheless it has been used as such since Holm-Hansen and Booth (1966) and Holm-Hansen (1970) showed that *in natural populations* ATP and POC frequently occur in nearly constant proportion, namely POC = 250 ATP (W/W). However, natural populations differ from batch culture, *inter alia*, in that cells disappear as quickly as they are produced (Hulburt, 1979), while in culture once nutrient uptake has ceased senescence sets in and ATP content diminishes. Furthermore ATP content varies greatly with the nature of the deficiency. Thus Karl (1980) has shown that nitrogen limitation has little enough effect, whereas with iron and especially phosphorus limitation the effect is very great, so much so that no reliability at all can be placed on ATP as a measure of biomass under nutrient-limiting conditions.

In addition to these so-called state measurements, rate measurements such as the rate of carbon assimilation have often been used to estimate "maximum growth" in batch cultures. This of course equates for the purpose the rate of a particular synthetic process with the accumulated product of all such processes, i.e. instantaneous standing crop. A few workers, however, (see Appendix Table, Chapter 5) have been naive enough to assume that such a rate measurement taken over a very short period (3 to 4 h) is necessarily a good measure of the biomass responsible for it. It is possible for instance that the same rate could be produced by a few very active cells or by a larger number of less active cells. On the other hand,

Healey (1979) has argued that a lag in the increase in photosynthetic rate will follow nutrient enrichment, and Lean and Pick (1981), Turpin (1982) and Quarmby et al. (1982) found evidence that nutrient-deficient populations seem to temporarily reduce, rather than enhance, photosynthesis when phosphorus or nitrogen (the respective actual limiting nutrients) becomes available. Nevertheless one can understand why carbon assimilation rate has been so popular; the great sensitivity possible with the ^{14}C method of Steemann Nielsen (1952), coupled with liquid scintillation spectrometry, offers incontestable advantages. Because of this sensitivity, some workers have believed that it could be used to identify and measure the rate-limiting factors (as opposed to that limiting AGP directly and have been able to reduce the duration of their experiments accordingly (Menzel and Ryther, 1961; Tranter and Newell, 1963; Barber and Ryther, 1969; Malewicz, 1975; Pojed and Kveder, 1977). But apart from this implied difference in principle, there are serious objections to this particular approach in the context of differential enrichments: the rate of photosynthesis of an alga responds differently to limitation by different nutrients and algae differ one from another. For example, with many algae, nitrogen or phosphorus limitation suppresses carbon assimilation so that the ^{14}C method might be expected to yield an exaggerated picture, while other algae e.g. the freshwater *Haematococcus pluvialis* continue to photosynthesize under severe N and P limitation. Some workers, conscious of these difficulties, and in an attempt to reduce the margin of error, prolonged their incubation to 24 h or even 48 h (Ryther and Guillard, 1959; Fournier, 1966; Glooschenko and Curl, 1971; Berland et al., 1978) or delayed adding the labelled bicarbonate until after a preliminary adjustment period in the enrichment samples (Charpy-Roubaud et al., 1978b). Longer incubation periods can cover more than a generation time and thus reduce the effects of the initial rapid absorption, if any, and also any that might be associated with circadian rhythms. Two methods are possible here, yet neither has been used so far. One is to add ^{14}C to the entire experimental bottle at the start of the experiment, then sample daily to establish a growth curve; in this case the labelled POC will plateau near maximum biomass, then drop because of excretion and cell lysis. The second method is to subsample daily and incubate briefly in the presence of ^{14}C, thus giving a more instantaneous rate of fixation which may indeed drop to zero as cultures become senescent. On the other hand, as Gieskes et al. (1979) remarked, prolonged incubation in a reduced volume (e.g. 30 ml) is harmful to cells and, at the very least, responsible for serious artefacts. According to these authors the incubated samples should be c. 4 litres in volume, which is quite impractical for bioassay work. The ^{14}C method does not therefore appear to be suited to the estimation of standing crop any more, incidentally, than does the O_2 method for carbon assimilation or the ^{15}N method for nitrogen assimilation (Neess et al., 1962). The last of these is of value in descriptive studies but is totally unsuited to bioassay work, for it requires 50 times as much biomass and takes very much longer than the ^{14}C measurement.

C. *Choice of the Best Measurement for Maximum Standing Crop*

Bioassays in batch culture involve two essentially different operations; daily monitoring the change in biomass to find the maximum and the accurate measurement of that maximum. The first requires a quick and sensitive method permitting a large number of analyses on a very small amount of material from each culture; the second can use the remaining material but then the analysis must be precise and of the variable most truly representative of biomass. Having regard for the different merits and failings of the various options discussed in the previous section, it is apparent that the two operations require a different analytical method.

Four variables meet daily monitoring requirement: *in vivo* fluorescence, chlorophyll *a* content, cell count, and carbon assimilation rate. The choice depends on whether natural mixed populations or unispecific cultures are being used.

With natural populations cell count is unsatisfactory because of the great size difference between species. For chlorophyll *a*, *in vivo* fluorescence has the advantage of speed and minimum preparation but is less sensitive than extracted fluorescence (Tunzi *et al.*, 1974), and is therefore perfectly suitable only for bioassays based on growth-rate variations (Sellner *et al.*, 1982). The ^{14}C method for carbon uptake is also simple and sensitive, but one must bear in mind that, unlike that of chlorophyll *a*, it is a rate measurement. The two however can be expressed in the same terms (as rates) if the fluorescence is read at the start and end of the ^{14}C incubation period.

$$\Delta\text{Chl } a = 3 \cdot 3[(\log_{10} \text{Chl } a \cdot \text{vol}^{-1})_t - (\log_{10} \text{Chl } a \cdot \text{vol}^{-1})_{t_0}]/(t - t_0)$$

$$\Delta\text{C} = 3 \cdot 3[(\log_{10} \text{C} \cdot \text{vol}^{-1})_t - (\log_{10} \text{C} \cdot \text{vol}^{-1})_{t_0}]/(t - t_0)$$

The end of the exponential growth phase is marked by a drop in ΔChl *a* or ΔC and, as was shown in Section VA, heralds the approach of the period of maximum biomass.

We favour the ^{14}C method as a first and the acetone extracted fluorescence as a second choice for monitoring cultures of natural populations. Since both methods require samples of *c*. 50 ml and at least five readings over the life of the culture and sufficient must remain for the final biomass measurement, a culture volume of 500 ml is indicated. Illumination should be moderate, for growth must not be so fast that it cannot be analysed by the methods used.

The same methods may be used for monitoring unialgal test cultures, but the speed of Coulter counting and the small volumes required (2 ml) offer overriding advantages. Monitoring by Coulter counter and subsequent biomass measurement only requires culture volumes of 100 ml.

The choice of the final biomass measurement is between a protein and a PON analysis, but failing either of these a Coulter-counter population volume is acceptable.

References

Allen, E. J. and E. W. Nelson (1910). *J. mar. biol. Ass. U.K.* **8**, 421–474.

Anderson, J. I. W. and W. P. Heffernan (1965). *J. Bact.* **90**, 1713–1718.

Anderson, M. A., F. M. M. Morel and R. R. L. Guillard (1978). *Nature (Lond.)* **276**, 70–71.

Anon. (1971a). "Algal Assay Procedure Bottle Test." National Eutrophication Research Program, Corvallis, Or. Environmental Protection Agency, 82 pp.

Anon. (1971b). "Provisional Algal Assay Procedures. Final report." Water Quality Office, 211 pp.

Anon. (1974). "Marine Algal Assay Procedure Bottle Test." Eutrophication and Lake Restoration Branch, National Environmental Research Center, Corvallis, Or. EPA-660/3-75-008, 43 pp.

Antia, N. J. and J. Y. Cheng (1970). *Phycologia* **9**, 179–183.

Antia, N. J. and V. Chorney (1968). *J. Protozool.* **15**, 198–201.

Antia, N. J., C. D. McAllister, T. R. Parsons, K. Stephens and J. D. H. Strickland (1963). *Limnol. Oceanogr.* **8**, 166–183.

Antia, N. J., B. R. Berland, D. J. Bonin and S. Y Maestrini (1975). *J. mar. biol. Ass. U.K.* **55**, 519–539.

Antia, N. J., B. R. Berland, D. J. Bonin and S. Y. Maestrini (1977). *Phycologia* **16**, 105–111.

Armstrong, F. A. J. (1965a). Phosphorus. *In* "Chemical Oceanography" (J. P. Riley and G. Skirrow, eds), Vol. I, pp. 323–364. Academic Press, London and New York.

Armstrong, F. A. J. (1965b). Silicon. *In* "Chemical Oceanography" (J. P. Riley and G. Skirrow, eds), Vol. I, pp. 409–432. Academic Press, London and New York.

Arnon, D. I. (1960). Some functional aspects of inorganic micronutrients in the metabolism of green plants. *In* "Perspectives in Marine Biology" (A. A. Buzzati-Traverso, ed.), pp. 351–383. Univ. Calif. Press, Berkeley, California.

Bagès, M., J. P. Dréno, E. Gonzalez-Rodriguez, S. Y. Maestrini and J. M. Robert (1978). *C. R. Acad. Sci. Paris*, sér. D. **287**, 1413–1415.

Banse, K. (1977). *Mar. Biol.* **41**, 199–212.

Banse, K., C. P. Falls and L. A. Hobson (1963). *Deep-Sea Res.* **10**, 639–642.

Barber, R. T. and J. H. Ryther (1969). *J. Exp. Mar. Biol. Ecol.* **3**, 191–199.

Barber, R. T., R. C. Dugdale, J. J. MacIsaac and R. L. Smith (1971). *Invest. Pesq.* **35**, 171–193.

Beardall, J., D. Mukerji, H. E. Glover and I. Morris (1976). *J. Phycol.* **12**, 409–417.

Berland, B. R., D. J. Bonin, R. A. Daumas, P. L. Laborde and S. Y. Maestrini (1970). *Mar. Biol.* **7**, 82–92.

Berland, B., D. Bonin, B. Coste, S. Maestrini and H. J. Minas (1973a). *Mar. Biol.* **23**, 267–274.

Berland, B. R., D. J. Bonin, S. Y. Maestrini and J. P. Pointier (1973b). *Int. Rev. Hydrobiol.* **58**, 203–220.

Berland, B. R., D. J. Bonin, S. Y. Maestrini and J. P. Pointier (1973c). *Int. Rev. Hydrobiol.* **58**, 401–416.

Berland, B. R., D. J. Bonin, S. Y. Maestrini and J. P. Pointier (1973d). *Int. Rev. Hydrobiol.* **58**, 473–500.

Berland, B. R., D. J. Bonin and S. Y. Maestrini (1974). *Ann. Inst. océanogr. (Paris)* **50**, 5–25.

Berland, B. R., D. J. Bonin and S. Y. Maestrini (1975). *Rapp. Sci. Techn. C.N.E.X.O.* **21**, 1–18.

Berland, B. R., D. J. Bonin and S. Y. Maestrini (1976). *Ann. Inst. océanogr. (Paris)* **52**, 45–55.
Berland, B. R., D. J. Bonin and S. Y. Maestrini (1978). *Int. Rev. Hydrobiol.* **63**, 501–531.
Berland, B. R., D. J. Bonin and S. Y. Maestrini (1980). *Oceanol. Acta* **3**, 135–141.
Bewers, J. M. and H. L. Windom (1982). *Mar. Chem.* **11**, 71–86.
Böhlen, P., S. Stein, W. Dairman and S. Udenfriend (1973). *Arch. Biochem. Biophys.* **155**, 213–220.
Bonin, D. J. and S. Y. Maestrini (1981). *Canad. Bull. Fish. Aquat. Sci.* **210**, 279–291.
Bonin, D. J., M. R. Droop and S. Y. Maestrini (in preparation). Physiological and Cultural Characteristics of Most-convenient Marine Algae for Bioassays.
Bonin, D. J., S. Y. Maestrini and J. W. Leftley (1981). *Canad. Bull. Fish. Aquat. Sci.* **210**, 310–322.
Borchardt, J. A. and H. S. Azad (1968). *J. Wat. Poll. Control Fed.* **40**, 1739–1754.
Brand, L. E., L. S. Murphy, R. R. L. Guillard and H. T. Lee (1981). *Mar. Biol.* **62**, 103–110.
Brewer, P. G. (1975). Minor elements in sea water. *In* "Chemical Oceanography" (J. P. Riley and G. Skirrow, eds), Vol. I, pp. 415–496. Academic Press, London and New York.
Broenkow, W. W. (1965)' *Limnol. Oceanogr.* **10**, 40–52.
Brown, L. M., B. T. Hargrave and M. D. MacKinnon (1981). *Canad. J. Fish. Aquat. Sci.* **38**, 205–214.
Bryan, G. W. (1976). Heavy metal contamination in the sea. *In* "Marine Pollution" (R. Johnston, ed.), pp. 185–302. Academic Press, London and New York.
Bumbu, Ya. V. (1976). In Russian. "Trace elements in the Life of Phytoplankton." 115 pp. Shiintsa, Kishinev, U.S.S.R.
Caperon, J. (1968). *Ecology* **49**, 866–872.
Carlucci, A. F. (1973a). Bioassay: biotin. *In* "Handbook of Phycological Methods. Culture Methods and Growth Measurements" (J. R. Stein, ed.), pp. 377–385. Cambridge University Press, Cambridge.
Carlucci, A. F. (1973b). Bioassay: cobalamin. *In* "Handbook of Phycological Methods. Culture Methods and Growth Measurements" (J. R. Stein, ed.), pp. 387–394. Cambridge University Press, Cambridge.
Castell, J. V., M. Cervera and R. Marco (1979). *Anal. Biochem.* **99**, 379–391.
Charpy-Roubaud, C. J., L. J. Charpy, S. Y. Maestrini and M. J. Pizarro (1978a). *C.R. Acad. Sci. Paris*, sér. D. **287**, 1031–1034.
Charpy-Roubaud, C. J., L. J. Charpy and S. Y. Maestrini (1978b). *C.R. Acad. Sci. Paris*, sér. D **287**, 539–542.
Charpy-Roubaud, C. J., L. J. Charpy and S. Y. Maestrini (1982). *Oceanol. Acta* **5**, 179–188.
Charpy-Roubaud, C. J., L. J. Charpy and S. Y. Maestrini (1983). *Mar. Ecol., P.S.Z.N.I. (Berlin)* **4**, 1–18.
Chiaudani, G. and M. Vighi (1978). *Quaderni dell' Istituto di Ricerca sulle Acque* **39**, 120 pp., Roma.
Collier, A., W. B. Wilson and M. Borkowski (1969). *J. Phycol.* **5**, 168–172.
Conover, S. A. M. (1956). *Bull. Bingham oceanogr. Coll.* **15**, 62–112.
Conway, H. L., P. J. Harrison and C. O. Davis (1976). *Mar. Biol.* **35**, 187–199.
Corredor, J. E. (1979). *Deep-Sea Res.* **26**, 731–741.
Culkin, F. (1965). The major constituents of sea water. *In* "Chemical Oceanography" (J. P. Riley and G. Skirrow, eds), Vol. I., pp. 121–161. Academic Press, London and New York.
Davies, A. G. (1976). *J. Mar. Biol. Ass. U.K.* **56**, 39–57.
De Valera, M. (1940). *Kgl. fysiogr. Sällsk. Lund Förh.* **10**, 52–58.

Dixon, P. S., J. Scherfig and C. A. Justice (1979). *Bull. Lake Sci.* **4**, 30–43.
Droop, M. R. (1955a). *J. Mar. Biol. Ass. U.K.* **34**, 233–245.
Droop, M. R. (1955b). *J. Mar. Biol. Ass. U.K.* **34**, 435–440.
Droop, M. R. (1957). *J. Gen. Microbiol.* **16**, 286–293.
Droop, M. R. (1961). *Bot. Mar.* **2**, 231–246.
Droop, M. R. (1966). *J. Mar. Biol. Ass. U.K.* **46**, 659–671.
Droop, M. R. (1968). *J. Mar. Biol. Ass. U.K.* **48**, 689–733.
Droop, M. R. (1973). *J. Phycol.* **9**, 264–272.
Droop, M. R. (1974). *J. Mar. Biol. Ass. U.K.* **54**, 825–855.
Droop, M. R. (1975). *J. Mar. Biol. Ass. U.K.* **55**, 541–555.
Dufour, P., J. L. Cremoux and M. Slepoukha (1981). *J. Exp. Mar. Biol. Ecol.* **51**, 247–267.
Dugdale, R. C. (1967). *Limnol. Oceanogr.* **12**, 685–695.
Dunstan, W. M. and K. R. Tenore (1974). *J. Appl. Ecol.* **11**, 529–536.
Eaton, A. D. and V. Grant (1979). *Limnol. Oceanogr.* **24**, 397–399.
Eide, I., A. Jensen and S. Melsom (1979). *J. Exp. Mar. Biol. Ecol.* **37**, 271–286.
Eppley, R. W. and E. H. Renger (1974). *J. Phycol.* **10**, 15–23.
Eppley, R. W. and J. D. H. Strickland (1968). Kinetics of marine phytoplankton growth. *In* "Advances in Microbiology of the Sea" (M. R. Droop and E. J. Ferguson Wood, eds), Vol. I., pp. 23–62. Academic Press, London and New York.
Eppley, R. W. and W. H. Thomas (1969). *J. Phycol.* **5**, 375–379.
Falkowski, P. G. (1977). *J. Exp. Mar. Biol. Ecol.* **27**, 37–45.
Falkowski, P. G. (1980). Light-shade adaptation in marine phytoplankton. *In* "Primary Productivity of the Sea" (P. G. Falkowski, ed.), Environmental Science Research Vol. 19, pp. 99–119. Plenum Press, New York.
Fiala, M., G. Cahet, G. Jacques, J. Neveux and M. Panouse (1976). *J. Exp. Mar. Biol. Ecol.* **24**, 151–163.
Filip, D. S. and E. J. Middlebrooks (1975). *Water Res.* **9**, 581–585.
Finney, D. J. (1952). "Statistical Method in Biological Assay." Charles Griffin, London, 611 pp.
Fitzgerald, G. P. (1975). "Factors Affecting the Algal Assay Procedure". Office of Research and Monitoring, Environmental Protection Agency, Washington, 31 pp.
Fitzgerald, G. P. and S. L. Faust (1967). *Limnol. Oceanogr.* **12**, 332–334.
Fleming, R. H. (1940). The Composition of Plankton and Units for Reporting Populations and Production. *Proc. 6th Pacific Science Congress* 1939, **3**, 535–540.
Fogg, G. E. (1965). "Algal Cultures and Phytoplankton Ecology." Univ. Wisconsin Press, Madison. 126 pp.
Ford, J. E., M. E. Gregory and E. S. Holdsworth (1955). *Biochem. J.* **61**, XXIII.
Forsberg, C., S. O. Ryding, A. Claesson and A. Forsberg (1978). *Mitt. Int. Ver. Limnol.* **21**, 352–363.
Fournier, R. O. (1966). *Chesapeake Sci.* **7**, 11–19.
Frey, B. E. and L. F. Small (1980). *J. Plankton Res.* **2**, 1–22.
Fuhs, G. W. (1969). *J. Phycol.* **5**, 312–321.
Gargas, E. and J. S. Pedersen (1974). "Algal Assay Procedure Batch Technique." Contr. Water Quality Institute. Danish Academy of Technical Science **1**, 48 pp.
Gavis, J., R. R. L. Guillard and B. L. Woodward (1981). *J. Mar. Res.* **39**, 315–333.
Gieskes, W. W. C. and G. W. Kraay (1975). *Neth. J. Sea Res.* **9**, 166–196.
Gieskes, W. W. C., G. W. Kraay and M. A. Baars (1979). *Neth. J. Sea Res.* **13**, 58–78.

Gilmartin, M. (1967). *Limnol. Oceanogr.* 12, 325–328.

Glooschenko, W. A. and H. Curl Jr (1971). *J. Fish. Res. Bd Can.* 28, 790–793.

Glover, H. E. (1978). *Limnol. Oceanogr.* 23, 534–537.

Goldberg, E. D. (1952). *Biol. Bull.* 102, 243–248.

Goldberg, E. D. (1963). The oceans as a chemical system. *In* "The Sea. Ideas and Observations on Progress in the Study of the Seas" (M. N. Hill, ed.), Vol. 2, pp. 3–25. John Wiley, New York.

Goldberg, E. D. (1965). Minor elements in sea-water. *In* "Chemical Oceanography" (J. P. Riley and G. Skirrow, eds), Vol. I, pp. 163–196. Academic Press, London and New York.

Goldman, C. R. and D. T. Mason (1962). *Science* 136, 1049–1050.

Gran, H. H. (1931). *Rapp. Cons. Explor. Mer* 75, 37–46.

Gran, H. H. (1933). *Biol. Bull.* 64, 159–181.

Granéli, E. (1978). *Vatten* 34, 117–128.

Granéli, E. (1981). *Kieler Meeresforsch.* 5, 82–90.

Grant, B. R., J. Madgwick and G. Dal Pont (1967). *Aust. J. Mar. Freshw. Res.* 18, 129–136.

Guillard, R. R. L. (1968). *J. Phycol.* 4, 59–64.

Guillard, R. R. L. and J. H. Ryther (1962). *Canad. J. Microbiol.* 8, 229–239.

Halldal, P. (1970). The photosynthetic apparatus of microalgae and its adaptation to environmental factors. *In* "Photobiology of Microorganisms" (P. Halldal, ed.), pp. 17–55. J. Wiley, London.

Hansen, A. L. and R. J. Robinson (1953). *J. Mar. Res.* 12, 31–42.

Harrison, P. J., H. L. Conway and R. C. Dugdale (1976). *Mar. Biol.* 35, 177–186.

Harrison, P. J., R. E. Waters and F. J. R. Taylor (1980). *J. Phycol.* 16, 28–35.

Haug, A., S. Myklestad and E. Sakshaug (1973). *J. Exp. Mar. Biol. Ecol.* 11, 15–26.

Hayward, J. (1968). *J. Mar. Biol. Ass. U.K.* 48, 295–302.

Hayward, J. (1969). *J. Mar. Biol. Ass. U.K.* 49, 439–446.

Healey, F. P. (1973a). *CRC Crit. Rev. Microbiol.* 3, 69–113.

Healey, F. P. (1973b). *J. Phycol.* 9, 383–394.

Healey, F. P. (1975). "Physiological Indicators of Nutrient Deficiency in Algae." Canada Fisheries and Marine Service Research and Development Directorate, Techn. Rep. No. 585, 30 pp.

Healey, F. P. (1978). *Mitt. int. Ver. Limnol.* 21, 34–41.

Healey, F. P. (1979). *J. Phycol.* 15, 289–299.

Hearn, M. T. W. and W. S. Hancock (1979). The role of ion-pair reversed phase HPLC in peptide and protein chemistry. *In* "Biological/Biomedical Applications of Liquid Chromatography" (G. L. Hawk, ed.), Chromatographic Science Series Vol. 12, pp. 243–271. Marcel Dekker, New York.

Heron, J. (1962). *Limnol. Oceanogr.* 7, 316–321.

Hitchcock, G. L. and T. J. Smayda (1977). *Limnol. Oceanogr.* 22, 132–139.

Hollaway, W. L., A. S. Bhown, J. E. Mole and J. C. Bennett (1979). HPLC in the structural studies of protein. *In* "Biological/Biomedical Applications of Liquid Chromatography" (G. L. Hawk, ed.), Chromatographic Science Series Vol. 10, pp. 163–176. Marcel Dekker, New York.

Holmes, R. W. (1966). *J. Phycol.* 2, 136–140.

Holm-Hansen, O. (1969). Environmental and Nutritional Requirements for Algae. Proc. Eutrophication-Biostimulation Assessment Workshop, Berkeley, 98–108.

Holm-Hansen, O. (1970). *Plant & Cell Physiol.* 11, 689–700.

Holm-Hansen, O. and C. R. Booth (1966). *Limnol. Oceanogr.* 11, 510–519.

Hood, D. and R. M. Pytkowicz (1974). Chemical oceanography. *In* "Handbook of Marine Science" (F. G. W. Smith, ed.), Vol. I, pp. 1–70. CRC Press, Cleveland.

Hulburt, E. M. (1979). *Mar. Sci. Comm.* 5, 245–268.

Huntsman, S. A. and W. G. Sunda (1980). The role of trace metals in regulating phytoplankton growth. *In* "The Physiological Ecology of Phytoplankton" (I. Morris, ed.), pp. 285–328. Blackwell, Oxford.

Ignatiades, L. and T. J. Smayda (1970). *J. Phycol.* 6, 357–364.

Jackson, G. A. and J. J. Morgan (1978). *Limnol. Oceanogr.* 23, 268–282.

Jacques, G., G. Cahet, M. Fiala and M. Panouse (1973). *J. Exp. Mar. Biol. Ecol.* 11, 287–295.

Jacques, G., M. Fiala, J. Neveux and M. Panouse (1976). *J. Exp. Mar. Biol. Ecol.* 24, 165–175.

Jenkins, D. (1967). *J. Wat. Poll. Control Fed.* 39, 159–180.

Jensen, A., B. Rystad and L. Skoglund (1972). *J. Exp. Mar. Biol. Ecol.* 8, 241–248.

Johnston, R. (1963). *J. Mar. Biol. Ass. U.K.* 43, 427–456.

Jones, G. E. (1967). *Limnol. Oceanogr.* 12, 165–167.

Jones, P. G. W. and C. P. Spencer (1963). *J. Mar. Biol. Ass. U.K.* 43, 251–273.

Jonge, V. N. de (1980). *Mar. Ecol. Prog. Ser.* 2, 345–353.

Karl, D. M. (1980). *Microbiol. Rev.* 44, 739–796.

Ketchum, B. H. (1939). *J. Cell. Comp. Physiol.* 13, 373–381.

Kochert, G. (1978). Protein determination by dye binding. *In* "Handbook of Phycological Methods. Physiological and Biochemical Methods" (J. A. Hellebust and J. S. Craigie, eds), pp. 96–97. Cambridge University Press, Cambridge.

Komarek, J. and O. Lhotský (1979). Review of algal assay strains. *In* "Algal Assays and Monitoring Eutrophication" (P. Marvan, S. Přibil and O. Lhotský, eds), pp. 103–118. E. Schweizerbart'sche, Stuttgart.

Komarek, J. and P. Marvan (1979). Selection and registration of strains of algae as assay organisms. *In* "Algal Assays and Monitoring Eutrophication" (P. Marvan, S. Přibil and O. Lhotský, eds), pp. 87–102. E. Schweizerbart'sche, Stuttgart.

Kossut, M. G. and S. Y. Maestrini (1977). *C.R. Acad. Sci. Paris*, sér. D 285, 393–396.

Krey, J. (1950). *Kieler Meeresforsch.* 7, 58–75.

Krey, J. (1964). *Kieler Meeresforsch.* 20, 18–29.

Kuenzler, E. J. and Ketchum, B. H. (1962). *Biol. Bull.* 123, 134–145.

Kuenzler, E. J. and J. P. Perras (1965). *Biol. Bull.* 128, 271–284.

Kuhl, A. (1974). Phosphorus. *In* "Algal physiology and biochemistry" (W. D. P. Stewart, ed.), Botanical Monographs Vol. 10, pp. 636–654. Blackwell, Oxford.

Kylin, A. (1941). *Kgl. fysiogr. Sällsk. Lund Förh.* 11, 217–232.

Lane, P. and R. Levins (1977). *Limnol. Oceanogr.* 22, 454–471.

Lännergren, C. (1978). *Int. Rev. Hydrobiol.* 63, 57–76.

Lean, D. R. S. and F. R. Pick (1981). *Limnol. Oceanogr.* 26, 1001–1019.

Leftley, J. W., D. J. Bonin and S. Y. Maestrini (1983). Problems in estimating marine phytoplankton growth, productivity and metabolic activity in nature: an overview of methodology. *Oceanogr. Mar. Biol. Ann. Rev.* 21, 23–66.

Lehman, J. T., D. B. Botkin and G. E. Likens (1975). *Limnol. Oceanogr.* 20, 343–364.

Leischman, A. A., J. C. Greene and W. E. Miller (1979). "Bibliography of Literature Pertaining to the Genus *Selenastrum.*" US EPA, Corvallis, Or. EPA-600/9-79-021, 191 pp.

Lewin, J. C. and C. H. Chen (1971). *Limnol. Oceanogr.* 16, 670–675.

Liebig, J. (1840). "Die Chemie in ihrer Anwendung auf Agrikultur und Physiologie." Taylor & Walton, London.

Lin, C. K. and C. L. Schelske (1981). *Canad. J. Fish. Aquat. Sci.* 38, 1–9.

Lindahl, P. E. B. and K. E. R. Melin (1973). *Oikos* 24, 171–178.

Lowry, H., N. J. Rosebrough, A. L. Farr and R. J. Randall (1951). *J. Biol. Chem.* 193, 265–275.

Lund, J. W. G. (1972). *Verh. int. Ver. Limnol.* 18, 71–77.

Macdonald, R. W. and F. A. McLaughlin (1982). *Water Res.* 16, 95–104.

Mackereth, F. J. (1953). *J. Exp. Bot.* 4, 296–313.

Maestrini, S. Y. and D. J. Bonin (1981). *Canad. Bull. Fish. Aquat. Sci.* 210, 264–278.

Maestrini, S. Y. and M. G. Kossut (1980). *J. Rech. Océanogr.* 5, 35–52.

Maestrini, S. Y. and M. G. Kossut (1981). *J. Exp. Mar. Biol. Ecol.* 50, 1–19.

Maestrini, S. Y. and J. M. Robert (1979). *J. Rech. Océanogr.* 4, 17–25.

Maestrini, S. Y. and J. M. Robert (1981). *Oceanol. Acta* 4, 13–21.

Malewicz, B. (1975). *Merentutkimuslait. Julk./Havsforskningsinst. Skr.* 239, 67–71.

Marker, A. F. H., E. A. Nusch, H. Rai and B. Riemann (1980). *Arch. Hydrobiol., Beih. Ergebn. Limnol.* 14, 91–106.

Marsot, P., R. Fournier and C. Blais (1981). *Canad. J. Fish. Aquat. Sci.* 38, 905–911.

Marvin, K. T., R. R. Proctor Jr and R. A. Neal (1972). *Limnol. Oceanogr.* 17, 777–784.

Matudaira, T. (1939). In Japanese. *Bull. Jap. Soc. Sci. Fish.* 8, 187–193.

McAllister, C. D., T. R. Parsons, K. Stephens and J. D. H. Strickland (1961). *Limnol. Oceanogr.* 6, 237–258.

McCarthy, J. J. (1972a). *J. Phycol.* 8, 216–222.

McCarthy, J. J. (1972b). *Limnol. Oceanogr.* 17, 738–748.

McCarthy, J. J. and J. C. Goldman (1979). *Science* 203, 670–672.

Melin, K. E. R. and P. E. B. Lindahl (1973). No. 7. *Oikos*, suppl. 15, 189–194.

Menzel, D. W. and J. H. Ryther (1960). *Deep-Sea Res.* 6, 351–367.

Menzel, D. W. and J. H. Ryther (1961). *Deep-Sea Res.* 7, 276–281.

Menzel, D. W. and J. P. Spaeth (1962). *Limnol. Oceanogr.* 7, 151–154.

Menzel, D. W., E. M. Hulburt and J. H. Ryther (1963). *Deep-Sea Res.* 10, 209–219.

Miller, W. E., J. C. Greene, E. A. Merwin and T. Shiroyama (1978). Algal bioassay techniques for pollution evaluation. *In* "Toxic Materials in the Aquatic Environment", pp. 9–16. Seminar Oregon State University, SEMINAR WR-024-78.

Monod, J. (1942). "Recherches sur la Croissance des Cultures Bactériennes." Hermann, Paris, 211 pp.

Monod, J. (1950). *Ann. Inst. Pasteur* 79, 390–410.

Morel, F. M. M., J. G. Rueter, D. M. Anderson and R. R. L. Guillard (1979). *J. Phycol.* 15, 135–141.

Morris, I. (1974). *Sci. Prog.* 61, 99–122.

Morris, I., C. M. Yentsch and C. S. Yentsch (1971). *Limnol. Oceanogr.* 16, 859–868.

Morris, I., J. Beardall and D. Mukerji (1978). *Mitt. int. Ver. Limnol.* 21, 174–183.

Murphy, L. S. (1978). *J. Phycol.* 14, 247–250.

Myklestad, S. (1974). *J. Exp. Mar. Biol. Ecol.* 15, 261–274.

Myklestad, S. and A. Haug (1972). *J. Exp. Mar. Biol. Ecol.* 9, 125–136.

Necas, J. (1979). Genetic variability and the resulting nonhomogeneity in algal populations.

In "Algal Assays and Monitoring Eutrophication" (P. Marvan, S. Přibil and O. Lhotský, eds), pp. 141–152. E. Schweizerbart'sche, Stuttgart.

Neess, J. C., R. C. Dugdale, V. A. Dugdale and J. J. Goering (1962). *Limnol. Oceanogr.* 7, 163–169.

North, B. B. (1975). *Limnol. Oceanogr.* 20, 20–27.

Novick, A. and L. Szilard (1950). *Science* 112, 715–716.

Officer, C. B. and J. H. Ryther (1980). *Mar. Ecol. Prog. Ser.* 3, 83–91.

O'Kelley, J. C. (1974). Inorganic nutrients. *In* "Algal Physiology and Biochemistry" (W. D. P. Stewart, ed.), pp. 610–635. Blackwell, Oxford.

Oswald, W. J. and C. G. Golueke (1966). *J. Wat. Poll. Control Fed.* 38, 964–975.

Parsons, T. R. (1973). Coulter Counter for phytoplankton. *In* "Handbook of Phycological Methods. Culture Methods and Growth Measurements" (J. R. Stein, ed.), pp. 345–358. Cambridge University Press, Cambridge.

Parsons, T. R., K. von Bröckel, P. Koeller, M. Takahashi, M. R. Reeve and O. Holm-Hansen (1977). *J. Exp. Mar. Biol. Ecol.* 26, 235–247.

Parsons, T. R., K. Stephens and M. Takahashi (1972). *Fish. Bull.* 70, 13–23.

Perkins, E. J. (1976). The evaluation of biological response by toxicity and water quality assessments. *In* "Marine Pollution" (R. Johnston, ed.), pp. 505–585. Academic Press, London and New York.

Perry, M. J. (1972). *Mar. Biol.* 15, 113–119.

Perry, M. J., M. C. Talbot and R. S. Alberte (1981). *Mar. Biol.* 62, 91–101.

Pojed, I. and S. Kveder (1977). *Rapp. Comm. Int. Mer Médit.* 24, 47–48.

Pratt, R. (1944). *Amer. J. Bot.* 31, 418–421.

Prézelin, B. B. (1981). *Canad. Bull. Fish. Aquat. Sci.* 210, 1–43.

Provasoli, L. (1963). Organic regulation of phytoplankton fertility. *In* "The Sea. Ideas and Observations on Progress in the Study of the Seas" (M. N. Hill, ed.), Vol. 2, pp. 165–219. J. Wiley, New York.

Provasoli, L. (1968). Media and prospects for the cultivation of marine algae. *In* "Cultures and Collection of Algae" (A. Watanabe and A. Hattori, eds), pp. 63–75. Proc. USA–Japan Conference, Hakone. Japanese Society of Plant Physiology.

Provasoli, L. and A. F. Carlucci (1974). Vitamins and growth regulators. *In* "Algal Physiology and Biochemistry" (W. D. P. Stewart, ed.), pp. 741–787. Blackwell, Oxford.

Provasoli, L., J. J. A. McLaughlin and M. R. Droop (1957). *Arch. Mikrobiol.* 25, 392–428.

Quarmby, L. M., D. H. Turpin and P. J. Harrison (1982). *J. Exp. Mar. Biol. Ecol.* 63, 173–181.

Rausch, T. (1981). *Hydrobiologia* 78, 237–251.

Redfield, A. C. (1934). On the proportions of organic derivatives in sea water and their relation to the composition of plankton. *In* "James Johnstone Memorial Volume" (R. J. Daniel, ed.), pp. 176–192. University Press, Liverpool.

Redfield, A. C. (1958). The inadequacy of experiment in marine biology. *In* "Perspectives in Marine Biology" (A. A. Buzzati-Traverso, ed.), pp. 17–26. Univ. Calif. Press, Berkeley, California.

Redfield, A. C. and A. B. Keys (1938). *Biol. Bull.* 74, 83–92.

Rhee, G. Y. (1972). *Limnol. Oceanogr.* 17, 505–514.

Rhee, G. Y. (1978). *Limnol. Oceanogr.* 23, 10–25.

Riley, G. A. (1938). *J. Mar. Res.* 1, 335–352.

Riley, J. P. (1965). Historical introduction. *In* "Chemical Oceanography" (J. P. Riley and G. Skirrow, eds), Vol. I, pp. 1–41. Academic Press, London and New York.

Rinne, I. and E. Tarkiainen (1975). *Merentutkimuslait. Julk./Havsforskningsinst. Skr.* **239**, 91–99.

Robarts, R. D. and G. C. Southall (1977). *Arch. Hydrobiol.* **79**, 1–35.

Robert, J. M., S. Y. Maestrini, M. Bagès, J. P. Dréno and E. Gonzalez-Rodriguez (1979). *Oceanol. Acta* **2**, 275–286.

Rosén, G. (1981). *Limnologica (Berlin)* **13**, 263–290.

Rueter, J. G., Jr and F. M. M. Morel (1981). *Limnol. Oceanogr.* **26**, 67–73.

Russel, F. S. (1939). *J. Mar. Biol. Ass. U.K.* **23**, 381–386.

Ryther, J. H. (1954). *Biol. Bull. (Lancaster)* **106**, 198–209.

Ryther, J. H. (1956). *Limnol. Oceanogr.* **1**, 72–84.

Ryther, J. H. and W. M. Dunstan (1971). *Science* **171**, 1008–1013.

Ryther, J. H. and R. R. L. Guillard (1959). *Deep-Sea Res.* **6**, 65–69.

Ryther, J. H. *et al.* (1975). *Aquaculture* **5**, 163–177.

Sakshaug, E. (1977). *J. Exp. Mar. Biol. Ecol.* **28**, 109–123.

Sakshaug, E. (1980). Problems in the methodology of studying phytoplankton. *In* "The Physiological Ecology of Phytoplankton" (I. Morris, ed.), Studies in Ecology, Vol. 7, pp. 57–91. Blackwell, Oxford.

Sakshaug, E. and O. Holm-Hansen (1977). *J. Exp. Mar. Biol. Ecol.* **29**, 1–34.

Sakshaug, E. and A. Jensen (1978). *Oceanogr. Mar. Biol. Ann. Rev.* **16**, 81–106.

Sakshaug, E. and S. Myklestad (1973). *J. Exp. Mar. Biol. Ecol.* **11**, 157–188.

Saldick, J. and Jadlocki, J. F., Jr (1978). *Mitt. Int. Ver. Limnol.* **21**, 50–56.

Sanders, J. G. (1978). *Mar. Environ. Res.* **1**, 59–66.

Schanz, F. (1977). *Hydrobiologia* **53**, 99–105.

Schanz, F. (1982). *Water Res.* **16**, 441–447.

Schreiber, E. (1927). *Wiss. Meeresunters. (Abt. Helgoland)* **16**, 1–34.

Schultz, J. S. and P. Gerhardt (1969). *Bact. Rev.* **33**, 1–47.

Scott, G. T. (1945). *J. Cell. Comp. Physiol.* **26**, 35–42.

Sellner, K. G., L. Lyons, E. S. Perry and D. B. Heimark (1982). *J. Phycol.* **18**, 142–148.

Sheldon, R. W. and T. R. Parsons (1967). "A Practical Manual on the Use of the Coulter Counter in Marine Science", 66 pp. Coulter Electronics Sales Company, Toronto.

Shiroyama, T., W. E. Miller and J. C. Greene (1976). Comparison of the algal growth responses of *Selenastrum capricornutum* Printz and *Anabaena flos-aquae* (Lyngb.) De Brébisson in waters collected from Shagawa Lake. *In* "Biostimulation and Nutrient Assessment" (E. J. Middlebrooks, D. H. Falkenborg and T. E. Maloney, eds), pp. 127–275. Proc. Workshop Utah State University, Logan, PRWG168-1.

Shuter, B. J. (1978). *Limnol. Oceanogr.* **23**, 1248–1255.

Simon, W., P. F. Sommer and G. H. Lyssy (1962). *Microchem. J.* **6**, 239–258.

Skipnes, O., I. Eide and A. Jensen (1980). *Appl. Environ. Microbiol.* **40**, 318–325.

Skirrow, G. (1975). The dissolved gases—carbon dioxide. *In* "Chemical Oceanography" (J. P. Riley and G. Skirrow, eds), Vol. II, pp. 1–192. Academic Press, London and New York.

Skulberg, O. M. (1964). Algal problems related to the eutrophication of European water supplies, and a bio-assay method to assess fertilizing influences of pollution on inland waters. *In* "Algae and Man" (D. F. Jackson, ed.), pp. 262–299. Plenum Press, New York.

Skulberg, O. M. (1966). Algal cultures as a means to assess the fertilizing influence of pollution. Proc. 3rd International Conference on water pollution research, sect. 1, paper 68, 1–15. Water Pollution Control Federation.

Skulberg, O. M. (1970). *Helgoländ. wiss. Meeresunters.* **20**, 111–125.

Smayda, T. J. (1964). Enrichment experiments using the marine centric diatom *Cyclotella nana* (clone 13-1) as an assay organism. *Proc. Symposium on marine ecology*, Graduate School of Oceanography, University of Rhode Island Occ. Publ. 2, 25–32.

Smayda, T. J. (1970). *Helgoländ. wiss. Meeresunters.* **20**, 172–194.

Smayda, T. J. (1971). Further enrichment experiments using the marine centric diatom *Cyclotella nana* (clone 13-1) as an assay organism. *In* "Fertility of the Sea" (J. D. Costlow, ed.), Vol. 2, pp. 493–511. Gordon and Breach, New York.

Smayda, T. J. (1973). *Norw. J. Bot.* **20**, 219–247.

Smayda, T. J. (1974). *Limnol. Oceanogr.* **19**, 889–901.

Smith, F. G. W. (1948). *Quart. J. Fla Acad. Sci.* **11**, 1–5.

Smith, F. G. W. and F. A. Kalber (1974). "Handbook of Marine Science" Vol. II, 390 pp. CRC Press, Cleveland.

Smith, W. O., Jr, R. T. Barber and S. A. Huntsman (1982). *J. Plankton Res.* **4**, 651–663.

Sorokin, C. (1973). Dry weight, packed cell volume and optical density. *In* "Handbook of Phycological Methods. Culture Methods and Growth Measurements" (J. R. Stein, ed.), pp. 321–343. Cambridge University Press, Cambridge.

Sournia, A. and J. Citeau (1972). *C.R. Acad. Sci. Paris*, sér. D **275**, 1299–1302.

Specht, D. T. (1975). Seasonal variation of algal biomass production potential and nutrient limitation in Yaquina Bay, Oregon. *In* "Biostimulation and Nutrient Assessment" (E. J. Middlebrooks, D. H. Falkenborg and T. E. Maloney, eds), pp. 149–174. Proc. Workshop Utah State University, Logan PRWG168-1.

Specht, D. T. and W. E. Miller (1974). Development of a standard marine algal assay procedure for nutrient assessment. Proc. Seminar on methodology for monitoring the marine environment. pp. 194–230 EPA-600/4-74-004.

Spencer, C. P. (1954). *J. Mar. Biol. Ass. U.K.* **33**, 265–290.

Spencer, C. P. (1975). The micronutrient elements. *In* "Chemical Oceanography" (J. P. Riley and G. Skirrow, eds), Vol. II, pp. 245–300. Academic Press, London and New York.

Steele, J. H. and I. E. Baird (1961). *Limnol. Oceanogr.* **6**, 68–78.

Steele, J. H. and D. W. Menzel (1962). *Deep-Sea Res.* **9**, 39–49.

Steemann Nielsen, E. (1952). *J. Cons. Explor. Mer* **18**, 117–140.

Steemann Nielsen, E. (1978a). *Mitt. int. Ver. Limnol.* **21**, 81–87.

Steemann Nielsen, E. (1978b). *Mar. Biol.* **46**, 185–189.

Stewart, W. D. P., G. P. Fitzgerald and R. H. Burris (1970). *Proc. Nat. Acad. Sci.* **66**, 1104–1111.

Stockner, J. G. and N. J. Antia (1976). *J. Fish. Res. Bd Can.* **33**, 2089–2096.

Stoermer, E. F., B. G. Ladewski and C. L. Schelske (1978). *Hydrobiologia* **57**, 249–265.

Strickland, J. D. H. and T. R. Parsons (1960). "A Manual of Sea Water Analysis" *Bull. Fish. Res. Bd Can.* **125**, 185 pp.

Strickland, J. D. H. and Parsons, T. R. (1972). "A Practical Handbook of Seawater Analysis" *Bull. Fish. Res. Bd Can.* **167**, 310 pp. (2nd edition).

Strickland, J. D. H. and L. D. B. Terhune (1961). *Limnol. Oceanogr.* **6**, 93–96.

Strickland, J. D. H., R. W. Eppley and B. R. De Mendiola (1969). *Bol. Inst. Mar Peru* **2**, 1–45.

Sunda, W. G., R. T. Barber and S. A. Huntsman (1981). *J. Mar. Res.* **39**, 567–586.

Sverdrup, H. U., M. W. Johnson and R. H. Fleming (1942). "The Oceans. Their Physics, Chemistry and General Biology, 1087 pp. Prentice Hall, Englewood Cliffs, N.J.

Syrett, P. J. (1953). *Ann. Bot. (London)* 17, 1–18.

Syrett, P. J. (1962). Nitrogen assimilation. *In* "Physiology and Biochemistry of Algae" (R. A. Lewin, ed.), pp. 171–188. Academic Press, New York and London.

Tarkiainen, E., I. Rinne and L. Niemistö (1974). *Merentutkimuslait. Julk./ Havsforskningsinst. Skr.* 238, 39–52.

Teixeira, C. and J. G. Tundisi (1981). *Bol. Inst. Oceanogr. S. Paulo* 30, 77–86.

Thayer, G. W. (1970). *Chesapeake Sci.* 11, 155–158.

Thomas, W. H. (1959). The culture of tropical oceanic phytoplankton. *In* "International Oceanographic Congress Preprints" (M. Sears, ed.), pp. 207–208. American Association for Advancement of Sciences, Washington.

Thomas, W. H. (1969). *J. Fish. Res. Bd Can.* 26, 1133–1145.

Thomas, W. H. (1979). *J. mar. Res.* 37, 327–335.

Thomas, W. H. and A. N. Dodson (1968). *Biol. Bull.* 134, 199–208.

Thomas, W. H., D. L. R. Seibert and A. N. Dodson (1974). *Estuar. Coast. Mar. Sci.* 2, 191–206.

Trainor, F. R. (1965). *Canad. J. Bot.* 43, 701–706.

Tranter, D. J. and B. S. Newell (1963). *Deep-Sea Res.* 10, 1–9.

Tunzi, M. G., M. Y. Chu and R. C. Bain Jr (1974). *Water Res.* 8, 623–635.

Turner, M. F. and M. R. Droop (1978). Culture media for algae. *In* "CRC Handbook Series in Nutrition and Food" (E. M. Rechcigl, ed.), Vol. 3, pp. 287–426. CRC Press, Cleveland.

Turpin, D. H. (1982). Interactions between photosynthesis and ammonium uptake in marine phytoplankton. Abstracts of papers for the 45th Meeting of The American Society of Limnology and Oceanography, Raleigh, pp. 68.

Udenfriend, S., S. Stein, P. Böhlen, W. Dairman, W. Leimgruber and M. Weigele (1972). *Science* 178, 871–872.

Utermöhl, H. (1958). *Mitt. int. Ver. Limnol.* 9, 1–138.

Vaccaro, R. F. (1965). Inorganic nitrogen in sea water. *In* "Chemical Oceanography" (J. P. Riley and G. Skirrow, eds), Vol. I, pp. 365–408. Academic Press, London and New York.

Van Slyke, D. D. and J. M. Neill (1924). *J. Biol. Chem.* 61, 523–573.

Vince, S. and I. Valiela (1973). *Mar. Biol.* 19, 69–73.

Welsch, M. (1947). "Le Dosage microbiologique des vitamines." 192 pp. Masson, Paris.

Wetzel, R. G. (1965). Nutritional aspects of algal productivity in marl lakes with particular reference to enrichment bioassays and their interpretation. *In* "Primary Productivity in Aquatic Environments." Proc. I.B.P. Symposium. *Mem. Ist. Ital. Idrobiol.* suppl. 18, 137–157.

Wheeler, P. A., B. B. North and G. C. Stephens (1974). *Limnol. Oceanogr.* 19, 249–259.

Whittaker, R. H., G. E. Likens and H. Lieth (1975). Scope and purpose of this volume. *In* "Primary Productivity of the Biosphere" (H. Lieth and R. H. Whittaker, eds), pp. 3–5. Springer, Berlin.

Wiessner, W. (1962). Inorganic micronutrients. *In* "Physiology and Biochemistry of Algae" (R. A. Lewin, ed.), pp. 267–286. Academic Press, New York and London.

Wilson, D. P. (1951). *J. Mar. Biol. Ass. U.K.* 30, 1–26.

Wilson, T. R. S. (1975). Salinity and the major elements of sea water. *In* "Chemical Oceanography" (J. P. Riley and G. Skirrow, eds), Vol. I, pp. 365–413. Academic Press, London and New York.

Wilson, W. B. (1966). *Fla Bd Conserv. Prof. Pap.* **7**, 1–42.

Yentsch, C. S. and D. W. Menzel (1963). *Deep-Sea Res.* **10**, 221–231.

Yoneshigue-Braga, Y., S. Y. Maestrini and E. Gonzalez-Rodriguez (1979). *C.R. Acad. Sci. Paris*, sér. D **288**, 135–138.

5 Test Algae as Indicators of Sea Water Quality: Prospects

Serge Y. Maestrini

*Centre de Recherche en Écologie
Marine et Aquaculture de
L'Houmeau, Case 5, 17137
Nieul-sur-Mer, France*

Michael R. Droop

*Dunstaffnage Marine Research Laboratory,
Oban, Argyll, Scotland*

Daniel J. Bonin

*Station Marine d'Endoume, Chemin de la Batterie des Lions, 13007
Marseille, France*

I. The Problems

Taking the field of marine bioassay as a whole, one is struck by the unsystematic attitude of the scientific community to the subject. First, both the principles and usefulness of bioassay have been contested (e.g. Holm-Hansen, 1969; Lane and Levins, 1977; Samuels *et al.*, 1979), and yet it is generally admitted that bioassays yield information that cannot be obtained in any other way (Berland *et al.*, 1972). Moreover, arguments accepted and valid for fresh waters (e.g. Wetzel, 1965; Fitzgerald, 1972; Maslin and Boles, 1978) apply equally well to the sea. Secondly, although the principle is old (Schreiber, 1927), it has until now only been put to use in a casual manner. Thirdly, the waters studied have been confined to a few circumscribed areas at privileged sites which are not necessarily typical of the rest of the world. Fourthly, the techniques employed have been very varied, in marked contrast to the general homogeneity of descriptive methods.

The three last points above reflect an immaturity in the approach to marine bioassays that has been absent from the freshwater field for the last decade and a half. Indeed the concept of a single principal limiting nutrient even now is only recognized by limnologists (Fuhs *et al.*, 1972). The development and use of standard method in the freshwater field has been in large part responsible for the general concern over eutrophication and accumulation of toxic wastes that has marked the last two decades. Political pressures have then led to the further

ALGAE AS ECOLOGICAL INDICATORS
ISBN 0 12 640620 0

simplification of methodology, with the result that the trophic status of any water can easily be assessed and predictions made of the probable dystrophic effects of any polluting influence. Bioassays are recognized as the only tool (e.g. Skulberg, 1964), and researches in methodology continue to multiply (see the reviews in Glass (1974), Middlebrooks *et al.* (1975), Skulberg (1978) and Marvan *et al.* (1979)). These led to the US Environmental Protection Agency's "Algal assay procedure: bottle test" (Anonymous, 1971), which, after improvement (Miller *et al.*, 1978) and a few minor adjustments, has become the universal method. That agency proposed a similar "Marine algal assay procedure: bottle test" (abbreviated MAAP; Anonymous, 1974), but to no avail: to our knowledge no one has subsequently used the method for sea water assays. This is regrettable for it was a step in the right direction. To redress this we propose a set of tested guidelines which we believe to be sufficiently well founded as to merit general adoption.

Neither current practice nor indeed the MAAP recommendations are altogether satisfactory. The batch methods recommend for estimating algal growth potential (AGP) (Oswald and Golueke, 1966) are in principle perfectly valid when used to assess fertility or even eutrophication due to pollution. They are based on biomass yield, which according to the Liebig law (1840) is solely proportional to concentration of the nutrient in shortest supply, so, there is not any real correlation between the growth rates of test algae calculated from short-term measurements and the concentrations of nutrients in the culture medium (Maloney *et al.*, 1972; Gerhart and Likens, 1975; O'Brien and Denoyelles, 1976). The logic of the yield method does not, as the MAAP recognized, extend to measuring the deleterious effects of pollution, since the primary action of any toxic compound is upon the rate of growth, effects upon yield being secondary and unpredictable. Accordingly, growth rate measurements (on batch cultures) were recommended. Rate measurements were also to be used for assessing the competitive potential of the members of natural populations in relation to the nutrient supply. Unfortunately batch cultures have very limited potential for growth rate measurement (for the reasons discussed in Section IIIA) and are not in our opinion generally to be recommended for this purpose. Either dialysis or continuous culture offers a better solution (see below).

II. Suggested Procedures for Marine Algal Bioassays

Many of the retained techniques from the MAAP report have already been discussed in detail in Chapter 4; others have not been included because they are dealt with fully elsewhere (e.g. Lund *et al.* (1971) on the preparation of glassware, Murray *et al.* (1971) and Chiaudani and Vighi (1978) on optimal irradiance for assays and Chiaudani and Vighi (1978) on optimal culture volumes and agitation). Otherwise we have followed the general protocols recommended by MAAP (Anonymous, 1974), by Gargas and Pedersen (1974) and by Chiaudani and Vighi (1978).

A. Cardinal Rules Regarding the Practice of Differential Enrichments

(*i*) The number of enrichments should not be fewer than 13 (which with the unenriched control, means 14 cultures) for each sea water sample. All have to be incubated in identical conditions favourable to algal growth.

(*ii*) The final concentrations of the nutrients in the enriched cultures should be of the same order as the highest concentrations observed in the water under study or at any rate very little higher. The complement of enrichments should embrace all nutrients essential to algal growth.

(*iii*) The increase in algal biomass should be followed with sufficient frequency to find the moment of maximum biomass.

(*iv*) Biomass should be estimated by a reliable measure.

(*v*) The volume of the test culture should be sufficient for wall effects to be negligible. Containers should neither liberate silicon nor toxic substances.

We suggest the procedure detailed below. It takes account of these rules and at the same time allows for the great variety of natural populations as well as for improvements effected by automation or by the increase in sensitivity of modern chemical and biochemical methods.

B. Suggested Procedure for Differential Enrichments Using Natural Populations

(*i*) Sea water is taken with non-metallic collecting bottles, and filtered immediately through a net of 300 μm mesh. If the assay cannot be performed at once the samples should be transferred to white polythene bottles made completely light-tight externally (or stored in a light-tight cabinet) and kept at a reduced temperature (4 –7 C).

(*ii*) The samples are apportioned into sterile Erlenmeyer flasks. These must be of transparent, autoclavable plastic (e.g. polycarbonate). After being sterilized ready for use the flasks have to be carefully acid-washed again, rinsed and resterilized. Detergents should not be used for washing. The best combination for the cultures is 500 ml in a litre flask, but failing this, 500 ml Erlenmeyers with 300 ml are possible.

(*iii*) Add the enrichment at the rate of 10 ml per 500 and then incubate the test cultures under identical conditions of temperature and light. One must obviously strive as far as possible to recreate natural conditions, always bearing in mind the tolerance of the populations under study. For example, in the laboratory phytoplankton from polar seas are cultivated at temperatures between 5 and 10°C, those from cold temperate seas between 10 and 15°C, those from the warm temperate seas between 15 and 20°C, and those from tropical waters are maintained at *c*. 20°C. A medium light intensity is used, *c*. 4000–5000 lux (i.e. about 80–100 $\mu E \cdot m^{-2}$ or 20–25 $W \cdot m^{-2}$), and is provided by fluorescent tubes designed for plant culture. When incubation is carried out in a pond or in a tank on deck, it is necessary to take great care never to expose the cultures to direct sunlight, by the use of neutral filters or metal screens. The most uniform illumination is obtained when each flask is surrounded by a black cylinder and is illuminated entirely from below. Complete uniformity of treatment of the cultures is not absolutely necessary if the objective is solely the estimation of maximum biomass produced by each culture.

(*iv*) Shake the cultures by hand twice a day.

(*v*) The time when growth stops must be determined as precisely as possible. For this it is necessary to take biomass readings at least daily: 50 ml from each culture for a chlorophyll *a* fluorescent measurement or ^{14}C uptake rate suffices. The duration and frequency of this monitoring depend on the richness of the water, notably respecting the nutrient that proves to be limiting, and on the activity of the phytoplankton. The experimenter must find the conditions which (in the majority of cases) do not prolong the incubation beyond 5 or 6 days. As soon as growth has stopped, as precise as possible a biomass estimation is carried out by several of the measurements normally used for this purpose (PON, proteins and cell volume; and, less acceptably, chlorophyll *a* and POC, and if it is desired to exploit the results to the limit, cells can be counted by the Utermöhl method), (an analogous estimation having been made just prior to setting up the cultures).

(*vi*) The results are presented diagrammatically as a series of vectors or sectors whose size is proportional to the maximum biomass reached by each culture. Growth rate is without significance in this method.

Note: Although replication of each test is theoretically a necessity, we believe that it is preferable to increase the number of different enrichments than to reduce them in favour of replication.

In addition to those listed for work with natural populations, three further rules apply to bioassays with cultured algae.

C. *Supplementary Rules to be Followed in Bioassay Work with Cultured Algae*

(*vii*) Planktonic organisms in the sample should be eliminated with as little other modification as possible.

(*viii*) The test species chosen should include at least one autotroph and one auxotroph, one capable of using a large number of organic substrates and one with a very narrow range, one capable of utilizing very low substrate concentrations and at least one ubiquitous and/or widely used strain and one species more endemic to the region.

(*ix*) The test algal inoculum must be in a similar physiological state to the natural population. For the most part this entails the absence of intracellular reserves, which would allow several cell divisions to take place independently of the nutrients in the water.

Taking account of these nine rules, we propose the following procedure.

D. *Suggested Procedure for Differential Enrichments Using Cultured Algae*

(*i*) Sea water is taken with non-metallic collecting bottles and filtered immediately through a washed 0·45 μm membrane. If the assay cannot be performed at once, the samples are stored in polythene vessels at −20°C.

(*ii*) The samples are apportioned to sterile 250 ml flasks, *c.* 100 ml to each flask. These should be of transparent plastic for diatoms, or of glass for unsilicified algae.

(*iii*) The enrichment mixtures are added aseptically at the rate of 5 ml per 100 ml; the concentrations at the end of the experiment should be of the same order as the mean of the waters being assayed.

(*iv*) Inoculations are made with 1·0 ml of stock culture on each test alga, suitably depleted of internal reserves but none the less completely viable, and of a convenient cell density, which, however, must be the same for all the cultures. At least three test species must be used.

(*v*) The cultures are incubated in conditions of temperature and light suitable for the test algae.

(*vi*) Algal biomass is monitored at least once a day by cell count, taking no more than 5·0 ml from the cultures.

(*vii*) When there is no more growth, as accurate as possible an estimation of biomass is made by several methods, population volume (Coulter counter), PON, proteins, etc., as above.

(*viii*) As before, the results are presented as vector or sector diagrams. Growth rate is also without significance in this method.

E. *Remarks*

This procedure differs in ten main points from that recommended by the US Environmental Protection Agency (Table 1), the remaining Agency recommendations being adopted in their entirety. The procedure does not allow examination of a large number of sea water samples. This is counter to current practice, but in our opinion it is a step forward, because a more profound knowledge of natural processes controlling primary producers is to be gained from the evidence of a small number of carefully chosen situations than by the accumulation of a mass of unconnected data. The proposed guidelines are self-consistent and probably represent the best overall solution, at least considering the analytical methods at our disposal, but in no way do we consider them perfect, nor indeed the only possible, for *there is no unique rational solution to the problems, only a series of compromises* between often conflicting constraints. For example, water is stored in polythene vessels because silicon dissolved from the walls of glass brings with it the problems of phosphorous adsorption by the plastic (Jenkins, 1967). Again, the choice of pore size for sample filtration presents problems: one wants to retain all organisms, including bacteria, while at the same time pass all nutrients including any adsorbed

Table 1 Differences in detail between the procedures for differential enrichment assays with cultured algal strains recommended respectively by ourselves and the US Environmental Protection Agency (Anonymous, 1974).

Methodological aspect	Maestrini et al. (this review)	US Environmental Protection Agency (Anonymous, 1974)
Removal of indigenous algae	Filtration	Filtration or autoclaving
Nutrients involved in nutrient mixtures	All nutrients needed for algal growth, except when inhibiting	Limited to the few which are likely to be limiting ($=$Si is not involved)
Final nutrient concentrations	Equal or just above the maximum in situ levels	High levels (e.g. 300 µg at-N·l$^-$1, 6·2 µg at-P·l^{-1})
Nutrient ratios in nutrient mixtures	Balanced levels (e.g. N/P $=$ 15)	Unbalanced levels (e.g. N/P $=$ 48)
Total number of enriching mixtures	Minimum number $=$ 13	Not clearly stated, but anyhow low
Recommended test-algal strains (listed according to the preference order)	At least 3 of the following species *Thalassiosira pseudonana* *Dunaliella tertiolecta* *Emiliania huxleyi* *Pavlova lutheri* *Phaeodactylum tricornutum* One or more endemic species	*Dunaliella tertiolecta* *Thalassiosira pseudonana*
Preparation of test cells prior to inoculation	Nutrient depletion	Cell washing
Volume of test cultures	100 ml to a minimum	40, 60 or 100 ml
Growth estimation	Maximum standing crop	Maximum standing crop and/or specific growth rate
Biomass estimation parameters	Cell density, total cell volume, chlorophyll-*a*, POC, PON and protein contents; use several parameters at a time	Dry weight (gravimetric measurement), cell density, chlorophyll-*a* and POC

on seston. Experiments indicated that the best compromise is a pore size of 0·8 μm (see Fig. 5, in Chapter 4) and this we would have adopted were it not that there is evidence that a large fraction of the bacteria pass 0·8 μm (Azam and Hodson, 1977; Chretiennot-Dinet and Vacelet, 1978) but are retained by 0·45 μm (Derenbach and Williams, 1974). Moreover, because 0·45 μm is the pore size conventionally adopted by oceanographers for separating "dissolved" from "particular" matter, we decided to adopt this size for pre-filtration of samples. The choice of biomass measurement is another problem without a really satisfactory solution. The methods proposed (see Chapter 4) have the merit of being sensitive and precise, but they do dictate the use of comparatively large culture volumes and consequently the processing of fewer samples. The number of treatments can only be increased at the expense of culture volume, so much so that the best way of following biomass increase would be the possibility of fluorometric measurement through the culture vessel without removal of the alga (Brand *et al.*, 1981).

Our guide-lines therefore do not relieve the researcher of his obligations to select his best compromise from many often conflicting requirements. We have only been able to draw attention to the main problems and discuss possible solutions, summarizing our conclusions in a set of provisional suggestions.

One should add that it is not always necessary to assay for all nutrients. Indeed the number diminishes as knowledge of an area increases, since those that are found always to be in plenty cease to be of interest. In the long run the batch assay could cease to be a useful approach in the few well-worked areas. On the other hand, as we have pointed out, the batch assay can only indicate AGP, the maximum algal biomass allowed by the nutrient status of the water, and identify the primary controlling nutrient. Any further information requires assay methods based on specific growth rate.

Growth rate-based assays are particularly suited to identifying the alga with the greatest potential for dominance in a particular situation—an important consideration in the field of aquaculture. Another area of application is in the assay of toxic materials and pollution, the measurement in this case being the depression in the growth rate of an assay organism. The remaining sections briefly discuss methods for growth rate based assays.

III. Estimation of Growth Rate by Dialysis Culture

A. Specific Growth Rate and the Limitations of Batch Cultures

Specific growth rate of any material (μ) is its fractional rate of increase. Thus

$$\mu = \frac{1}{x}\frac{dx}{dt}, \text{ or } \frac{1}{n}\frac{dn}{dt}, \text{ or e.g. } \frac{1}{N_c}\frac{dN_c}{dt} \tag{1}$$

according to whether the material considered is biomass (x), cell number (n) or e.g. cell nitrogen (N_c). Only when cells are in a true steady state of growth will all the possible measures of μ be the same. *This condition is never attained in practical batch cultures* because the uptake of the various cell materials does not in general have time to become mutually coupled before depletion sets in (Droop, 1975). Total biomass is accepted as the best basis for the expression of specific growth rate, but the coupling between biomass and cell numbers is usually sufficiently good for the latter to be used. Its use is however not without pitfalls since the coupling is sometimes sensitive to nutrients and other conditions (e.g. Burmaster, 1979; Droop, 1979; Eppley, 1980; Redalje and Laws, 1981).

The measurement of μ in a batch culture requires at least two determinations of biomass separated by time intervals and thus can only properly be made if μ is constant over the period required to measure it, namely during the so-called log phase of the culture. The specific growth rate is then given by

$$\mu = \frac{\ln[x_{t_2}/x_{t_1}]}{t} \text{ or } \frac{2\cdot3 \log [x_{t_2}/x_{t_1}]}{t} \tag{2}$$

When cell number is used as a measure of biomass, growth rate is often expressed as number of cell division per unit time, which in an organism multiplying by binary division is greater than μ by a factor $1/\ln2$, or $1\cdot44$.

If, for the sake of argument, we adopt Monod kinetics, the relation between specific growth rate and concentration (S) of a rate-controlling substrate is

$$\mu = \frac{\mu_{r} S}{K_s + S} \tag{3}$$

In *early* batch culture the substrate concentration S would have an initial value S_R, while by the time growth ceased due to nutrient exhaustion one would find the relation

$$x = Y_x S_R \tag{4}$$

Y_x being the (maximum) yield coefficient. Thus according to this model, in a batch culture either μ or x appears as a function of S_R according to the phase of the culture. Equation (4) is of course the basis of the AGP bioassay.

Unfortunately the rate-controlling range of nutrient concentrations indicated by the value of K_s for growth is usually at the limit of or beyond practical measurement. Moreover, unless the inoculum is vanishingly small, the demand of the population is such that cultures starting with nutrient concentrations within this range will not exhibit sufficient growth for the slope of the growth curve to be ascertained, whereas if the initial concentration is outside that range the growth rate will be maximal in any case (Droop, 1961).

Furthermore, initial very rapid rates of nutrient uptake, the so-called luxury consumption (which is not confined to phosphorus) (Kuenzler and Ketchum, 1962; Droop, 1975), and consequent depletion of the medium before growth has even started ensure that *in a practical batch culture there can be no guarantee of constant observable relation between S and μ*, and that Monod kinetics cannot be applied. The conclusion is inevitable that generally in a practical batch culture the only rate that can be meaningfully measured is $μ_{,.}$, the maximum specific growth rate under the physical conditions prevailing.

These limitations do not apply to dialysis and continuous cultures, both of which can be operated in the physiological concentration range. The technique of dialysis (or cage) culture has been developed and described by Jensen and co-workers (Jensen *et al.*, 1972; Jensen and Rystad, 1973; Sakshaug and Jensen, 1978).

B. Measurement of μ in the Physiological Range of Substrate Concentration by Dialysis Culture

A dialysis culture is essentially a batch fermentor/continuous reservoir system. The principle has been fully described by Schultz and Gerhardt (1969) and need only be outlined here.

(*i*) The rate of molecular diffusion across a membrane is expressed as

$$Dρ = P_{,.}A_{,.}Δ_s$$

Where $Dρ$ is the rate of diffusion (dimensions: mass·t^{-1}); P_m, the permeability coefficient of the membrane (dimensions: length·t^{-1}) and is a function of the pore area/membrane area ratio and of pore length; $A_{,.}$, the membrane surface area (dimension: length^2); $Δs$, the concentration difference across the membrane (dimensions: moles·vol^{-1}). Other symbols used in the treatment following are: x, biomass (dimensions: mass·vol^{-1}); Y_x, yield coefficient; V_i, reactor volume; M, specific excretion rate (t^{-1}); U_L, specific rate of uptake:

$$(1/x)(d\frac{x}{Y_x}/dt)(\text{dimensions: } t^{-1})$$

(*ii*) Typically dialysis cultures pass through three phases—exponential, linear and asymptotic—the first of which is of most concern to us.

(a) *Exponential phase.* So long as both Mx and U_Lx are small compared with D/V_i, $μ$ will be a function (according to Monod kinetics) merely of $μ_{max}$ and S and growth will be exponential. During this phase, provided $V_i/A_{,.}$ is not too large, the diffusion rate is sufficient to maintain the internal and external concentrations virtually identical at all times.

(b) *Linear phase.* When biomass has increased to such an extent that $(1/Y_x)U_Lx$ nearly equals $Dρ/V_i$ we have

$$\frac{dx}{dt} = Y_x\frac{Dρ}{V_i} - Mx \qquad (6)$$

Thus growth rate dx/dt is virtually linear while both the specific rates of uptake (U_L) and growth $(1/x)(dx/dt)$ decrease exponentially.

(c) *Asymptotic phase*. This sets in when biomass has increased to an extent that Mx approaches the value of $Y_x(D\rho/V_i)$. By this time carbon limitation is likely to have taken over, so that equation (6) will now be in terms of carbon/light rather than the previously limiting nutrient.

Dialysis culture has the advantage over simple batch culture in that the duration of the exponential phase, during which time specific growth rate is controlled by the concentration of the limiting nutrient, is greatly increased, owing to the fact that the rate of diffusion is large compared to the rate of utilization and the external reservoir is inexhaustible. Moreover, many of the algal excretory products pass through the membranes and do not accumulate as they do in simple batch cultures. Quite high cell densities can be achieved before the onset of the linear phase. Jensen and Rystad (1973), who first applied the dialysis system to algae, report $6 \cdot 5 \times 10^9$ cell·1^{-1} for *Phaeodactylum tricornutum*, 6×10^9 cell·1^{-1} for *Thalassiosira pseudonana* and $2 \cdot 5 \times 10^9$ cell·1^{-1} for *Skeletonema costatum* in water containing no more than $0 \cdot 2$ µg at·1^{-1} NO_3-N.

C. Practical Employment of Dialysis Cultures

There are two main ways of using dialysis culture: (1) with simple dialysis sacks suspended in their entirety in the water under study or (2) with special apparatus in which the water is either pumped or through the membrane. Category (1) can be used in the field while category (2) is only suitable for laboratory use.

1. Simple sacks immersed in their entirety   These can be used in one of three ways (Fig. 1): simple immersion in the sea, immersion in a tank through which sea water is pumped, and immersion in an isolated tank. Simple immersion has the advantage of simplicity but the sacks are vulnerable to destruction by wave action and moreover are not easily accessible. Land-based tanks and pumping avoid these drawbacks and are consequently the most popular (Jensen *et al.*, 1972; Prakash *et al.*, 1973; Sakshaug, 1977; Yoder, 1979). Tanks could be shipboard operated, although this has not yet been attempted. Owens *et al.* (1977) immersed sacks in an isolated tank of limited volume without renewing the sea water. This can provide some useful information: for example, by suspending several sacks with different species in the same tank and manipulating the nutrient composition of the water, one could learn something about the competitive ability of the several cultured species.* In general, the volume of an isolated tank should be large in comparison with the total volume of dialysis sacks. Moreover low cell densities have to be used, which raises the problem of a precise numeration.

Regenerated cellulose dialysis tubing, usually 16 mm diameter, seldom greater, forms a suitable sack. This diameter allows one to obtain a sufficiently large surface/volume ratio. Mechanical agitation as described by Jensen *et al.* (1972) is employed by most authors to increase diffusion (Jensen and Rystad, 1973; Prakash *et al.*, 1973; Jensen and Rystad, 1974; Eide *et al.*, 1979, Yoder, 1979). Jensen's

* See Note Added in Proof (p. 188).

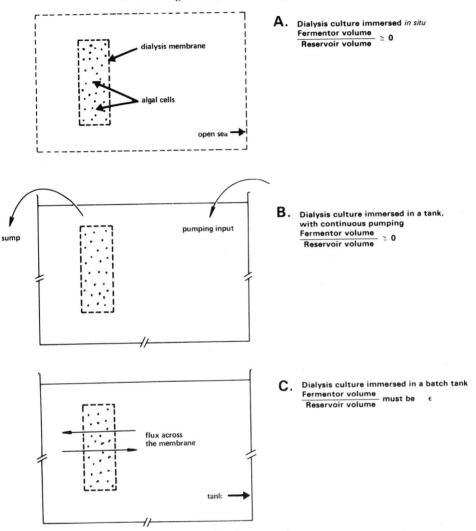

A. Dialysis culture immersed *in situ*
$$\frac{\text{Fermentor volume}}{\text{Reservoir volume}} \cong 0$$

B. Dialysis culture immersed in a tank, with continuous pumping
$$\frac{\text{Fermentor volume}}{\text{Reservoir volume}} \cong 0$$

C. Dialysis culture immersed in a batch tank
$$\frac{\text{Fermentor volume}}{\text{Reservoir volume}} \text{ must be } \epsilon$$

Fig. 1   Possible uses of dialysis culturing with simple sacks immersed in their entirety.

method (see Chapter 4) used both circular and up and down motion; glass beads enclosed in the sack agitate the culture. More complex dialysis chambers have been used by some authors, for instance those of Owens *et al.* (1977) which consist of a 16 ml plexiglass chamber with parallel 5–9 mm windows covered with millipore membranes of various porosity. Originally used for *in situ* bacterial cultures by McFeters and Stuart (1972), it was reported to be entirely suitable for cultures of such planktonic algae as the dinoflagellates *Gymnodinium nelsoni* and *Prorocentrum mariæ-lebouriae.* There is the great practical benefit of direct access to the reactor interior by hypodermic needles mounted in the plexiglass support. By contrast, dialysis tubing requires to be lifted and opened, which is difficult during the course of a culture especially with *in situ* experiments. As yet there is no entirely satisfactory answer to this technical problem. More research is needed, since it is essential to follow biomass during the course of growth. McFeters and Stuart's solution is attractive but does not in our view permit sufficient agitation within the culture and is also inapplicable *in situ.* Any solution for *in situ* "cage cultures" will have to overcome the difficulties encountered by Maestrini and Kossut (1980). In any case it is unlikely that a single recommendation applicable both to tank and *in situ* cultures could ever be made, as the problems are different. With tank cultures the main problem is agitation; with *in situ* cultures it is accessability. For sampling the latter Baskett and Lulves (1974) suggest the use of vertically orientated dialysis tubes each terminated at the top by glass tubing held above water by a foam-plastic float and capped by a glass stopper.

2. *Methods not involving the immersion of the whole culture*   In these systems the culture vessel is traversed by a membrane, on one side of which are the growing algae and on the other the circulating water under study. This system can be just as efficient as fully immersed sacks. Jensen *et al.* (1972) recommend an upper culture chamber of 150 ml and lower circulation chamber of 500 ml receiving filtered sea water. Dodson and Thomas (1977) used a perforated 850 ml chamber surrounded by a silk gauze; this was bathed in a 3-litre reservoir through which sea water circulated, the whole surrounded by a temperature-controlled water jacket. Marsot *et al.* (1981) have devised the first equipment separating the growth reactor from the dialysis walls; the haemodialyser units (hollow dialysis fibres) they used provide a large surface/volume ratio and a large algal biomass growing in a 10-litre chamber. With the exception of the apparatus of Skoglund and Jensen (1976), which corresponds less to a dialysis type of culture to two batch cultures separated by a membrane, these are the only examples of method not involving immersion of the whole culture. That of Skipnes *et al.* (1980) is discussed in Section IVD.

   The apparatus is delicate and expensive and can only be used in the laboratory. Moreover it is open to all the usual criticisms directed at all purely laboratory methods, notably regarding temperature, light and turbulence, all of which can affect nutrient absorption (see Bonin *et al.,* 1981) and which are variable in nature but constant in the laboratory. *In situ* cage culture is the only method of estimating growth rate and at the same time taking account of all the natural variability. On that count therefore, *in vitro* dialysis culture would appear to be a backward step, and in our opinion, effort should be directed towards *in situ* cage cultures and to eliminating the technical difficulties of that method.

D. *Application and Limitations of Cage Cultures*

1. Reproduction of natural conditions   Growth rate values measured by dialysis cultures are as near to the "natural" ones as it is possible to achieve. The qualification is important on account of three natural conditions that are either reduced or eliminated in the cages; turbulence of the water surrounding the cells, mobility of the cells, and patchiness. Although these factors are of lesser importance than temperaure, light and nutrient concentration, they are by no means negligible. Thus Munk and Riley (1952) showed that turbulence surrounding the cells increased absorption of nutrients, which, according to Pasciak and Gavis (1974, 1975), is due to the increased diffusion coefficient across the cell membrane. Cell mobility has the same net effect in this respect as turbulence. Although the small-scale effects of movement and turbulence can be well reproduced in dialysis sacks by efficient agitation and inclusion of glass beads, larger-scale turbulence and active migration of algal populations (e.g. Gran, 1929; Eppley *et al.*, 1968; Eppley and Harrison, 1975; Staker and Bruno, 1980) cannot be reproduced. For this reason results obtained with cage cultures cannot be applied to algae such as dinoflagellates capable of large-scale vertical migration. Non-motile algae are less affected, although those whose buoyancy alters with nutrient status should be included (Kahn and Swift, 1978). Keeping the algae at a defined depth in the water column creates an artificial situation that prevents the profitable interactions generated by vertical mixing and fast algal photosynthetic adaptation (Nelson and Conway, 1979; Falkowski, 1980). The effect of patchiness *vis à vis* the cages is less easy to visualize. Water appearing homogeneous on the evidence of widely used hydrological methods is in fact patchy respecting nutrients. The extent of the patches can be measured in km (Armstrong *et al.*, 1967), in m (Strickland, 1968) or in mm (Shanks and Trent, 1979). Goldman *et al.* (1979) suggested that because diatoms such as *Thalassiosira pseudonana* can obtain sufficient nutrients for a single cell division from a high nutrient pulse of duration only 1/30 of a generation time, they could maintain a much higher rate of division by virtue of their encountering plumes of zooplankton excretion than could be supported by the mean nutrient concentration of the water. Similarly Turpin *et al.* (1981) found that lower saturation constants are measured if nutrients are pulsed. Williams and Muir (1981) however, have proved the plume hypothesis to be untenable on microhydrodynamic grounds, so that the absence of zooplankton from experimental cages will not on that count bias results.

Nevertheless it would be rash to consider even *in situ* cage cultures to be totally representative of the natural situation of a growing population of the same species. Apart from anything else the pore diameter of the membrane determines which size of molecules or particles can freely pass into the cage. The membrane should therefore be chosen with great care.

2. Properties of the dialysis membrane The regenerated cellulose used for dialysis membranes has pores of the order 25–50 nm, which allows through molecules up to about 12 000. All mineral ions and organic nutrient molecules pass through easily, as do indeed many, but by no means all, algal excretory products and "ectocrines". Among those held back to the possible detriment of dialysis cultures are the vitamin-B_{12}-binding proteins (Droop, 1968). Likewise all sestonic material is held back together with the nutrients inevitably adsorbed on it. Add to all this the fact that the membranes are impervious to bacterial cells and once more we see a conflict of requirements, for we need to expose the cultures to *all* the natural chemical and physical actions of the environment (Lee, 1973; Goldman, 1978) with the single exception of grazing. Increasing the pore size is only a partial answer, for it introduces other problems. Prakash *et al.* (1973), Skoglund and Jensen (1976) and Owens *et al.* (1977) used 1·0 μm membranes and Dodson and Thomas (1977) a silk gauze of 7 μm mesh; hence the term "cage cultures" coined by Sakshaug and Jensen (1978) to embrace barriers of all porosities. The chief drawback of materials of larger porosity is that, being in sheet form, they require support structures, which until now (McFeters and Stuart (1972) excepted) have not been immersible and are therefore not suited to *in situ* cage cultures.

3. Summary of requirements

(*i*) Cage culture walls should be as transparent as possible. Sakshaug and Jensen (1978) suggest three possible materials: regenerated cellulose dialysis tubing (absorbing 3·7% of incident irradiance), Nuclepore 0·40 μm filters (absorbing 9·8%) and nylon gauze (10 μm Nytal, absorbing 12%). All other possible materials are more opaque, ranging from 27 to 64% absorption.

(*ii*) The walls must be able to withstand the exigencies of immersion in the sea. There is little to go on here as most experiments have been carried out on land. Our experience is that dialysis tubing does not stand up to a rough sea. Given suitable supports Nuclepore membranes and especially nylon gauze should be able to weather many days immersion.

(*iii*) The material should be heat-sterilizable, strong enough to withstand brushing to remove epiphytic algae, and resistant to bacterial action. All the materials mentioned in (*i*) above are autoclavable and can be brushed, but cellulose tubing will only withstand bacterial action for a few days (Vargo *et al.*, 1975). Nylon and Nucleopore are more resistant.

(*iv*) Pore diameter should be as large as possible, but according to Sakshaug and Jensen (1978) not greater than a quarter of the cell diameter of the cultured alga, and in any case not greater than 10 μm, having regard for the size of the smallest herbivores. Having made the choice of pore size below this limit, one has still to decide whether the algal cells are to be influenced solely by dissolved compounds or by a combination of dissolved and particulate, whether living or inert, and in particular whether bacteria are to be included or excluded.

4. Some possible solutions

(*i*) Algal populations in contact solely with dissolved nutrients. Since in this case all particulates, including bacteria are to be excluded, the algal source culture must be axenic and the culture vessel (e.g. dialysis tube) must be autoclaved and the medium sterile. Aseptic handling techniques must be employed, which raises the question of accessibility for frequent sampling. The method of access is of prime importance but has not been given very much attention. A simple knot, as most frequently used in the past, only allows one sampling at the end of the incubation (e.g. Jensen *et al.*, 1972, fig. 2; Maestrini and Kossut, 1981). Eide *et al.* (1979) closed the tubes with a rubber penicillin cap, which allowed frequent sampling with a hypodermic syringe, but the method is not very convenient, especially when practised at sea.

Dialysis cage cultures require continuous agitation and frequent cleaning and are only practical in sheltered sites such as fjords (Eide *et al.*, 1979) or tanks simulating natural conditions (Yoder, 1979).

Dialysis culture does not suit all algae. Sakshaug and Jensen (1978) note that species that do not undergo cell enlargement (e.g. *Phaeodactylum tricornutum* and *Thalassiosira pseudonana*) thrive well in 1·6 to 2·6 cm diameter tubing, while those that do (e.g. *Chaetoceros*) will not grow at all in this type of culture. Filamentous species, such as *Oscillatoria agardhi*, will only grow in large diameter tubing. Yet others, dinoflagellates in particular (e.g. *Gonyaulax spinifera* and *Prorocentrum micans*), do not tolerate the movement of double agitation.

Nucleopore membranes require a rigid holder-cum-culture chamber, similar to that developed by McFeters and Stuart (1972) but in Pyrex glass and sterilizable. Furthermore, the dialysing faces on such a chamber would require careful placing in order that the agitating movement affected both faces efficiently; otherwise severe fouling would be a problem. A porosity of 0·45 μm is suitable for excluding bacteria (see Section IIE).

(*ii*) Algal populations in contact with all elements of the environment other than herbivores. The large range of sizes of possible subjects for this type of culture permits an equal range of pore sizes. Nucleopore membranes can be selected from the range 1 to 8 μm to match most algae (adopting the convention that the pore size should be one quarter that of the alga). Larger species such as *Coscinodiscus* could be cultured with 10 to 20 μm membranes.

From the point of view of strength nylon gauze is the most suitable material for membranes in this range, and the same support chambers are suitable for nylon as for polycarbonate membranes. Owens *et al.* (1977) used McFeters and Stuart's (1972) for both. One could imagine better support designs, without for instance the disadvantage of the flattened chamber which has to be orientated with its principal face towards the light to avoid shading by the frame (a difficult operation in view of the requirement for agitation and the equal treatment of all membranes). A perforated spherical glass vessel covered with nylon gauze might be a solution were it not that the sphere has the least favourable surface to volume ratio.

Mesh sizes of the finest nylon gauzes quoted by the manufacturer bear little relation to the maximum porosity. Owens et al. (1977) for instance found that 100% of small flagellates passed a '1 μm' nylon mesh while only 1% passed a Nucleopore filter of the same specification. This should be taken account of when selecting a suitable mesh size to use.

Algal species might in principle be cultivated and studied in situ. But it is necessary to recognize that for instance a small species with a membrane of small porosity is in a different situation from a large species with a membrane of equal relative porosity. The small species e.g. Pavlova lutheri with a 1 or 2 μm membrane exists in virtual isolation, whereas a larger species e.g. Thalassiosira rotula, with an 8 or 10 μm membrane, coexists with all organisms below this size. In the extreme case, e.g. Coscinodiscus concinnus (160–200 μm) the pores would be large enough to admit zooplankton predators, which could modify the coexisting flora. Against this one might argue that in the circumstances the mobility of dissolved molecules is such that it is of little account whether the coexisting organisms are inside or outside the chamber. All the same, we should never forget that the growth rate we are interested in must reflect the organism's ability to perform in the face of its natural competitors, and to exclude the latter might favour the subject unnaturally. Certainly no one screen will be at once equally suitable for all species at all times and moreover much might be learned by observing the differences produced.

5. Measuring growth rate in polluted water   Jensen and associates (Jensen and Rystad, 1974; Jensen et al., 1976; Eide et al., 1979; Jensen, 1980) have demonstrated the advantage cage culture has over batch for monitoring the effects of and detecting toxic materials, including heavy metals. In batch culture in effect metal ions are added only once and the effective concentration is modified by uncontrollable factors, namely adsorption to the vessel walls, precipitation during autoclaving and above all sequestration by chelating materials in the medium. Being pH-sensitive, all these effects (in a bicarbonate-buffered medium such as sea water) are influenced by the CO_2 status of the water and hence eventually by the algae themselves. It is virtually impossible to know what the active concentration is at any time. Eide et al. (1979) suggest 30% inactivation. With the single addition the progress of surviving algae are subject to less stress than their parents who will have removed much of the metal from the medium. In effect a batch culture can only predict the result of a massive pulse of toxic materials to the environment and can give no indication of the harm that a continuous small input might have.

Cage cultures, on the other hand, are well adapted for observing the effects of small continuous additions, and the results are sometimes surprising. Thus Jensen and Rystad (1974) and Jensen et al. (1976) found that the diatoms Phaeodactylum tricornutum, Skeletonema costatum and Thalassiosira pseudonana, in dialysis cultures suspended in tanks with constant concentrations of copper or zinc, accumulated these ions in quantity from concentrations so low as to have no effect on the division rate. Following this up, Eide *et al.* (1979) immersed dialysis sacks *in situ* at varying distances from a source of pollution and confirmed the accumulation and presumed transmission up the food chain of toxic ions from concentrations too low to affect cell division.

It is, however, by the effect on cell division that the action of heavy metal ions is most easily detected. The sensitivity of the cage culture makes it a most valuable environmental tool, especially as chronic rather than acute pollution is the rule and largely goes undetected until too late.

The major drawback to the use of cage cultures is the length of the response time, which is in effect the time needed for an accurate growth rate measurement (11 to 15 days according to Jensen and associates). This greatly reduces the utility of the method when the objective is monitoring with a view to prevention of, for example, an agreed threshold of pollution. That would need a response time of days or even hours rather than weeks, which brings us to continuous culture with its potential for a much shorter response time.

IV. Continuous Culture

A. Principle

The essential of a continuous culture is a constant volume reactor with inlet and outlet through which culture medium is caused to pass at a controlled rate. The rate of cell increase due to growth is μx, while the rate that cells wash out is $(f/V)x$ (f being the flow rate and V the vessel volume). f/V is known as the dilution rate (symbol D). Thus the change in biomass is given by

$$\frac{dx}{dt} = \mu x - Dx \qquad (7)$$

When μ and D are equal, there is a steady state and the biomass in the reactor remains constant. To see how this is achieved it is necessary to take account of the rate of nutrient supply and its effect on growth rate. A reasonable assumption is that for a given condition (e.g. growth rate) the rates of biomass increase (dx) and nutrient consumption $(-dS)$ are in constant proportion (Monod, 1942). The yield coefficient Y_x is thus given by

$$Y_x = -\frac{dx}{dt} \cdot \frac{dt}{dS} = -\frac{dx}{dS} \qquad (8)$$

Now, the rate of change in the concentration of the controlling nutrient in the reactor is the difference between the rate of controlling nutrient output and utilization. Therefore, in the steady state

$$\frac{dS}{dt} = 0 = DS_R - DS - \frac{\mu x}{Y_x} \tag{9}$$

(S_R being the concentration in the inflow, S that in the outflow). Note that the specific rate of nutrient uptake is μ/Y_x.

It can be shown formally (Spicer, 1955) that provided (1) the dilution rate does not exceed the maximum specific growth rate of the organism and (2) the derivative of the function relating to μ to S, i.e. $d\mu/dS$, is positive, that a system fully described by equations (7) and (9) tends to a stable equilibrium (steady state) with μ and D equal and x and S constant.

In the steady state equation (9) becomes

$$x = Y_x(S_R - S) \tag{10}$$

and, adopting Monod (1942) kinetics (equation (3)), substituting D for μ and rearranging,

$$S = \frac{K_s D}{\mu_m - D} \tag{11}$$

Thus by substitution of equation (11) into (10)

$$x = Y_x(S_R - \frac{K_s D}{\mu_m - D}) \tag{12}$$

This equation relates the steady-state biomass to dilution rate and it is easily seen that at washout (zero biomass) the controlling nutrient concentration is equal to the input concentration and that the specific growth rate at washout is less than the maximum specific growth rate by a factor $S_R/(K_s + S_R)$.

Since μ in a continuous culture depends on S and not S_R, it follows that to observe the growth rate potential of any medium or water sample the apparatus would need to be operated as near to the washout point (i.e. with as low a biomass) as possible.

From equation (12) it follows that control of a continuous culture can be effected either by controlling the dilution rate (chemostat) or by controlling the steady-state biomass (turbidostat). Chemostat operation merely requires an adjustable metering pump in the inflow, whereas the turbidostat requires a biomass-monitoring device to control the metering pump. Although the mathematical description of the steady state is independent of the method of control, the efficacy of control is not. The chemostat works better than the turbidostat at low dilution rates, whereas at high dilution rates the turbidostat is the better.

The control device in an algal turbidostat is a photoelectric device monitoring the optical density of the culture whose output is used to switch on the pump in the medium flow line when the optical density falls below a predetermined value.

The literature on microbial continuous culture is voluminous: reviews of the theory and microbial application will be found in Tempest (1970), Evans *et al.*

(1970) and Munson (1970); for algal application see, for example, Myers and Clark (1944), Phillips and Myers (1954), Fogg *et al.* (1959), Maddux and Jones (1964), Droop (1966) and Rhee (1980); accounts of the control systems for algal turbidostats will be found in Maddux and Jones (1964), Droop (1975), Premazzi *et al.* (1978) and Skipnes *et al.* (1980).

B. *Response Time*

Of primary importance in any bioassay work is the response time of the assay, for in the final analysis it determines the cost. In batch assay a yield or growth rate measurement requires several days, and the same applies to the simple dialysis assays (Jensen *et al.*, 1972; Jensen and Rystad, 1973; Saskshaug and Jensen, 1978; Maestrini and Kossut, 1981). However, in those applications in which a photosynthesis measurement is acceptable, the response time may be reduced to hours for natural assemblages and to minutes for cultured populations.

The response time of continuous culture assays should also be much shorter than that of other batch growth rate assay methods (including dialysis). In 1974 one of the authors (MRD) was asked to report to the ICES Plankton Committee on the feasibility of using continuous culture for shipboard water quality assay. Time did not allow any practical evidence to be presented, but enough was known about the nutrient kinetics of one possible assay organism (*Pavlova lutheri*) for a realistic simulation study to be undertaken. The results were clear and unequivocal in favour of the turbidostat over all other forms of growth assay. We summarize them below.

The simulations were designed to compare the sensitivity and response times obtained with turbidostat and chemostat control with those of batch assays. Nutrient (vitamin B_{12}) concentration was taken as example of water quality and the Cell Quota model (Droop, 1974, 1975) was used for the simulations, using the measured parameters. The model is summarized in Fig. 2. A chemostat differs from a turbidostat merely in the point where control is exerted. In the chemostat the dilution rate is set and the biomass equilibrates, while in the turbidostat biomass is

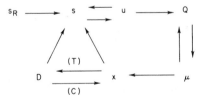

Fig. 2 Cell quota model applied to continuous culture. Key: S_R, Input substrate concentration; S, output substrate concentration; U, rate of uptake per unit biomass; Q, cell quota ($\equiv 1/Y_x$), cell nutrient per unit biomass; μ, specific growth rate; x, biomass per unit volume; D, dilution rate; (*T*) Turbidostat control; (*C*), Chemostat control.
The diagram is to be interpreted, *e.g.* as $S = F(S_R, U, x, D)$. For details of the functions (*F*) and associated parameters see Droop (1975).

set and the growth rate equilibrates. Thus, in a chemostat bioassay the dilution rate would be set at a convenient value and the biomass would be monitored; in a turbidostat assay the apparatus would be set to maintain a convenient biomass and the dilution rate would be monitored.

Both chemostat and turbidostat simulations were run. The range of substrate levels used (2·0 and 0·02 fM·ml^{-1} of vitamin B$_{12}$) were realistic for coastal waters. The effect of step increases or decreases in concentration varying in magnitude and duration was observed. Figure 3 shows examples of chemostat and Fig. 4 of turbidostat performance. It is clear that the turbidostat is vastly superior to the chemostat in response time, especially when the change is in the upward direction (Table 2). Figure 5 relates the size of the maximum turbidostat response to that of a step pulse of 4 day duration. The curve is S shaped, but the change in dilution rate varies as the log of the change in input over a 30 fold range of input concentrations. The slope around the origin is 0·67, i.e. a change in dilution range of 0·08 (20%) would be produced by a 1·3 fold (30%) change in input concentration.

It is clear from these simulations that the turbidostat has considerable potential for on-line bioassay of "water quality"; the response is sensitive and fast. Probably undefined water quality assay would be the immediate application, but by the use of appropriate enrichments, the method could be used to assay concentrations of a particular nutrient. This however would involve a knowledge of the parameters of the Monod equation (equation 3) for the nutrient in question. Alternatively, since in the steady state the input concentration is the sum of the output concentration and

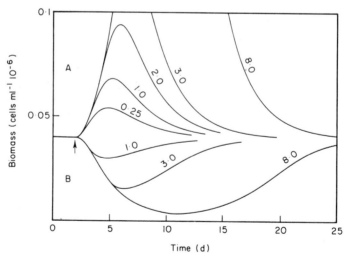

Fig. 3 Chemostat response to tenfold step-pulses in nutrient level of duration as indicated (days). Dilution rate setting, 0·5 vol d^{-1}.
A, upward step-pulses, 0·2 to 2·0 fM vitamin B$_{12}$ ml^{-1}.
B, downward step-pulses, 0·2 to 0·02 fM vitamin B$_{12}$ ml^{-1}.

Table 2 Time taken for a 20% response to a tenfold substrate change, in chemostat and turbidostat.

		Time (days)
Chemostat	{ upward step	1·5–3
	{ downward step	1·3–1·5
Turbidostat	{ upward step	0·2
	{ downward step	0·5–0·6

cell concentrations, while the output concentration is likely to be vanishingly small, it follows that the former is approximately the product of the cell quota $(1/Y)$ and biomass

$$S_R \approx \frac{x}{Y_x} \qquad (13)$$

The method would of course involve a cell quota measurement (x/Y_x).

The other great advantage the turbidostat has over the chemostat is in automation, for its output is easily linked to a chart recorder, which is not the case with the chemostat. The drawback is the extra complexity of the apparatus.

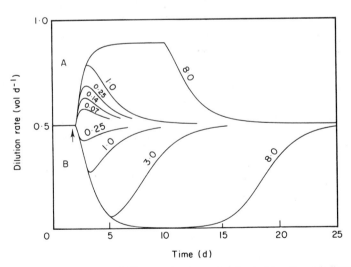

Fig. 4 Turbidostat response to tenfold step-pulses in nutrient level of duration as indicated (days). Biomass setting $0·04 \times 10^6$ cells ml^{-1}.
A, upward step-pulses, 0·2 to 2·0 fM vitamin B$_{12}$ ml^{-1}.
B, downward step-pulses, 0·2 to 0·02 fM vitamin B$_{12}$ ml^{-1}.

S. Y. Maestrini et al.

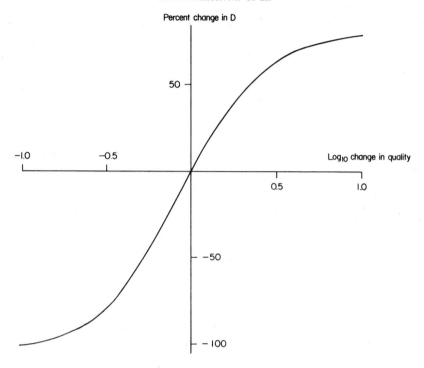

Fig. 5   Maximum response to a step-pulse of duration $4/\mu_{m}$ days as a function of the magnitude of the step. Initial dilution rate, $\mu_{m}/2$.

C. *Application of Continuous Culture to Bioassay*

Barlow and colleagues were probably the first to apply continuous methods to bioassay problems (Barlow *et al.*, 1973, 1975). They used filtered lake or pond water on the natural populations and compared the population changes with those taking place in turbidostats and chemostats. Chemostats have been employed for laboratory bioassays by Steyn *et al.* (1974), Reynolds *et al.* (1975) and Kayser (1976). Trotter and Hendricks (1976) and Lin (1977) describe flow systems for bioassay with sessile algae. A turbidostat was used by Premazzi *et al.* (1978) for laboratory toxicity assays. The paper contains a useful description of the control system.

D. *The Cage Culture Turbidostat*

Skipnes *et al.* (1980) have described an ingenious cage culture turbidostat. The additional complexity of this apparatus over the simple turbidostat is well compensated for by the fact that it can utilize unprocessed medium in the inflow.

This is without doubt potentially the most important development in this field and therefore warrants description.

The apparatus is represented diagrammatically in Fig. 6. A cage (of volume V_i) containing the culture is formed by two membranes and culture medium is pumped through the cage while culture is removed (pumped) from it via a port to maintain constant biomass (turbidostat operation).

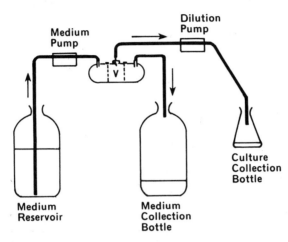

Fig. 6 Cage culture turbidostat, diagrammatic. By permission from Skipnes *et al.* (1980).

In the apparatus the equivalent of equation (7) in the steady state is

$$\frac{dx}{dt} = 0 = x(\mu - W) \qquad (14)$$

(WV_i being the flow rate out of the culture port). The conservation equation (9) however is applicable to the cage (DV_i being the flow rate into the cage).

When Michaelis kinetics for nutrient uptake are assumed

$$\frac{\mu}{Y_x} = \frac{U_m S}{K_m + S} \qquad (15)$$

and substituted into equation (9), a rather complex second-degree expression is obtained and it is difficult to visualize the general relation between μ/Y_x and S_R, D and x. However three special cases afford easy solution and are instructive:

(*i*) When the biomass setting is so large that $S \ll K_m$. Then equation (15) becomes (on rearrangement)

$$S \approx \frac{\mu}{Y_x} \cdot \frac{K_m}{U_m} \qquad (16)$$

S. Y. Maestrini et al.

which, when substituted into equation (9), gives

$$\frac{\mu}{Y_x} \approx \frac{D\,S_R}{x+(DK_m/U_m)} \approx \frac{DS_R}{x}$$

(since $x \gg DK_m/U_m$). In other words, uptake at a given biomass setting is dependent on the product of dilution rate and input concentration. This is the special case envisaged by Skipnes *et al.* (1980), being equivalent to the linear phase of a normal dialysis culture.

(*ii*) At the other extreme, when the biomass setting approaches zero and $S \approx S_R$. Then equation (15) becomes

$$\frac{\mu}{Y_x} \approx \frac{U_m\,S_R}{K_m+S_R} \tag{17}$$

and uptake is independent of the dilution rate (D) and depends only on the input concentration.

(*iii*) When $S_R \ll K_m$. Then simply

$$\frac{\mu}{Y_x} \approx \frac{U_m}{K_m}S_R \tag{18}$$

and there will be no alternative but to work with a very low biomass setting.

As in any continuous culture, since uptake (and hence specific growth rate) always depends directly and solely on S and not S_R, and since the bioassay interest lies in S_R, not S, the cage should be run with as low a biomass as the apparatus will allow (the constraint being the sensitivity of the biomass-monitoring device). Furthermore, since it is physically impossible for WV_i to be greater than DV_i there is a lower safe limit to the permitted flow, namely $\mu_m V_i$ (or $Y_x U_m V_i$). On the other hand, a practical upper limit to the flow will depend on the porosity of the membranes used. Skipnes *et al.* (1980) suggest membranes with 2 to 3 μm pore diameter, and the reader is referred to the original paper for practical details. Their work was carried out with the marine diatoms *Thalassiosira pseudonana*, *Skeletonema costatum* and *Phaeodactylum tricornutum*.

V. Appendix Table

Main technical details of batch bioassays used with the aim of determining the magnitude (= "fertility") and the nature of nutrient limiting the "Algal Growth Potential" of sea water. All usually cited papers are listed, but papers reporting pollution bioassay are not involved here.

(*a*) Estu = estuarine, oc = oceanic, upw = upwelling.

(*b*) Storage duration is: d = day, m = month.

(*c*) Elements as trace metals, vitamins and chelating substances, for whom concentrations used are not given here but were used at balanced levels with regard to N, P and Si; + = element added, but at unpublished concentration.

(*d*) Number of mixtures denotes number of different enrichments added to aliquots taken from the same seawater sample. Such a "mixture" may contain only one element; unenriched control is not included; duplicated chemical forms of an element (e.g. NO_3, NH_4, urea) or different concentrations are not considered here.

(*e*) Condition of growth denotes: deck RV = deck research vessel—test cultures are immersed in sea water (tank), receive natural light and are equilibrated with *in situ* temperature by cooling or heating with running-pumped sea water; tank = *idem*, except the tank is on dry land; incub = incubator; cooling with running sea water or fan; lab = laboratory culture room facilities; *in situ* = test cultures, in glass bottle or dialysis sack, are immersed *in situ*.

(*f*) Initial biomass of test cultures is usually given as number of cells per litre, except for few cases herein mentioned by Chl-*a* = chlorophyll *a* content or dw = dry weight.

(*g*) Duration denotes period to the start of biomass measurements; it may differ between different samples of sea water belonging to the same study or between different aliquots of a sample enriched with different mixtures.

(*h*) ^{14}C = rate of carbon uptake; ^{14}C-*x* h = rate of carbon uptake measurement with an *x* hours incubation period; O_2 = rate of photosynthetic oxygen production; POC = particulate organic carbon, PON = particulate organic nitrogen; Chl-*a* = chlorophyll *a*; ATP = adenosine triphosphate; cell density refers either to maximum cell number or division rate.

(*i*) Value computed or read by the reviewers from author's figures or tables.

(*j*) Two of the mixtures were glycine (10 μg-at-N·l^{-1}) and zooplankton homogenate freshly prepared; EDTA was added at 27 μM·l^{-1}.

(*k*) Two chelating substances were added: EDTA and Deferriferrioxamine B, a ferric-specific high-stability chelator available under the brand name of Desferal, from Ciba Pharmaco Co.; Fe was added at 1 μg-at·l^{-1}; mixtures were: + EDTA, + Fe + metal mix, + EDTA + Fe + metal mix; enrichments were made with charcoal-treated or untreated sea water.

(*l*) Cells were not depleted in poor sea water, but for preconditioning were grown in an aliquot of the sea water to be tested.

(*m*) Counts were made every 48 hours on each test culture until the stationary phase was reached; therefore growth duration was variable, but the range is not indicated.

(*n*) Filtration was not carried out prior storage deep-frozen but after thawing; filter used was a glass fibre No. 984 H Ultra from Reeves and Angel Co.

(*o*) The parent culture was preconditioned by culturing in post-bloom water containing 4·5 μg-at-N·l^{-1}, 0·41 μg-at-P·l^{-1} and 6·0 μg-at-Si·l^{-1}.

(*p*) Phytoplankton was collected at one station, then filtered through a 100 μm mesh and centrifuged; 10 mg dry weight $\cdot l^{-1}$ was suspended in enriched test samples.

(*q*) Authors wrote: "The quantity of algal culture inoculated was $0.014\ \mu l \cdot l^{-1}$"; since a regular drop from a disposable sterile Pasteur pipette has usually a volume of 50 μl, a mistake had probably occurred during printing or the parent culture was diluted and this operation was not reported.

(*r*) Tests cells were grown *in situ* in dialysis sacks without any nutrient enrichment; author's statement is based on relative cell component concentrations.

$+$ = operation made, but usual information not given; $-$ = information neither clearly given by the author nor found by the reviewers.

No	Reference	Seawater sample			Filtration (μm)	Storage		Enrichments						Total number of mixtures used[d]
		Origin						Elements or substances involved (μg-at·l^{-1})[c]						
		Area	Nature[a]	Depth (m)		t°C	Duration[b]	N	P	Si	M	Chel	Vit	
1	Agius and Jaccarini (1982)	Marsaxlokk Bay, Malta	Ner	0	250	None	None	2–32	0·3–4·8	0	0	0	0	3
2	Aïd et al. (1981)	Bay of Alger, Algeria	Ner	5	200	None	None	10	1	10	Fe-EDTA	←	B_1, B_{12} Biotin	9
3a	Anonymous (1974)	Any origin			0·45	4	—	Δ	Δ	Δ	B, Co, Cu, Fe, Mn, Mo, Zn	EDTA	B_1, B_{12} Biotin	Δ
3b	Anonymous (1974)	Yakina Bay, Oregon	Ner	0	0·45	—	—	71	1·7	0	0	0	0	2
4	Bages et al. (1978), Robert et al. (1979)	Oyster pond; Bay of Bourgneuf, Atlantic French coast	Ponds ner	0	1·2	−20	0–14 m	0	0	0	0	0	0	0
5	Barber and Ryther (1969)	Cromwell current	Oc upw	0–500	None	None	None	50	5	50	Fe-EDTA	←	B_1, B_{12}	8[j]
6	Barber et al. (1971)	Peru upwelling	Oc upw	0–75	None	None	None	0	0	0	Fe	EDTA[k]	0	3[k]
7	Berland et al. (1973a)	Northwestern Mediterranean Sea	Oc	0	GF/C 0·22	−20	—	0	0	0	0	0	0	0

No	Enriched test samples		Test algae		Growth				Remarks
	Volume (ml)	No. of replicata	Species (eventually strain)	Depletion	Condition[e]	Starting[f] cell density (10^6 cell·l^{-1}) or biomass	Duration[g]	Estimation[h]	
1	20 000	2	Nat. pop.	None	Lab	0·5 µg·l^{-1}	7 d	Cell density Chl-a; NO_3 and PO_4 disappearance	N and P were added singly at 3 different levels and in a mixture at 3 other levels. Daily estimation of biomass
2	10 000	1	Nat. pop.	None	Lab	0·1–0·4 µg·l^{-1} Chl-a	—	Cell density Chl-a; ^{14}C-3 h	Cell count for each species; nutrient consumption also recorded
3a	40, 60 or 100	3	Dunaliella tertiolecta / Thalassiosira pseudonana	Washed twice	Lab	0·1	12–20 d	Dry weight from cell density	Describes a proposed standard method: MAAP-bottle test. An example of utilization is given, but main field data are in Specht and Miller (1974), Specht (1975) and Baumgartner et al. (1977). Δ = design dictated by the actual situation
3b	—	3	D. tertiolecta / T. pseudonana	Washed twice	Lab	0·1	—	Dry weight from cell density	
4	150	1	Navicula ostrearia / Phaeodactylum tricornutum / Skeletonema costatum	2 d in charcoal treated poor sea water	Lab	1·5 / 5 / 5	7–10 d / 8 d / 6 d	Chl-a; cell density; nutrient uptake	Measurement of AGP. Sea water sampled over 14 months bioassayed all at a time. Daily counting for cell test cultures of Navicula ostrearia and 10 cultures (over 90) of P. tricornutum and S. costatum
5	1000	—	Nat. pop.	None	Incub	—	72 h	^{14}C	^{14}C measurements are made at 12 h intervals; glycine and zooplankton extract also involved in mixtures added
6	500	1 / –	Nat. pop.	None	Deck RV	—	7 d	^{14}C	4 or 5 ^{14}C analyses during the 7 d period
7	15	3	Chaetoceros lauderi / Chlamydomonas palla (=C. magnusii) / P. tricornutum	2–5 d in charcoal treated poor sea water	Lab	1 / 15 / 15	27 d	^{14}C-4 h; cell density	Measurements were made on days 2, 4, 7, 11, 19 and 27

No	Reference	Seawater sample						Enrichments						Total number of mixtures used[d]
		Origin		Nature[a]	Depth (m)	Filtration (μm)	Storage	Elements or substances involved ($\mu g\text{-}at \cdot l^{-1}$)[c]						
		Area					$t^{\circ}C$	Duration[b]	N	P	Si	M	Chel	Vit	
8	Berland et al. (1973c)	Rhône outlet area, Villefranche Bay		Ner-estu Ner	0–75	0·45	−25	—	588[i]	29[i]	528[i]	B, Co, Cu, Fe, Mn, Zn	EDTA	B₁, B₁₂	11
9	Berland et al. (1974)	Northwestern Mediterranean Sea		Oc	0	0·45	−18	—	588[i]	29[i]	528[i]	B, Co, Cu, Fe, Mn, Zn	EDTA	B₁, B₁₂ Biotin	7 or 10
10	Berland et al. (1978)	Calanque d'En-Vau, Marseilles, Mediterranean Sea		Ner	0	0·45	None		49	4·6	57	Fe: 1·4	0	B₁, B₁₂ Biotin	16
11	Charpy-Roubaud et al. (1978a, 1982)	Gulf of San Jose, Gulf Nuevo, Camarones, Patagonia coast, Argentina		Ner	0 5 10	0·45	−20	0–5 m	0	0	0	0	0	0	0
12a	Charpy-Roubaud et al. (1978b, 1982)	Gulf of San Jose, Patagonia coast, Argentina		Ner	0 5 10	None	None		25	2·3	26	B, Co, Cu, Fe, Mn, Mo, Zn	EDTA	B₁, B₁₂ Biotin	10
12b	Charpy-Roubaud et al. (1978b, 1982)	Gulf of San Jose, Patagonia coast, Argentina		Ner	0 5 10	0·45	None		25	2·3	26	B, Co, Cu, Fe, Mn, Mo, Zn	EDTA	B₁, B₁₂ Biotin	10

No	Enriched test samples		Test algae			Growth			Remarks
	Volume (ml)	No. of replicata	Species (eventually strain)	Depletion	Condition[e]	Starting cell density (10^6 cell·l^{-1}) or biomass[f]	Duration[g]	Estimation[h]	
8	10	2	C. lauderi P. tricornutum S. costatum	2–7 d in poor sea water	Lab	1·5 15 22	7 d 7 d 5 d	Cell density	0·22 μm filtration after thawing; depletion duration previously studied (Berland et al., 1973b) Algae are local isolates
9	10	2	C. lauderi P. tricornutum	2–7 d in poor sea water	Lab	0·6 & 1·8 15 & 17	10 d	Cell density	Samples treated in two series; 0·22 μm filtration after thawing. Algae are local isolates
10	33	2	Nat. pop. Asterionella japonica C. lauderi Cylindrotheca closterium Hemiselmis virescens Pavlova lutheri Pavlova pinguis P. tricornutum Prorocentrum micans S. costatum T. rotula	in situ, in dialysis sacks (15 d)	Lab	–[i] 0·8 0·4 1·0 4·0 0·4 0·4 2·0 0·2 1·0 0·4	1 and 2 d	^{14}C-24 h and 48 h	Natural phytoplankton and cultured test cells were enclosed in dialysis sacks and left in situ for two weeks; enrichments were done after this period. Starting cell densities of enriched test cultures are given in Berland et al. (1976). Algae are local isolates
11	25	2	C. palla Coscinodiscus lineatus Monallantus salina P. tricornutum Tetraselmis striata	6 d in charcoal treated poor sea water	Lab	10 – 16 8 5·8	max 8 d	Cell density; Chl-a in vivo	Measurement of AGP. Samples collected over 5 months bioassayed all at a time. Daily count of all testcultures
12a	125	2	Nat. pop.	None	Tank	0·2–1	0·8 d	Cell density; Chl-a in vivo	^{14}C added 16 h after nutrient enrichments. Daily count of 3 test-cultures for each species; not all species used at one time
12b	20	2	C. palla C. sp. Monallantus salina P. tricornutum Platymonas suecica Tetraselmis striata	6 d in charcoal treated poor sea water	Lab	8 1 8 8 1 1	max 8 d	^{14}C-4 h	

No	Reference	Seawater sample					Enrichments							Total number of mixtures used[a]
		Origin			Filtration (μm)	Storage	Elements or substances involved (μg-at·l⁻¹)[c]							
		Area	Nature[a]	Depth (m)		t°C	Duration[b]	N	P	Si	M	Chel	Vit	
13	Charpy-Roubaud et al. (1983)	Gulf of San Jose, Patagonia coast, Argentina	Ner	0	None	None		25	2·3	26	B, Co, Cu, Fe, Mn, Mo, Zn	EDTA	B_1, B_{12} Biotin	12-14
14	Chiaudani and Vighi (1978)	Ligurian Sea, Tiber estuary	Ner	2	0·45	4 <0	<1-2 d >2	3-23	0·2-0·7	0	Co, Cu, Fe, Mn, Mo, Zn	EDTA	0	4
15	Chiaudani and Vighi (1982) Chiaudani et al. (1980a, 1980b)	Northern Adriatic Sea	Ner	1	150 0·45	4	<1-2 d	+ + +	+	+ 0 0	- 0 0	- 0 0	B_1, B_{12} Biotin 0 0	5+1
16	Collier et al. (1969)	West Florida coast	Ner Oc	-	0·45	Deep frozen		10 000 1000 0·22	0·06	0	Ba, Cu, Fe, Mn, Ti, V, Zn, Zr	EDTA	B_1, B_{12} Biotin	10
17	Conover (1956)	Long Island Sound	Ner	-	-	None		20	2	0	Fe, Mn	Citric acid	Soil extract	11
18	Corredor (1979)	Caribbean upwelling	Oc	10	None	None		0	0	0	0	EDTA	B_1, B_{12} Biotin	1
19	Davies and Sleep (1981)	English Channel	Ner	0	170	None		0·4-10	0·15-1·5	0	Cu, Mn, Mo, Zn	EDTA	B_1, B_{12}	8
20	Dufour and Slepoukha (1981); Dufour et al. (1981)	Ebrie lagoon (Ivory Coast)	Mar. Pond	0	200	None		588	29	0	B, Co, Cu	EDTA	B_1, B_{12}	-

No	Enriched test samples		Test algae			Starting[f] cell density (10^6 cell·l^{-1}) or biomass	Growth		Remarks
	Volume (ml)	No. of replicata	Species (eventually strain)	Depletion	Condition[e]		Duration[g]	Estimation[h]	
13	20 000	1	Nat. pop.	None	Tank	0·4– 0·6 µg·l^{-1} Chl-a	14 d	^{14}C-4 h; Chl-a; POC; PON; cell density; nutrient uptake	Daily sampling and analysis. Two experiments which involved different enrichment mixtures. Cell counts for each species
14	100	2	D. tertiolecta P. tricornutum	Cell washing + 1 d in poor sea water	Lab	50 200	6–8 d	Cell density (daily counting)	Mainly contains a proposed standard method for marine bioassays; data reported illustrate its use. NO_3 and PO_4 added at variable levels, i.e. no more than 3 times the natural values, for oligotrophic sea water
15	1 000	2	Nat. pop. D. tertiolecta P. tricornutum	None	Lab	– 50 200	5–9 d	^{14}C; Chl-a in vivo; cell density	Sewage also involved as an enrichment; concentrations as in Chiaudani and Vighi (1978)
16	10^i	5	Gymnodinium breve	–	Lab	–	28 d	Cell density	CO_3^{2-} and S also involved in enriching mixture
17	125	–	Nat. pop.	None	In situ	–	5, 7 or 8 d	Cell density; Chl-a; nutrient uptake	Probably surface water
18	50	–	Nat. pop.	None	Deck RV	–	20 h	^{14}C-4 h	EDTA and vitamins added together
19	1000	1	Nat. pop.	None	Lab	1·3– 4·8 µg·l^{-1} Chl-a	18 h	^{14}C-5 h	N added as NO_3, NH_4 and urea; P added as PO_4 and glucose-6-PO_4; N and P were added at varied concentrations within the same sets of enrichments
20	300	3 or 4	Nat. pop.	None	Tank	–	–	Chl-a in vivo	C (as $HNaCO_3$) also involved in the enrichments at 2700 µg-at-C·l^{-1}.

No	Reference	Seawater sample			Filtration (μm)	Storage		Enrichments						Total number of mixtures used[d]
		Origin						Elements or substances involved (μg-at·l^{-1})[c]						
		Area	Nature[a]	Depth (m)		t°C	Duration[b]	N	P	Si	M	Chel	Vit	
21	Dunstan and Tenore (1974)	Woods Hole, Massachusetts	Ner	1	100		None	100	16·4[i]	24	Co, Cu, Mn, Fe-EDTA, Mo, Zn	↓	B_1, B_{12} Biotin	6
22	Fedorov et al. (1967)	Chupa Bay, White Sea	Ner	5	—			None	55 110 220	10 20 40	300 600 900	Fe Fe Fe	0 0 0	0 0 0	16 16 16
23	Fedorov and Semin (1970)	Chupa Bay, White Sea	Estu ner	0·5	—	—	—	NO_3:6 NH_4:1	0·7	18	Fe	0	0	15
24	Fedorov et al. (1972)	Chupa Bay, White Sea	Ner	0	Net		None	90	20	500	Fe	0	0	16
25	Fiala et al. (1976)	Golfe du Lion, Mediterranean Sea	Ner	20	—		None	10	1	10	Fe-EDTA	↓	B_1, B_{12} Biotin	1
26	Fournier (1966)	York River estuary, Virginia	Estu ner	0·6 3·0	— —		None	1280	160	90	Fe-EDTA, B, Co, Cu, Mn, Mo, Zn	↓	B_1, B_{12} Biotin	6
27	Frey and Small (1980)	Yaquina Bay, Oregon	Ner	0·5	0·45	—	2 d	150	7·3	21	Co, Cu, Mn, Mo, Zn, Fe-EDTA	EDTA Humic acid	B_1, B_{12} Biotin	9

No	Enriched test samples		Test algae		Growth				Remarks
	Volume (ml)	No. of replicata	Species (eventually strain)	Depletion	Condition[e]	Starting[f] cell density (10^6 cell·l^{-1}) or biomass	Duration[g]	Estimation[h]	
21	400 000	–	Nat. pop.	None	Tank	–	4–7	Cell density; POC	After 4–7 days of batch culturing; semi-continuous culturing was established, up to 10 days. Species counts were made.
22	250	1	Nat. pop.	None	In situ	–	12 h	^{14}C-12 h	10% nitrogen provided as NH_4-N. See also Federov and Maksimov (1967).
23	250	–	Nat. pop.	None	In situ	–	8 h or 5 d	^{14}C-8 h	NO_3 and NH_4 treated as separate nutrients. Other details in Fedorov et al. (1967)
24	1500	1	Nat. pop. Chaetoceros compressus Chaetoceros constrictus Chaetoceros wighamii Nitzschia delicatissima Thalassionema nitzschioides Thalassiosira nordenskioldi S. costatum	None	In situ	– 9900 1400 760 1900 900 800 13460	5 d	Cell density, daily counting	Size mesh in Soviet standard; 4 replicata for unenriched reference; listed test algae are dominant species of natural population. See also Fedorov and Kol'tsova (1973)
25	10 000	1	Nat. pop.	None	Lab	0.5–153	10 d	^{14}C; Chal-a; cell density	Growth measured every 2 days, Fe added as 23 µg-at·l^{-1}
26	300	4	Nat. pop.	None	Lab	–	1 d	O_2-24 h	First 4 experiments, only 6 h growth. EDTA added with M
27	50 000	3	Nat. pop.	None	Tank	–	14 d	Cell density; in vivo fluorescence; nutrient uptake	Charcoal-treated and charcoal-untreated samples were used. Initial cell density was very low. Daily fluorescence measurements were made; species counts were made; DOC was analysed

No	Reference	Seawater sample				Storage		Enrichments						Total number of mixtures used[d]
		Origin						Elements or substances involved (μg-at·l[-1])[c]						
		Area	Nature[a]	Depth (m)	Filtration (μm)	t °C	Duration[b]	N	P	Si	M	Chel	Vit	
28	Gargas and Pedersen (1974)	Any origin			GF/C	4	—	Δ	Δ	0	Fe	0	0	2
29	Glooschenko and Curl (1971)	Northeast and central Pacific Ocean	Oc ner	10 or 15	None	None		28·2	2·9	0	Fe-EDTA	—	0	3
30	Glover (1978)	Maine coast	Ner	0-10	None	None		18	0	0	Fe	0	0	2
31	Goldman et al. (1973)	Woods Hole coastal waters	Ner	0	0·45	None		+	0	0	0	0	0	1
32	Goldman (1976)	Massachussetts and R.I. coastal waters	Ner	0	1·2	None		+	+	0	0	0	0	1
33a	Gran (1931)	Romsdalsfjord, Norwegian coast	Ner	0	None	None		+	0	0	Fe-saccharate	Soil extr. oxy-haem	↑	5
33b	Gran (1931)	Puget Sound, Washington	Ner	0	None	None		+	+	0	Fe, Mn, Zn	Soil extr.	↑	4
34	Gran (1933)	Gulf of Maine	Ner-oc	0	None	None		92[i]	5·1[i]	0	Fe, Mn	Soil extr.	↑	4
35	Graneli (1978)	Öresund, Baltic Sea	Ner	0	100	None		600	18·7	0	Al, B, Br, Cd, Co, Cr, Cu, Fe, J, Mg, Mo, Ni, V, W, Zn	EDTA	0	12
36	Graneli (1981)	Falsterbo Channel, Baltic Sea	Ner	0	None	None		25-36 +1·2-11	0·8-1·2 +0·1-0·4	0	0	0	0	3

No	Enriched test samples		Test algae			Starting[f] cell density (10^6 cell·l^{-1}) or biomass	Growth		Remarks
	Volume (ml)	No. of replicata	Species (eventually strain)	Depletion	Condition[e]		Duration[g]	Estimation[h]	
28	1000	–	P. tricornutum	Washed	Lab	1	15 d	Cell density; cell volume; POC; PON; POP	Technical paper which describes a proposed standard method also valuable for freshwaters by the use of Selenastrum capricornutum as test alga. Δ = design dictated by the actual situation
29	120	1	Nat. pop.	None	In situ	–	1 d	^{14}C-24 h	
30	–	–	Nat. pop.	None	–	≃1 µg l^{-1} i	4 h	^{14}C-4 h	Experiments were done when Dinoflagellates bloomed
31	250	4	Mixture of species, mainly Chaetoceros simplex	None	Lab	3·5	6 d	Chl-a in vivo fluorescence	Sea water was enriched with waste water and/or N-NH$_4$
32	2750 or 1000	1	P. tricornutum	None	Lab	–	7–10 d	Dry weight (ash free); POC; POP	Sea water enriched with waste water; bioassays conducted in continuous culture (0·5 volume by day).
33a	250	3	Nat. pop. dominated by S. costatum	None	In situ	4·10^{-4}	6 d	Cell density	Oxy-haem = oxy-haemoglobin
33b	250	3	Nat. pop. dominated by Nitzschia seriata S. costatum T. condensata	None	In situ	0·001 0·007 0·025	5 d	Cell density	
34	150	3 or 4	Nat. pop.	None	Lab	0·09	3, 5 or 12 d	Cell density	
35	300	–	Nat. pop. P. tricornutum	None	Lab	0·5–1·5	2–4 d	^{14}C-2 h; Chl-a; cell density; nutrient uptake	Species identification was made at the beginning and at the end of the experiments
36	150 000	1	Nat. pop.	None	In situ	0·7–1·7 µg l^{-1} Chl-a	7–15 d	^{14}C-2 h; Chl-a; nutrient uptake; specific plasma volume	P and N were added at the highest values on day 0, and added daily at the lowest ranges

No	Reference	Seawater sample			Filtration (μm)	Storage		Enrichments						Total number of mixtures used[d]
		Origin						Elements or substances involved (μg-at·l⁻¹)[c]						
		Area	Nature[a]	Depth (m)		t°C	Duration[b]	N	P	Si	M	Chel	Vit	
37	Hitchcock and Smayda (1977)	Narragansett Bay, R.I.	Ner	4 or 8	GF	Frozen	1 m	70	3	8	Fe-EDTA Co, Cu, Mn, Mo, Zn	↓	B_1, B_{12} Biotin	8
38	Ignatiades and Smayda (1970)	Narragansett Bay, R.I.	Ner	0	GFⁿ 984 H	Deep-frozen	1–6 m	50	5	50	Fe-EDTA Co, Cu, Mn, Mo, Zn	↓	B_1, B_{12} Biotin	14
39	Jacques et al. (1973)	Golfe du Lion, Mediterranean Sea	Ner	10–40	–	None		10	1	10	Fe-EDTA	↓	B_1, B_{12} Biotin	5 or 13
40	Jacques et al. (1976)	N.W. African upwelling	Upw Oc	10	–	None		40	2·5	25	Fe-EDTA Co, Cu, Mn, Mo, Zn	↓	B_1, B_{12} Biotin	7
41	Johnston (1963; 1964)	North Sea	Oc	0	None paper	None	–	588	37	264	Fe-EDTA Co, Cu, Mn, Zn	↓ 21 sub	B_1, B_{12}	6
42	Kabanova (1967)	Black Sea	Ner	1	None	None		100	5	178	Al, Co, Cu, Fe, Mn, Mo, V, Zn	0	B_{12}	15?
43	Kabanova and Domanov (1979)	Caribbean Sea	–	–	None	None		21	1·7	18	Al, Co, Cu, Fe, Mn	0	B_{12}	25
44	Lanigan (1972; cited in Taylor, 1973, as Miss Nola)	Hauraki gulf, N.Z.	Ner	0	None 0·45	None		500	50	250	Co, Cu, Fe, Mo, Mn, Zn	EDTA	B_1, B_{12} Biotin	12

No	Enriched test samples		Test algae		Growth				Remarks
	Volume (ml)	No. of replicata	Species (eventually strain)	Depletion	Condition[c]	Starting[b] cell density (10^6 cell·l^{-1}) or biomass	Duration[g]	Estimation[h]	
37	50	2	*S. costatum*	6 d in assayed water[f]	Lab	0·5	Δ'''	Cell density	Incubation at 2°C; waters collected at 4 m and 8 m are combined
38	10	3	*Rhizosolenia fragilissima*	4 days in spring water[a]	Lab	0·4	7 d	Cell density	
39	10 000	1	Nat. pop.	None	Lab	0·04-2·0	12 d	^{14}C-6 h; Chl-*a*; pheophytin	Growth daily measured; Fe added as 23 µg-at·l^{-1}
40	2000	1	Nat. pop.	None	Deck RV	1·2-6·8 µg Chl-*a* l^{-1}	6 d	^{14}C-6 h; Chl-*a*; nutrient uptake	Daily measurements; no clear statement on nutrient limiting AGP
41	—	2	Nat. pop. *Chaetoceros teres* *Peridinium trochoideum* *S. costatum*	None	Lab	—	5-12 d	Cell density; optical density	The number of enrichment mixtures is left unclear; IAA, PAB and cystein also involved in enrichments. Cell counting is made after 5-8 days. 1964 paper involves 21 chelating substances studied for effect on algal growth through metal complexes
42	—	—	Nat. pop. dominated by *Exuviella* sp. *Thalassionema nitzschioides*	None	—	0·001	5 d	^{14}C; O$_2$; cell density	?= information remained uncertain; article in Russian
43	—	—	Nat. pop. dominated by *Asterionella notata* *Licmophora abbreviata* *Navicula* sp. *S. costatum*	None	*In situ*	0·015 0·004 0·038 0·010	6 d	^{14}C; cell density	
44	10	3	Nat. pop. *S. costatum*	None 18 h in poor sea water	Lab	0·1-10 500-600	7 d 6 d	Cell density	Results of Lanigan (1972) are unfortunately not published, but her statements appear in Taylor (1973). With natural populations, cell counting was made for every species

No	Reference	Seawater sample						Enrichments						
		Origin			Filtration (μm)	Storage		Elements or substances involved ($\mu g \cdot at \cdot l^{-1}$)[c]						Total number of mixtures used[d]
		Area	Nature[a]	Depth (m)		t°C	Duration[b]	N	P	Si	M	Chel	Vit	
45	Lännergren (1978)	Lindåspollene Fjord, Norway	Ner	0	None	None		14·2[i]	1·3[i]	17·4[i]	0	0	0	4
46	Lännergren (1980)	Hjeltefjorden, Norwegian coast	Ner	0	None	None		14	1·3	14	Fe-EDTA Ca, Co, Cu, Mn, Mo, Zn	←	B_1, B_{12} Biotin	6
47	Lindahl and Melin (1973) Melin and Lindahl (1973)	Stockholm archipelago, Baltic Sea	Ner	0	GF/C	−20		500	50	0	B, Co, Cu, Fe, Mn, Mo, Zn	Na_2 Vernesol	0	2 or 7
48	Maestrini and Kossut (1981)	Ligurian Sea	Ner	1	0·45	None		50	5	50	B, Co, Cu, Fe, Mn, Mo, Zn	EDTA	B_1, B_{12} Biotin	27 18 or 24 at a time
49	Malewicz (1975)	Southern Baltic Sea	Ner	0–40[i]	0·45	4	–	1000[i]	23[i]	58[i]	B, Co, Cu, Fe, Mn, Mo, V, Zn	0	Vitamin pool (9 substances)	3
50	Matudaira (1939)	Tita peninsula, Japan	Ner	Δ	None	None		3775	2666	83	Fe	0	0	1?
51a	Menzel and Ryther (1961)	Sargasso Sea	Oc	0	None	None		50	5	0	Fe-EDTA Co, Cu, Fe, Mn, Mo, Zn	←	B_1, B_{12} Biotin	15
51b	Menzel and Ryther (1961)	Bermuda–West Indies	Oc	0	None	None		50	5	0	Fe	0	B_1, B_{12} Biotin	3
52	Menzel et al. (1963)	Sargasso Sea	Oc	0	None	None		50	5	50	Fe-EDTA Cu, Co, Mn, Mo, Zn	←	B_1, B_{12}	6

No.	Enriched test samples		Test algae		Growth				Remarks
	Volume (ml)	No. of replicata	Species (eventually strain)	Depletion	Condition[c]	Starting[b] cell density (10^6 cell·l^{-1}) or biomass	Duration[g]	Estimation[h]	
45	100 000 200	1	Nat. pop. Nat. pop.	None None	*In situ* Lab	–	7–9 d	^{14}C-4 h; Chl-*a*; turbidity	For *in situ* exp: culture were made in 160 l polyethylene bag; nutrient uptake was also measured
46	200	–	Nat. pop.	None	Lab	–	7 d	^{14}C-4 h	Measurements were made on days 2, 5 and 7. Few bioassays also made with varying-Co concentrations
47	200	–	*Aphanizomenon flos-aquae* *Oscillatoria agardhii*	Cells washed	Lab	0·5–2 mg dw·l^{-1}	8–12 d	Dry weight, Chl-*a*	*A. flos-aquae* probably fixes N_2
48	15	2	*Cylindrotheca closterium* *Hemiselmis virescens* *P. tricornutum* *Platymonas suecica* *S. costatum* *T. pseudonana*	7–13 d *in situ* culturing	Lab	2 5 2 1 2 2	7 d	Cell density	*Cylindrotheca closterium* is a wild contaminant which replaced the inoculated algae and gave several unialgal populations
49	1 000	2, 3 or 7	Nat. pop. *Chlorella vulgaris* *Scenedesmus quadricauda*	None	Lab	–[p] 1 1	18 h 5 d 5 d	^{14}C-6 h, dry weight	Cultured strains used with 2 samples only; NaCl added to ensure $S‰ = 10$. Vitamin pool $= B_1 + B_6 + B_{12} + Ca$-pantothenate + biotin + folic acid + inositol + nicotinic acid + thymine
50	100	?	Nat. pop.	None	Lab	0·009–0·05	4–7 d	Cell density	? = information remained unclear; article in Japanese
51a	1000	–	Nat. pop.	None	Incub	–	1–3 d	^{14}C-4 h	
51b	1000	–	Nat. pop.	None	Incub	–	1 d	^{14}C-4 h	
52	10 000	–	Nat. pop.	None	Lab	0·1[i]	3 d 7 d 9 d	^{14}C-4 h	Final populations resulting from enrichment in terms of species was also surveyed

No	Reference	Seawater sample						Enrichments						Total number of mixtures used[d]
		Origin (Area)	Nature[a]	Depth (m)	Filtration (µm)	Storage t°C	Duration[b]	Elements or substances involved (μg-at·l^{-1})[c] N	P	Si	M	Chel	Vit	
53	Perry (1972)	Subtropical central north Pacific Ocean	Oc	0–150	None	None	—	8.5^i	0.85^i	8.5^i	Fe-EDTA Co, Cu, Mn, Mo, Zn	↓	B_1, B_{12} Biotin	7
54	Pojed and Kveder (1977)	Northern Adriatic	Ner, Oc	0	—	—	—	+	+	+	+	+	0	—
55	Riley (1938)	Gulf of Mexico, Tortugas region	Oc	1–153	Net 20	None	—	0.7 7.1	0.3 3.3	0	Fe	0	0	4
56	Rinne and Tarkiainen (1975)	Helsinki area, Baltic Sea	Ner, Oc	0	0.45	Deep frozen	—	100 & 285	2, 3.3, 26	0	B, Co, Cu, Fe, Mn, Mo, Zn-EDTA	↓	0	7
57	Ryther (1954)	Long Island Bay	Ner	0	—	—	—	+	+	0	0	0	0	2
58	Ryther in Ketchum et al. (1958)	Caribbean Sea	Oc	0	None	None	—	100	6.6	0	0	0	0	1
59	Ryther and Guillard (1959)	From New York to Bermuda	Oc	0	None	None	—	50	5	50	Fe-EDTA, Co, Cu, Mn, Mo, Zn	↓	B_1, B_{12} Biotin	8
60a	Ryther and Dunstan (1971)	Long Island Bay	Ner	0	+	None	—	100	10	0	0	0	0	2
60b	Ryther and Dunstan (1971)	From New York to 200 milles	Ner, Oc	0	+	Frozen	—	100	10	0	0	0	0	2
61	Sakshaug and Myklestad (1973)	Trondheims fjord	Ner	0–50	GF/C	Deep frozen	—	176^i	7.2^i	21^i	Fe-EDTA, Cu, Co, Mn, Mo, Zn	↓	B_1, B_{12}[i] Biotin	4
62	Sakshaug (1977)	Narragansett Bay	Ner	6	0	None	—	0^i	0	0	0	0	0	0
63a	Skulberg (1970)	Oslofjord	Ner	1	GF/A		—	+	+	0	Fe	—	0	—
63b	Skulberg (1970)	Baltic Sea	Ner, Oc	0	GF/A		—	+	+	0	Fe	—	0	—

No	Enriched test samples		Test algae		Condition[e]	Starting[f] cell density (10^6 cell·l^{-1}) or biomass	Growth		Remarks
	Volume (ml)	No. of replicata	Species (eventually strain)	Depletion			Duration[g]	Estimation[h]	
53	150	–	Nat. pop.	None	Incub	–	3 d	[14]C	Concentrations of added elements are given in Eppley et al. (1967)
54	2000	–	–	–	Lab	–	2 d	[14]C; Chl-a	Test algae are probably natural populations
55	4000	–	Nat. pop.	None	Tank	–	7 d	Chl-a; O_2	N added at 4 concentrations, P added at 3 different concentrations
56	30	3	Chlorella sp. Oscillatoria agardhii	–	Lab	–[q]	14 d 24 d	Turbidity	N added as NH_4; sewage also added in 2 experiments
57	300	–	Stichococcus sp. poss. Stichococcus cylindricus	–	Lab	–	10-15 d	Cell density	
58	–	–	Nat. pop.	None	Deck RV	–	3 d	[14]C; O_2	
59	–	–	Nat. pop.	None	Incub	–	1 d	[14]C-24 h	
60a	50	1	Nannochloris atomus	–	Lab	–	7 d	Cell density	Review of previous data
60b	50	1	S. costatum	Washing 2 d in poor sea water	Lab	–	5 d	Cell density	
61	50	–	Chaetoceros curvicetus Chaetoceros debilis Gonyaulax tamarensis S. costatum	10-14 d in unenriched or test-enriched sea water	Lab	1·3-5·7 (diatoms) 0·2-1·2 (G. tamarensis)	14 d	Cell density	Algae are local isolates
62	130	2 4	Asterionella japonica S. costatum	In situ dialysis culturing	In situ	10-30[i] 3-9	6-21 d	Cell density; Chl-a; ATP; POC; PON	Test algae are in situ cultures without any nutrient enrichment
63a	1000	–	S. costatum	–	Lab	–	14 d	Cell density	Concentration of added nutrient not given. Experiments also made with freshwater samples and algae
63b	1000	–	S. costatum	–	Lab	–	14 d	Cell density	

No	Reference	Seawater sample						Enrichments							
		Origin			Filtration (μm)	Storage		Elements or substances involved (μg-at·l⁻¹)[c]							Total number of mixtures used[d]
		Area	Nature[a]	Depth (m)		t°C	Duration[b]	N	P	Si	M	Chel	Vit		
64	Smayda (1964)	Section from R.I. to Bermuda	Oc	0·5 100	GF		None	50	5	56	Fe-EDTA, Cu, Co, Mn, Mo, Zn	↓	B₁, B₁₂ Biotin		12
65	Smayda (1964)	Sargasso Sea	Oc	30	GF/A		None	50	5	59	Fe-EDTA, Co, Cu, Mn, Mo, Zn	↓	B₁, B₁₂ Biotin		17
66	Smayda (1971)	From R.I. to Puerto Rico	Oc	0·5	GF/A	—	—	50	5	50	Fe-EDTA, Co, Cu, Mo, Mn, Zn	↓	B₁, B₁₂ Biotin		13
67	Smayda (1973)	Narragansett Bay, R.I.	Ner	0	Ne. 10		None	8·8	3·6	52·8	Fe-EDTA, Co, Cu, Mo, Mn, Zn	↓	0		1-6
68	Smayda (1974)	Narragansett Bay, R.I.	Ner	0	0·1		None	50	5	50	Fe-EDTA, Co, Cu, Mo, Mn, Zn	↓	B₁, B₁₂ Biotin		15
69	Smith et al. (1982)	N.W. Africa	Upw	2-5	None and 0·45		None	0	0	0	Cu	EDTA	0		2
70	Specht and Miller (1974) Specht (1974, 1975)	Yaquina Bay, Oregon	Ner-estu	0	0·45	4°	1 d	71·4	1·6	0	0	0	0		3

No	Enriched test samples		Test algae			Growth			Remarks
	Volume (ml)	No. of replicata	Species (eventually strain)	Depletion	Condition[e]	Starting[f] cell density (10^6 cell·l^{-1}) or biomass	Duration[g]	Estimation[h]	
64	25	—	Cyclotella nana (clone 13-1) later named Thalassiosira pseudonana	None	Incub	0·4–1·6	7 d	Cell density	Cell measurements were also made to calculate cell volume
65	15	—	Bacteriastrum hyalinum Chaetoceros lauderi Rhizosolenia setigera S. costatum T. pseudonana (clone 13-1) T. rotula	Poor sea water	Incub	1·8	5 d	Cell density	Other species used to check the representativity of Bacteriastrum hyalinum, Chaetoceros lauderi, Rhizosolenia setigera, S. costatum, T. pseudonana, T. rotula
66	15	—	T. pseudonana (=Cyclotella nana)	Poor sea water	Lab RV	5–7	5 d	Cell density	Probably no storage and no duplicates; duration of depletion incubation not given
67	2500 or 3000	—	S. costatum	None	Tank	0·02–0·07	2–14 d	Cell density	Enriched mixtures variable with experiments
68	15	3	T. pseudonana (clone 13-1)	5 d in Poor sea water	Lab	1·0	—	Cell density	Styrene flask for Si; for other elements glass vessel. Poor sea water: collected from surface water of Sargasso Sea
69	500	1	Nat. pop.	None	Deck RV	1/10 natural population	6 d	^{14}C	Sea water with suspended phytoplankton is diluted with filtered sea water of same sample. Cu and EDTA are added at various concentrations. Daily measurement of carbon uptake
70	100	3	D. tertiolecta Selenastrum capricornutum T. pseudonana	Washed	Lab	0·1	7–10 d	Cell density	Cell density data give calculated dry weight

| No | Reference | Seawater sample | | | | | | Enrichments | | | | | | |
| | | Origin | | | Filtration (μm) | Storage | | Elements or substances involved (μg-at·l^{-1})[c] | | | | | | Total number of mixtures used[d] |
		Area	Nature[a]	Depth (m)		t°C	Duration[b]	N	P	Si	M	Chel	Vit	
71	Subba Rao (1981)	St Margaret's Bay, N.S.	Ner	0	100 0·8	—	—	1760	73·6	107	Co, Cu, Fe, Mn, Mo	EDTA	B_1, B_{12} Biotin	1
72	Sunda et al. (1981)	Off North Carolina coast	Oc	800	None	4 4	1 d 1–4 d	0	0	0	Cu, Fe Mn	EDTA NTA	0	10[i]
73	Tarkiainen et al. (1974)	Farö Deep, Baltic proper	Oc	0–145	0·45	Deep frozen	5 m	100	2	0	Fe-EDTA, B, Co, Cu, Mo, Mn, Zn	←	0	7
74	Teixeira and Vieira (1976)	Victoria Island coast, Brazil	Ner	0	0·45	None		50·	5	56	Fe-EDTA, Co, Cu, Mo, Mn, Zn	←	B_1, B_{12} Biotin	10
75	Teixeira and Tundisi (1981) Ubatuba region, Brazil		Ner	0	None	None		50	5	0	0	0	0	3
76	Thayer (1974)	North Carolina coast	Ner estu	0	Net 10	None		50	5	50	Fe-EDTA, Co, Cu, Mo, Mn, Zn	←	B_1, B_{12} Biotin	7
77	Thomas (1959)	Costa Rica dome	Oc-upw	10	None	None		100	10	10	Fe, M	Soil extr.	←	11

No	Enriched test samples		Test algae		Condition[e]	Growth			Remarks
	Volume (ml)	No. of replicata	Species (eventually strain)	Depletion		Starting[f] cell density (10^6 cell·l^{-1}) or biomass	Duration[g]	Estimation[h]	
71	250 50	3	Nat. pop. D. tertiolecta Chaetoceros septentrionalis P. tricornutum Rhizosolenia sp. Thalassiosira sp. S. costatum	None	Lab	–	14 d	Optical density Cell density	Trace metals were added at 4 concentration levels. Results are expressed as growth-rate
72	500	1[i]	Nat. pop. Chaetoceros socialis	Cultured in deep sea water	Lab	6 ng·l^{-1} Chl-a 4 ng·l^{-1} Chl-a	6–11 d	^{14}C-6–11 d; Chl-a	All nutrients were added singly, with one of them at varied concentrations; natural population inocula were taken in near-shore area; ^{14}C was added to the whole experimental bottle, after a 24 h equilibration delay, in Huntsman and Barber (1975)
73	30	–	Chlorella sp. local strain	–	Lab	–[q]	14 d	Turbidity	N was added as NH_4 and NO_3; including these duplicates, total number of mixtures is 12
74	100	1	P. tricornutum	–	Lab	15–25	10 d	^{14}C-4 h; Chl-a	
75	10 000 or 25 000	–	Nat. pop.	None	In situ	0·6– 2·4 µg·l^{-1} Chl-a	5–8 d	^{14}C; Chl-a	Daily estimations of Chl-a content and C uptake rate
76	12 000	1	Nat. pop.	None	Tank	–	0·1–0·5 2 and 3 d	^{14}C-2 h	Nutrient analysis also made at the start and the end of each experiment. S also added (8 µg-at l^{-1})
77	50	–	Nat. pop.	None	Lab	–	15 d	Cell density	M: Arnon's solution. Fe: Rhode's solution

No	Reference	Seawater sample Origin (Area)	Nature[a]	Depth (m)	Filtration (μm)	Storage t°C	Duration[b]	N	P	Si	M	Chel	Vit	Total number of mixtures used[d]
78	Thomas (1967)	Gulf of Tehuantepec, Costa Rica dome	Ner, Oc upw	0	None	None		25	2·5	5	Fe-EDTA, Co, Cu, Mo, Mn, Zn	→	B_1, B_{12} Biotin	9
79	Thomas (1969, 1977)	Baja California (tropical sea water)	Ner, Oc upw	0	None	None		25	2·5	5	Fe-EDTA, Co, Cu, Mo, Mn, Zn	EDTA	B_1, B_{12} Biotin	9 or 15
80	Thomas et al. (1974)	Southern California	Ner	0	None	None		25	2·5	5	Fe-EDTA, Co, Cu, Mo, Mn, Zn	→	B_1, B_{12} Biotin	15
81	Tranter and Newell (1963)	N.W. Australia	Ner, Oc	160	None	None		28·5[i]	3·3[i]	0	Fe-EDTA	→	0	5
82	Vedernikov and Sapozhnikov (1978)	Antarctic S.W. Pacific	Oc	0	None	None		26	1·6	26	Co, Fe, Mn	EDTA	0	15
83	Vince and Valiela (1973)	Vineyard Sound, Massachusetts	Ner	0	300	None		25 50 75	3·3 5·0 10·0	0	0	0	0	3
84a	Wilson (1966) exp. on 1956	West coast South Florida	Ner	—	0·45	Deep frozen	—	10[i] 18	3·7[i]	0	Mg	EDTA Soil extr.	B_1, B_{12} Biotin	5
84b	Wilson (1966) exp. on 1964	Tampa bay, Florida	Ner	—	0·45	Deep frozen	—	18 20	7·4	0	Fe-EDTA, Ba, Cu, Mn, V, Ti, Zn, Zr	→ Tris	B_1, B_{12} Biotin	20
85	Yoneshigue-Braga et al. (1979)	Cabo Frio, Rio de Janeiro, Brazil	Ner	50, 89 120	0·45	−20	6 m	+	+	+	+	+	+	13

No	Enriched test samples		Test algae				Growth		Remarks
	Volume (ml)	No. of replicata	Species (eventually strain)	Depletion	Condition[e]	Starting[f] cell density (10^6 cell·l^{-1}) or biomass	Duration[g]	Estimation[h]	
78	20 000	1	Nat. pop.	None	Deck RV	—	4 d	Chl-a in vivo	Starting biomass: 11 µg Chl-a l^{-1}
79	17 500	1	Nat. pop.	None	Lab	—	4 or 5 d	^{14}C-3 h or Chl-a in vivo	Thomas (1977)'s paper summarizes all author's previous articles
80	4000	1	Nat. pop.	None	Deck RV	—	4 or 5 d	Chl-a	Sewage added in two mixtures
81	500	1	Nat. pop.	None	Incub	—	20 h	^{14}C-4 h	Fe added as 1·8 µg-at l^{-1}
82	250 20 000	3	Nat. pop.	None	Tank Deck	—	3–10 d	Chl-a; ^{14}C-6 h	
83	2000	2 or 4	Nat. pop.	None	Deck RV	1·9	6 d	Cell density	N added as NH_4; daily estimation of biomass
84a	10	—	Gymnodinium breve	—	Lab	0·01	10–56 d	Cell density	N added as 10 µg-at·N-NO_3·l^{-1} and/or 18 N-NH_4. Also added CO_3^{2-} and S. Cell density was estimated every 3–5 days.
84b	10	—	Gymnodinium breve	—	Lab	0·005 0·02	28 d	Cell density	N added as 20 µg-at l^{-1} N-NO_3 and/or 18 µg-at l^{-1} N-NH_4 CO_3^{2-} and S also added
85	20	2	P. tricornutum Platymonas sp.	None	Lab	15	6 d	Cell density	Cell density numeration made at 2nd, 4th and 6th days

References

Agius, C. and V. Jaccarini (1982). *Hydrobiologia* **87**, 89–96.

Aid, F., G. Gaumer and F. L. Samson-Kechacha (1981). *C.R. Acad. Sci. Paris*, sér. D **293**, 435–437.

Anon. (1971). "Algal Assay Procedure Bottle Test". National Eutrophication Research Program, Corvallis, Or. Environmental Protection Agency, 82 pp.

Anon. (1974). "Marine Algal Assay Procedure Bottle Test". Eutrophication and Lake Restoration Branch, National Environmental Research Center, Corvallis, Or. EPA-660/3-75-008, 43 pp.

Armstrong, F. A. J., C. R. Stearns and J. D. H. Strickland (1967). *Deep-Sea Res.* **14**, 381–389.

Azam, F. and R. E. Hodson (1977). *Limnol. Oceanogr.* **22**, 492–501.

Bagès, M., J. P. Dréno, E. Gonzalez-Rodriguez, S. Y. Maestrini and J. M. Robert (1978). *C.R. Acad. Sci. Paris*, sér. D **287**, 1413–1415.

Barber, R. T. and J. H. Ryther (1969). *J. Exp. Mar. Biol. Ecol.* **3**, 191–199.

Barber, R. T., R. C. Dugdale, J. J. MacIsaac and R. L. Smith (1971). *Invest. Pesq.* **35**, 171–193.

Barlow, J. P., B. J. Peterson and A. E. Savage (1973). Continuous-flow studies of phosphorus as a limiting nutrient for Cayuga Lake phytoplankton. *In* "Proc. 16th Conf. Great Lakes Res." 1973, 7–14.

Barlow, J. P., W. R. Schaffner, F. de Noyelles Jr and B. J. Peterson (1975). Continuous flow nutrient bioassays with natural phytoplankton populations. *In* "Bioassay Techniques and Environmental Chemistry" (G. E. Glass, ed.), pp. 299–319. Ann Arbor Science Publ., Ann Arbor, MI.

Baskett, R. C. and W. J. Lulves (1974). *J. Fish. Res. Bd Can.* **31**, 372–374.

Baumgartner, D. J., D. W. Schults, S. E. Ingle and D. T. Specht (1977). Interchange of nutrients and metals between sediments and water during dredged material disposal in coastal waters. *In* "Management of bottom sediments containing toxic substances" (S. A. Peterson and K. K. Randolph, ed.), pp. 229–245. Corvallis Environmental Research Laboratory, Corvallis, Or. EPA-600/3-77-083.

Berland, B. R., D. J. Bonin, S. Y. Maestrini and J. P. Pointier (1972). *Int. Rev. Hydrobiol.* **57**, 933–944.

Berland, B., D. Bonin, B. Coste, S. Maestrini and H. J. Minas (1973a). *Mar. Biol.* **23**, 267–274.

Berland, B. R., D. J. Bonin, S. Y. Maestrini and J. P. Pointier (1973b). *Int. Rev. Hydrobiol.* **58**, 203–220.

Berland, B. R., D. J. Bonin, S. Y. Maestrini and J. P. Pointier (1973c). *Int. Rev. Hydrobiol.* **58**, 473–500.

Berland, B. R., D. J. Bonin and S. Y. Maestrini (1974). *Ann. Inst. océanogr. (Paris)* **50**, 5–25.

Berland, B. R., D. J. Bonin and S. Y. Maestrini (1976). *Ann. Inst. océanogr. (Paris)* **52**, 45–55.

Berland, B. R., D. J. Bonin and S. Y. Maestrini (1978). *Int. Rev. Hydrobiol.* **63**, 501–531.

Bonin, D. J., S. Y. Maestrini and J. W. Leftley (1981). *Canad. Bull. Fish. Aquat. Sci.* **210**, 292–309.

Brand, L. E., R. R. L. Guillard and L. S. Murphy (1981). *J. Plankton Res.* **3**, 193–201.

Burmaster, D. E. (1979). *J. Exp. Mar. Biol. Ecol.* **39**, 167–186.

Charpy-Roubaud, C. J., L. J. Charpy, S. Y. Maestrini and M. J. Pizarro (1978a). *C. R. Acad. Sci. Paris*, sér. D **287**, 1031–1034.

Charpy-Roubaud, C. J., L. J. Charpy and S. Y. Maestrini (1978b). *C.R. Acad. Sci. Paris*, sér. D **287**, 539–542.

Charpy-Roubaud, C. J., L. J. Charpy and S. Y. Maestrini (1982). *Oceanol. Acta* **5**, 179–188.

Charpy-Roubaud. C. J., L. J. Charpy and S. Y. Maestrini (1983). *Mar. Ecol., P.S.Z.N.I. (Berlin)* **4**, 1–18.

Chiaudani, G. and M. Vighi (1978). *Quaderni dell' Istituto di Ricerca sulle Acque* **39**, 120 pp (*Roma*).

Chiaudani, G. and M. Vighi (1982). *Water Res.* **16**, 1161–1166.

Chiaudani, G., G. F. Gaggino and M. Vighi (1980a). *Vèmes Journées Etud. Pollutions*, Cagliari, CIESM, 383–390.

Chiaudani, G., R. Marchetti and M. Vighi (1980b). *Prog. Wat. Tech.* **12**, 185–192.

Chrétiennot-Dinet, M. J. and E. Vacelet (1978). *Oceanol. Acta* **1**, 407–413.

Collier, A., W. B. Wilson and M. Borkowski (1969). *J. Phycol.* **5**, 168–172.

Conover, S. A. M. (1965). *Bull. Bingham Oceanogr. Coll.* **15**, 62–112.

Corredor, J. E. (1979). *Deep-Sea Res.* **26**, 731–741.

Davies, A. G. and J. A. Sleep (1981). *J. Mar. Biol. Ass. U.K.* **61**, 551–563.

Derenbach, J. B. and P. J. Le B. Williams (1974). *Mar. Biol.* **25**, 263–269.

Dodson, A. N. and W. H. Thomas (1977). *J. Exp. Mar. Biol. Ecol.* **26**, 153–161.

Droop, M. R. (1961). *Bot. Mar.* **2**, 231–246.

Droop, M. R. (1966). *J. Mar. Biol. Ass. U.K.* **46**, 659–671.

Droop, M. R. (1968). *J. Mar. Biol. Ass. U.K.* **48**, 689–733.

Droop, M. R. (1974). *J. Mar. Biol. Ass. U.K.* **54**, 825–855.

Droop, M. R. (1975). *J. Mar. Biol. Ass. U.K.* **55**, 541–555.

Droop, M. R. (1979). *J. Exp. Mar. Biol. Ecol.* **39**, 203.

Dufour, P. and M. Slepoukha (1981). *Rev. Hydrobiol. Trop.* **14**, 103–114.

Dufour, P., J. L. Cremoux and M. Slepoukha (1981). *J. Exp. Mar. Biol. Ecol.* **51**, 247–267.

Dunstan, W. M. and K. R. Tenore (1974). *J. Appl. Ecol.* **11**, 529–536.

Eide, I., A. Jensen and S. Melsom (1979). *J. Exp. Mar. Biol. Ecol.* **37**, 271–286.

Eppley, R. W. (1980). Estimating phytoplankton growth rates in the central oligotrophic oceans. *In* "Primary productivity in the sea" (P. G. Falkowski, ed.), pp. 231–242, Environmental Science Research Vol. 19. Plenum Press, New York.

Eppley, R. W. and W. G. Harrison (1975). Physiological ecology of *Gonyaulax polyedra* a red water dinoflagellate of Southern California. *In* "Proceedings of the First International Conference on Toxic Dinoflagellate Blooms" (V. R. LoCicero, ed.), pp. 11–22. The Massachusetts Science and Technology Foundation, Wakefield, Mass.

Eppley, R. W. and J. D. H. Strickland (1968). Kinetics of marine phytoplankton growth. *In* "Advances in Microbiology of the Sea" (M. R. Droop and E. J. Ferguson Wood, eds), Vol. I, pp. 23–62. Academic Press, London and New York.

Eppley, R. W., R. W. Holmes and J. D. H. Strickland (1967). *J. Exp. Mar. Biol. Ecol.* **1**, 191–208.

Evans, C. G. T., D. Herbert and D. W. Tempest (1970). The continuous cultivation of micro-organisms. 2. Construction of a chemostat. *In* "Methods in Microbiology" (J. R. Norris and D. W. Ribbons, eds) Vol. 2, pp. 277–327. Academic Press, London and New York.

Falkowski, P. G. (1980). Light–shade adaptation in marine phytoplankton. *In* "Primary productivity in the Sea" (P. G. Falkowski, ed.), pp. 99–119. Environmental Science Research Vol. 19, Plenum Press, New York.

Fedorov, V. D. and T. I. Kol'tsova (1973). *Oceanology* 13, 63–70.

Fedorov, V. D. and V. N. Maksimov (1967). In Russian. *Nauchn. dokl. vysshey shkoly. Biol Nauki* 10, 132–142.

Fedorov, V. D. and V. A. Semin (1970). *Oceanology* 10, 242–253.

Fedorov, V. D., T. I. Kol'tsova, K. A. Kokin and T. V. Khlebovich (1972). In Russian. *Bot. Zh.* 57, 482–495.

Fedorov, V. D., V. A. Semin, V. N. Maksimov and V. G. Bogorov (1967). In Russian. *Dokl. Akad. Nauk. S.S.S.R.* 175, 220–223.

Fiala, M., G. Cahet, G. Jacques, J. Neveux and M. Panouse (1976). *J. Exp. Mar. Biol. Ecol.* 24, 151–163.

Fitzgerald, G. P. (1972). Bioassay analysis of nutrient availability. *In* "Nutrients in Natural Waters" (H. E. Allen and J. R. Kramer eds), pp. 147–169. J. Wiley, New York.

Fogg, G. E., W. E. E. Smith and J. D. A. Miller (1959). *Biochem. Microbiol. Technol. Engineer.* 1, 59–76.

Fournier, R. O. (1966). *Chesapeake Sci.* 7, 11–19.

Frey, B. E. and L. F. Small (1980). *J. Plankton Res.* 2, 1–22.

Fuhs, G. W., S. D. Demmerle, E. Canelli and M. Chen (1972). Characterization of phosphorus-limited plankton algae (with reflections on the limiting-nutrient concept). *In* "Nutrients and Eutrophication" Special symposia (G. E. Likens, ed.), Vol. I, pp. 113–132. Amer. Soc. Limnol. Oceanogr.

Gargas, E. and J. S. Pedersen (1974). "Algal Assay Procedure Batch Technique". Contr. Water Quality Institute. Danish Academy of Technical Science 1, 48 pp.

Gerhart, D. Z. and G. E. Likens (1975). *Limnol. Oceanogr.* 20, 649–653.

Glass, G. E. (ed.) (1974). "Bioassay techniques and environmental chemistry". Ann Arbor Science, Ann Arbor, MI, 2nd printing, 499 pp.

Glooschenko, W. A. and H. Curl, Jr (1971). *J. Fish Res. Bd Can.* 28, 790–793.

Glover, H. E. (1978). *Limnol. Ocenogr.* 23, 534–537.

Goldman, C. R. (1978). *Mitt. int. Ver. Limnol.* 21, 364–371.

Goldman, J. C. (1976). *Wat. Res.* 10, 97–104.

Goldman, J. C., J. J. McCarthy and D. G. Peavey (1979). *Nature (Lond.)* 279, 210–215.

Goldman, J. C., K. R. Tenore and H. I. Stanley (1973). *Science* 180, 955–956.

Gran, H. H. (1929). *Rapp. Cons. Explor. Mer* 56, 1–112.

Gran, H. H. (1931). *Rapp. Cons. Explor. Mer* 75, 37–46.

Gran, H. H. (1933). *Biol. Bull.* 64, 159–181.

Granéli, E. (1978). *Vatten* 34, 117–128.

Granéli, E. (1981). *Kieler Meeresforsch.* 5, 82–90.

Hitchcock, G. L. and T. J. Smayda (1977). *Limnol. Oceanogr.* 22, 132–139.

Holm-Hansen, O. (1969). Environmental and nutritional requirements for algae. *In* Proc. Eutrophication-Biostimulation Assessment Workshop, Berkeley, June 1969, 98–108.

Huntsman, S. A. and R. T. Barber (1975). *J. Phycol.* 11, 10–13.

Ignatiades, L. and T. J. Smayda (1970). *J. Phycol.* 6, 357–364.

Jacques, G., G. Cahert, M. Fiala and M. Panouse (1973). *J. Exp. Mar. Biol. Ecol.* 11, 287–295.

Jacques, G., M. Fiala, J. Neveux and M. Panouse (1976). *J. Exp. Mar. Biol. Ecol.* 24, 165–175.

Jenkins, D. (1967). *J. Wat. Poll. Control Fed.* **39**, 159–180.

Jensen, A. (1980). *Rapp. Cons. Explor. Mer* **179**, 306–309.

Jensen, A. and B. Rystad (1973). *J. Exp. Mar. Biol. Ecol.* **11**, 275–285.

Jensen, A. and B. Rystad (1974). *J. Exp. Mar. Biol. Ecol.* **15**, 145–157.

Jensen, A., B. Rystad and L. Skoglund (1972). *J. Exp. Mar. Biol. Ecol.* **8**, 241–248.

Jensen, A., B. Rystad and S. Melsom (1976). *J. Exp. Mar. Biol. Ecol.* **22**, 249–256.

Johnston, R. (1963). *J. Mar. Biol. Ass. U.K.* **43**, 427–456.

Johnston R. (1964). *J. Mar. Biol. Ass. U.K.* **44**, 87–109.

Kabanova, Yu. G. (1967). In Russian. *Okeanologija* **7**, 495–503.

Kabanova, Yu. G. and M. M. Domanov (1979). In Russian. *Okeanologija* **19**, 692–698.

Kahn, N. and E. Swift (1978). *Limnol. Oceanogr.* **23**, 649–658.

Kayser, H. (1976) *Mar. Biol.* **36**, 61–72.

Ketchum, B. H., J. H. Ryther, C. S. Yentsch and N. Corvin (1958). *Rapp. Cons. Explor. Mer* **144**, 132–140.

Kuenzler, E. J. and B. H. Ketchum (1962). *Biol. Bull.* **123**, 134–145.

Lane, P. and R. Levins (1977). *Limnol. Oceanogr.* **22**, 454–471.

Lanigan, K. S. (1972). Nutrients influencing phytoplankton growth in the Jellicoe Channel. MS. Thesis, University of Auckland, 111 pp.

Lännergren, C. (1978). *Int. Rev. Hydrobiol.* **63**, 57–76.

Lännergren, C. (1980). *Sarsia* **65**, 287–299.

Lee, G. F. (1973). *Wat. Res.* **7**, 1525–1546.

Liebig, J. (1840). "Die Chemie in ihrer Anwendung auf Agrikultur und Physiologie." Taylor & Walton, London.

Lin, C. K. (1977). *J. Phycol.* **13**, 267–271.

Lindahl, P. E. B. and K. E. R. Melin (1973). *Oikos* **24**, 171–178.

Lund, J. W. G., G. H. M. Jaworski and H. Bucka (1971). *Acta Hydrobiol.* **13**, 235–249.

Maddux, W. S. and R. F. Jones (1964). *Limnol. Oceanogr.* **9**, 79–86.

Maestrini S. Y. and M. G. Kossut (1980). *J. Rech. Océanogr.* **5**, 35–52.

Maestrini S. Y. and M. G. Kossut (1981). *J. Exp. Mar. Biol. Ecol.* **50**, 1–19.

Malewicz, B. (1975). *Merentutkimuslait. Julk/Havsforskningsinst. Skr.* **239**, 67–71.

Maloney, T. E., W. E. Miller and T. Shiroyama (1972). Algal responses to nutrient additions in natural waters. I. Laboratory assays. *In* "Nutrients and Eutrophication" Special symposia (G. E. Likens ed.), Vol. I, pp. 134–140+154–156. Amer. Soc. Limnol. Oceanogr.

Marsot, P., R. Fournier and C. Blais (1981). *Can. J. Fish. Aquat. Sci.* **38**, 905–911.

Marvan, P., S. Přibil and O. Lhotský (1979). "Algal Assays and Monitoring Eutrophication", 253 pp. E. Schweizerbart'sche, Stuttgart.

Maslin, P. E. and G. L. Boles (1978). *Hydrobiologia* **58**, 261–269.

Matudaira, T. (1939). In Japanese. *Bull. Jap. Soc. Sci. Fish.* **8**, 187–193.

McFeters, G. A. and D. G. Stuart (1972). *Appl. Microbiol.* **24**, 805–811.

Melin, K. E. R. and P. E. B. Lindhal (1973). No 7. *Oikos*, suppl. **15**, 189–194.

Menzel, D. W. and J. H. Ryther (1961). *Deep-Sea Res.* **7**, 276–281.

Menzel, D. W., E. M. Hulburt and J. H. Ryther (1963). *Deep-Sea Res.* **10**, 209–219.

Middlebrooks, E. J., D. H. Falkenborg and T. E. Maloney (ed.) (1975). "Biostimulation and Nutrient Assessment". Proc. Workshop Utah State University, Logan, PRWG168-1, 390 pp.

Miller, W. E., J. C. Greene and T. Shiroyama (1978). "The *Selenastrum capricornutum* Printz algal assay bottle test. Experimental design, application, and data interpretation protocol." Environmental Research Laboratory, Corvallis, Or. EPA-600/9-78-018, 125 pp.

Monod, J. (1942). "Recherches sur la Croissance des Cultures Bactériennes". Hermann, Paris. 211 pp.

Munk, W. H. and G. A. Riley (1952). *J. Mar. Res.* 11, 215–240.

Munson, R. J. (1970). Turbidostats. *In* "Methods in Microbiology". (J. R. Norris and D. W. Ribbons, eds) Vol. 2, pp. 349–376. Academic Press, London and New York.

Murray, S., J. Scherfig and P. S. Dixon (1971). *J. Wat. Poll. Control Fed.* 43, 1991–2003.

Myers, J. and L. B. Clark (1944). *J. Gen. Physiol.* 28, 103–112.

Nelson, D. M. and H. L. Conway (1979). *J. Mar. Res.* 37, 301–318.

O'Brien, W. J. and F. Denoyelles Jr (1976). *Hydrobiologia* 49, 65–76.

Oswald, W. J. and C. G. Golueke (1966). *J. Wat. Poll. Control Fed.* 38, 964–975.

Owens, O. V. H., P. Dresler, C. C. Crawford, M. A. Tyler and H. H. Seliger (1977). *Chesapeake Sci.* 18, 325–333.

Pasciak, W. J. and J. Gavis (1974). *Limnol. Oceanogr.* 19, 881–888.

Pasciak, W. J. and J. Gavis (1975). *Limnol. Oceanogr.* 20, 604–617.

Perry, M. J. (1972). *Mar. Biol.* 15, 113–119.

Phillips, J. N. and J. Myers (1954). *Plant Physiol. (Lancaster)* 29, 148–152.

Pojed, I. and S. Kveder (1977). *Rapp. Comm. Int. Mer Médit.* 24, 47–48.

Prakash, A., L. Skoglund, B. Rystad and A. Jensen (1973). *J. Fish Res. Bd Can.* 30, 143–155.

Premazzi, G., O. Ravera and A. Lepers (1978). *Mitt. Int. Ver. Limnol.* 21, 42–49.

Redalje, D. G. and E. A. Laws (1981). *Mar. Biol.* 62, 73–79.

Reynolds, J. H., E. J. Middlebrooks, D. B. Porcella and W. J. Grenney (1975). Comparison of semi-continuous and continuous flow bioassays. *In* "Biostimulation and Nutrient Assessment" (E. J. Middlebrooks, D. H. Falkenborg and T. E. Maloney, eds) pp. 241–265. Proc. Workshop Utah State University, Logan PRWG168-1.

Rhee, G. Y. (1980). Continuous culture in phytoplankton ecology. *In* "Advances in Aquatic Microbiology" (M. R. Droop and H. W. Jannasch eds.), Vol. 2, pp. 151–203. Academic Press, New York and London.

Riley, G. A. (1938). *J. mar. Res.* 1, 335–352.

Rinne, I. and E. Tarkiainen (1975). *Merentutkimuslait. Julk/Havsforskningsinst. Skr.* 239, 91–99.

Robert, J. M., S. Y. Maestrini, M. Bagès, J. P. Dréno and E. Gonzalez-Rodriguez (1979). *Oceanol. Acta* 2, 275–286.

Ryther, J. H. (1954). *Biol. Bull. (Lancaster)* 106, 198–209.

Ryther, J. H. and W. M. Dunstan (1971). *Science* 171, 1008–1013.

Ryther, J. H. and R. R. L. Guillard (1959). *Deep-Sea Res.* 6, 65–69.

Sakshaug, E. (1977). *J. Exp. Mar. Biol. Ecol.* 28, 109–123.

Sakshaug, E. and A. Jensen (1978). *Oceanogr. Mar. Biol. Ann. Rev.* 16, 81–106.

Sakshaug, E. and S. Myklestad (1973). *J. Exp. Mar. Biol. Ecol.* 11, 157–188.

Samuels, W. B., A. Uzzo and R. Nuzzi (1979). *Hydrobiologia* 64, 233–237.

Schreiber, E. (1927). *Wiss. Meeresuntersuch. (Abt. Helgoland)* 16, 1–34.

Schultz, J. S. and P. Gerhardt (1969). *Bact. Rev.* 33, 1–47.

Shanks, A. L. and J. D. Trent (1979). *Limnol. Oceanogr.* 24, 850–854.

Skipnes, O., I. Eide and A. Jensen (1980). *Appl. Environ. Microbiol.* 40, 318–325.

Skoglund, L. and A. Jensen (1976). *J. Exp. Mar. Biol. Ecol.* 21, 169–178.

Skulberg, O. M. (1964). Algal problems related to the eutrophication of European water supplies, and a bio-assay method to assess fertilizing influences of pollution on inland waters. *In* "Algae and Man" (D. F. Jackson, ed.), pp. 262–299. Plenum Press, New York.

Skulberg, O. M. (1970). *Helgoländ. wiss. Meeresunters.* **20**, 111–125.

Skulberg, O. M. (ed.) (1978). "Symposium on Experimental use of Algal Cultures in Limnology" *Mitt. int. Ver. Limnol.* **21**, 607 pp.

Smayda, T. J. (1964). Enrichment experiments using the marine centric diatom *Cyclotella nana* (clone 13-1) as an assay organism. *In* Proc. Symposium on Marine Ecology. Graduate School of Oceanography, University of Rhode Island occ. publ. **2**, 25–32.

Smayda, T. J. (1971). Further enrichment experiments using the marine centric diatom *Cyclotella nana* (clone 13-1) as an assay organism. *In* "Fertility of the Sea" (J. D. Costlow, ed.), Vol. 2, pp. 493–511. Gordon and Breach, New York.

Smayda, T. J. (1973). *Norw. J. Bot.* **20**, 219–247.

Smayda, T. J. (1974). *Limnol. Oceanogr.* **19**, 889–901.

Smith, W. O., Jr, R. T. Barber and S. A. Huntsman (1982). *J. Plankton Res.* **4**, 651–663.

Specht, D. T. (1974). The use of standardized marine algal bioassays for nutrient assessment of Oregon coastal estuaries. *In* "Proc. Fourth Annual Technical Conference on Estuaries of the Pacific Northwest". Oregon State University, Corvallis, Or. Circular No. 50, 15–31.

Specht, D. T. (1975). Seasonal variation of algal biomass production potential and nutrient limitation in Yaquina Bay, Oregon. *In* "Biostimulation and Nutrient Assessment" (E. J. Middlebrooks, D. H. Falkenborg and T. E. Maloney, eds.) pp. 149–174 Proc. Workshop Utah State University, Logan PRWG168-1.

Specht, D. T. and W. E. Miller (1974). Development of a standard marine algal assay procedure for nutrient assessment. *In* "*Proc. Seminar on Methodology for Monitoring the Marine Environment*". pp. 194–230. US EPA. EPA-600/4-74-004.

Spicer, C. C. (1955). *Biometrics* **11**, 225–230.

Staker, R. D. and S. F. Bruno (1980). *Bot. Mar.* **23**, 167–172.

Steyn, D. J., D. F. Toerien and J. H. Visser (1974). *South African J. Sci.* **70**, 277–278.

Strickland, J. D. H. (1968). *Limnol. Oceanogr.* **13**, 388–391.

Subba Rao, D. V. (1981). *Bot. Mar.* **24**, 369–379.

Sunda, W. G., R. T. Barber and S. A. Huntsman (1981). *J. Mar. Res.* **39**, 567–586.

Takahashi, M. and N. Fukazawa (1982). *Mar. Biol.* **70**, 267–273.

Tarkiainen, E., I. Rinne and L. Niemistö (1974). *Merentutkimuslait. Julk./Havsforskningsinst. Skr.* **238**, 39–52.

Taylor, F. J. (1973). Phytoplankton and nutrients in the Hauraki gulf approaches. *In* "Oceanography of the South Pacific" (R. Fraser, ed.), pp. 485–492. New Zealand National Commission for UNESCO, Wellington.

Teixeira, C. and J. G. Tundisi (1981). *Bolm Inst. Oceanogr. S. Paulo* **30**, 77–86.

Teixeira, C. and A. A. H. Vieira (1976). *Bolm Inst. Oceanogr. S. Paulo* **25**, 29–42.

Tempest, D. W. (1979). The continuous cultivation of micro-organisms. I. Theory of the chemostat. *In* "Methods in Microbiology" (J. R. Norris and D. W. Ribbons eds.), Vol. 2, pp. 259–276. Academic Press, London and New York.

Thayer, G. W. (1974). *Oecologia (Berlin)* **14**, 75–92.

Thomas, W. H. (1959). The culture of tropical oceanic phytoplankton. *In* "International Oceanographic Congress Preprints" (M. Sears, ed.), pp. 207–208. Amer. Ass. Adv. Sci., Washington.

Thomas, W. H. (1967). The nitrogen nutrition of phytoplankton in the northeastern tropical Pacific Ocean. *In* "*Proc. International Conference on Tropical Oceanography*" **5**, 280–289, Stud. trop. Oceanogr., Miami.

188   S. Y. Maestrini et al.

Thomas, W. H. (1969). *J. Fish. Res. Bd Can.* **26**, 1133–1145.
Thomas, W. H. (1977). *Bull. Inter-Amer. Trop. Tuna Comm.* **17**, 173–212.
Thomas, W. H., D. L. R. Seibert and A. N. Dodson (1974). *Estuar. Coast. Mar. Sci.* **2**, 191–206.
Tranter, D. J. and B. S. Newell (1963). *Deep-Sea Res.* **10**, 1–9.
Trotter, D. M. and A. C. Hendricks (1976). *Wat. Res.* **10**, 913–917.
Turpin, D. H., J. S. Parslow and P. J. Harrison (1981). *J. Plankton Res.* **3**, 421–431.
Vargo, G. A., P. E. Hargraves and P. Johnson (1975). *Mar. Biol.* **31**, 113–120.
Vedernikov, V. I. and V. V. Sapozhnikov (1978). *Trans. P.P. Shirsov Institute of Oceanology* **112**, 76–82.
Vince, S. and I. Valiela (1973). *Mar. Biol.* **19**, 69–73.
Wetzel, R. G. (1965). Nutritional aspects of algal productivity in marl lakes with particular reference to enrichment bioassays and their interpretation. *In* "Primary Productivity in Aquatic Environments". Proc. I.B.P. Symposium. *Mem. Ist. Ital. Idrobiol.* suppl. **18**, 137–157.
Williams, P. J. L. and I. R. Muir (1981). Diffusion as a constraint on the biological importance of microzones in the sea. *In* "Ecohydrodynamics" (J. C. Nihoul, ed.), pp. 209–218. Elsevier, Amsterdam.
Wilson, W. B. (1966). *Fla Bd Conserv. Prof. Pap.* **7**, 42 pp.
Yoder, J. A. (1979). *Limnol. Oceanogr.* **24**, 97–106.
Yoneshigue-Braga, Y., S. Y. Maestrini and E. Gonzalez-Rodriguez (1979). *C. R. Acad. Sci. Paris*, sér. D **288**, 135–138.

Note Added in Proof

A paper relevant to the topic reviewed here appeared after going to press. Takahashi and Fukazawa (1982) have indeed used the approach within Section C1 (as depicted by Fig. 1C), namely dialysis cultures immersed in a differentially enriched water tank, but their tanks were too small to yield all the advantages we expected from this method.

As expected, this approach provided some interesting information on the behaviour of the various algae. Thus, the diatoms as a whole appeared to be favoured by conditions of macronutrient (N and P) sufficiency. The dominant species behaved differently only when nutrients became scarce; for example *Skeletonema costatum* was still able to maintain an active growth rate, whereas *Thalassiosira sp.* was not. In N- and P-exhausted conditions two flagellates, *Gymnodinium sp.* and *Heterosigma sp.*, took the upper hand but they required the presence of some micronutrients (Fe and vitamin B_{12} for the *Gymnodinium*, and Mn and vitamin B_{12} for the *Heterosigma*). Curiously, the flagellate *Eutreptiella sp.* behaved like a diatom, although the addition of Mn and vitamins did enhance its growth rate.

6 The Use of Seaweeds for Monitoring Coastal Waters

H. G. Levine

Marine Sciences Research Center
State University of New York, Stony Brook
Stony Brook, New York, USA

I. Introduction

The use of seaweeds as monitors of pollution has increased in recent years. Towards this end, the macrophytic algae have several intrinsic advantages. They are, for the most part, sessile in nature and can therefore be used to characterize one location over time. Seaweeds are easily collected in abundance at many coastal localities and they readily accumulate compounds present within the waters of their environment.

These plants have often been used as indicators of pollution at the community level, with the occurrence and abundance of taxa taken as an indication of environmental conditions (Grenager, 1957; Bellamy *et al.*, 1967; Widdowson, 1971; Borowitzka, 1972; Edwards, 1972, 1975; Mathieson and Fralick, 1973; Littler and Murray, 1975, 1977; Murray and Littler, 1978; Klavestad, 1978; Wilkinson and Tittley, 1979; Thom, 1980; Schneider, 1981). Although community structure investigations can be extremely revealing, there are disadvantages inherent in this approach. Sampling, separation and enumeration of organisms is time consuming and requires expertise. Also, adequate keys and checklists are not always available (Stein and Denison, 1967). The ultimate interpretation of results is difficult because cause and effect relationships for species distributions are intricately tied up with natural processes. As a result, detecting the impact of pollution on community structure often requires a long-term investigation.

Alternatively, approaches exist for appraising polluted environments employing individual species as indicators. The use of organisms which respond to environmental conditions in characteristic ways can be termed biodetection (Baker, 1976). Algal growth often changes as some function of the level of pollution. Size or productivity measurements can therefore give a relatively simple method for evaluating a polluted condition and one which is sensitive to grades of effects (Oglesby, 1967; Kusk, 1980; Kindig and Littler, 1980). Biodetectors can be used in

ALGAE AS ECOLOGICAL INDICATORS
ISBN 0 12 640620 0

laboratory bioassay studies or in field deployment investigations. The former involves laboratory tests in either static or flow-through systems. In the latter, organism growth varies as some function of distance from a pollution source.

Seaweeds can also be used as bioaccumulators, i.e. as organisms that accumulate compounds from the environment in which they have grown. In this capacity they function as continuous sampling monitors for pollutants. Tissue analyses minimize the difficulties associated with obtaining representative samples of compounds in coastal waters. The degree of pollutant accumulation is a complex function of numerous factors, but if these are understood and taken into consideration when interpreting results, a meaningful comparison of water quality conditions within and between coastal environments is possible.

This chapter focuses on procedures for the use of seaweeds as bioindicators of pollution at the species level. It is meant as a guide to the variety of approaches that have been developed. In view of the wide range of pollutants that have been studied, a review of the results obtained by way of seaweed methodologies exceeds the scope of this presentation.

II. Life History Considerations

The diverse life history types among the algae offer different attractions for coastal monitoring. Seaweeds with an annual life cycle are valuable as they reflect environmental conditions over a well-defined time interval. Their rapid growth rates also make them appealing for laboratory investigations. Alternatively, species of perenniating algae exist from which water quality information can be extracted pertaining to conditions up to several years past.

The various stages in algal life cycles make a range of within-species assays possible. Different stages can be expected to have different susceptibilities and suitabilities for biomonitoring. Of particular interest are the reproductive cells, which often constitute the most vulnerable link in the algal life cycle.

The selection of assay organisms involves seemingly conflicting considerations. The species should be hardy, because it is desirable to be able to propagate and subject it to laboratory manipulations, but it must be sensitive enough to respond to levels of pollution found in the environment. The degree of response should be reproducible within the strain and life stage under consideration. A ubiquity of occurrence is advantageous both for ease of procurement and for intra- and inter-regional comparisons. Lastly, results generally have a wider appeal if the organism is either of ecological importance or economic interest.

Brown algae (Phaeophyceae) have been frequently employed for coastal monitoring. They often constitute the dominant members of the seaweed community both in the littoral and sublittoral zones. Some species are economically important, their polysaccharide constituents being used in a wide range of applications. Members of the Fucales (*Fucus, Ascophyllum*) and Laminariales (*Laminaria, Macrocystis*) have received the most attention for pollution assessment. Some knowledge of their life cycles is requisite for comprehension of experimental protocols that have been developed.

The life cycle of *Laminaria* involves an alternation between haploid gametophytes and a diploid sporophyte. Meiosis occurs in localized patches of sporogenous tissue called sori. Upon release, the motile meiotic products (zoospores) settle and give rise to microscopic male and female gametophytes. Fertilization is oogamous with the sporophyte emerging from the oogonium and frequently overgrowing the female gametophyte.

In *Fucus*, the gametes are the only haploid cells in the life cycle. Fertile receptacles contain pits (conceptacles) in which gametes are produced directly as a result of meiosis. The reproductive cells (oospheres, spermatozoids) are exuded into the seawater where fertilization takes place. Zygotes settle, germinate and develop directly into the adult plants.

Red algae (Rhodophyceae) have also been used in pollution assessment studies. In terms of species composition, reds usually make up the largest component of seaweed communities. Although they rarely constitute the ecological dominants as determined by biomass in intertidal localities, red algae can be the major biomass component in the subtidal community (*Chondrus, Phyllophora*). The polysaccharide constituents of many Rhodophyceae are also of economic importance.

The generalized life cycle of a red alga involves three plant types: gametophytes, carposporophytes, and tetrasporophytes. Male and female gametophytes are haploid and give rise to gametes (spermatia and carpogonia respectively) by mitotic divisions. After syngamy, the zygote develops into a diploid carposporophyte which remains attached to the female gametophyte. Carposporophytes give rise to carpospores via mitotic divisions. Carpospores are dispersed into the environment and germinate into diploid tetrasporophytes. Meiosis occurs on these plants within tetrasporangia, giving rise to haploid tetraspores which are dispersed and develop into the male and female gametophytes. Any or all of these stages can be used for the investigation of pollution effects.

Green seaweeds (Chlorophyceae) have received the least attention in marine pollution applications. Although some species can be ecological dominants (*Ulva, Enteromorpha*), members of this class find their greatest preponderance in freshwater environments and are currently of little economic importance. Some green algae however, have potential for coastal water monitoring due to advantageous structural and ecological characteristics.

The green seaweed *Ulva* has a flat blade-like thallus two cell layers thick. Its life cycle consists of an alternation between haploid gametophytes and diploid sporophytes which are not always readily distinguishable in the field. Gametophytes produce gametes (by mitotic divisions) which fuse to form zygotes that develop into sporophytes. Meiosis occurs in the peripheral tissues of mature sporophytes, producing zoospores which settle and develop into gametophytes.

Departures from this basic scheme are possible. Of particular interest is the ability of gametes to develop parthenogenetically, i.e. without fusion. This results in the production of genotypically identical plants which can be used to minimize data variation attributable to genetic differences between experimental organisms.

III. Seaweeds as Biodetectors

Plants grow by cell division and elongation. The rate and magnitude of these processes can be measured in several ways. Increase in dimensions is the simplest means of assessment. Changes in rates of primary productivity and biomass determinations constitute alternative approaches.

Both laboratory and field studies of this nature have been undertaken. Laboratory investigations are useful in revealing cause and effect relationships, but many important modifying circumstances and/or synergistic relationships can best be uncovered by field studies, so field and laboratory investigations complement one another.

A. Size Measurements

The most frequently used measurement for pollution assessment studies with seaweeds is size. Exactly what is measured, and the sensitivities involved, vary greatly.

1. Kelp growth studies   The localized meristematic regions of some seaweeds have been capitalized on for growth determinations. This approach has been widely applied to the kelps, in which growth is primarily restricted to the base of the blades. The usual procedure is to punch a hole within the blade at a predetermined location relative to the meristematic zone. After the growth interval, another hole is punched at the same location, and the distance between holes taken as the growth measurement (Parke, 1948; Sundene, 1964; Mann, 1973; Kain, 1976; Lüning, 1979).

Markham *et al.* (1979) have devised a continuous flow culture system to assess the effects of pollutants on *Laminaria saccharina* growth as measured by the hole-punching method. Laboratory maintained strains were used, permitting a standardization with respect to genetic composition. Male and female gametophytes were cultured separately under red light conditions which prevented fertility induction. When required, samples of both sexes were removed, briefly ground together with mortar and pestle in a small amount of medium, and the resulting suspension was poured on to substrata and maintained in white light for two days. This procedure induced gametophyte fertility and subsequent fertilization. The substrata, with developing sporophytes attached, were transferred to a culture system, the plants thinned at periodic intervals and cultured to the desired size. The sporophytes produced were of comparable size, and genetic variation was minimized. Plants were subsequently transferred to a continuous-flow culture system, holes punched at the base of the blades, and the effects of toxicants on growth measured. Sporophytes suitable for study (50–60 mm long) were produced within seven weeks. Assays typically lasted six days.

Eutrophication of natural waters is perhaps the most visible category of pollution. Accelerating rates of input for nitrogen and phosphorus have resulted in the overproduction of organic matter, its decay, and subsequent degradation of water quality (Goldman and Ryther, 1976). North (1977) capitalized on the rapid growth rate of the giant kelp *Macrocystis* sp. to characterize the capacity of different waters to promote primary productivity, and thus generated a qualitative measure of eutrophication. Juvenile plants were maintained in 40-litre aquaria receiving vigorous agitation and frequent changes of seawater from different localities. Similar studies with *Laminaria saccharina* have been performed (Burrows and Pybus, 1970, 1971; Burrows, 1971).

In an approach which permits utilization of earlier developmental stages, Steele and Hanisak (1979) have collected sporogenous tissue from *Laminaria saccharina* plants in the field and induced the release of zoospores which were cultured in the presence of toxicant solutions. Increase in gametophyte dimensions, development of sexual structures, and growth of sporophytes were scored. Pybus (1973) has also worked with *L. saccharina* assays in which zoospores were exposed to a toxicant solution and growth of the resulting gametophytes measured. Growth of young sporophytes cultured in the presence of pollutants were measured as well.

2. Fucoid growth studies Perhaps the most precise method for measuring the effects of pollutants on seaweed growth is that developed by Strömgren (1975, 1977a,b; 1979a,b; 1980a,b). The effects of heavy metals on members of the Fucales have been studied in this fashion. The procedure was initiated by collecting plants from the field and cutting off the growing apices (about 20 mm). The tips were then individually attached to plexiglass frames with tape and covered by plexiglass plates so that the most apical portion of the tips protruded over a hole within the glass plate (Fig. 1). Directly opposite to the apex was an adjustable diffraction edge. Together, the diffraction edge and the apex formed a slit through which monochromatic light from a low energy laser was passed, resulting in the production of a diffraction pattern made up of spots which were photographed. Apical growth decreased the slit width, which was calculated based upon the distance between two spots in the diffraction pattern.

The units with their attached apices were preconditioned in flowing seawater and exposed to experimental treatments. Growth was then measured at given intervals by the laser diffraction method. Measurements required less than 30-second emersions of the experimental unit for handling, and less than two-second laser

194   *H. G. Levine*

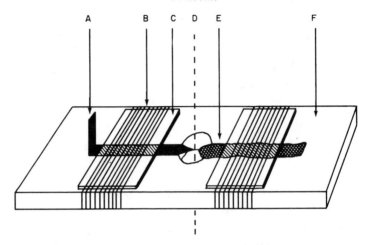

Fig. 1 Growing tip mounted on plexiglass frame. A: adjustable polyethylene diffraction edge. B: glass fibre tape. C: thin plexiglass plate. D: laser beam. E: growing tip of alga. F: plexiglass frame. (From Stromgren (1977) *J. Exp. Mar. Biol. Ecol.* **29**, 181.)

beam exposures. The technique was sensitive enough to detect growth at five minute intervals. Daily measurements have been made over the course of three weeks without injurious effects to the apices.

Steele and Hanisak (1979) have developed procedures for the use of *Fucus edentatus* in more traditional bioassay investigations. Fertile plants were collected, and gamete release followed by fertilization was induced. The zygotes were then cultured in toxicant solutions and the resulting plants scored for length increases after 12 days.

3. Red algal growth studies   Boney and co-workers (1959a,b; 1962; 1963a,b) established procedures for using the following red algae in bioassay investigations: *Plumaria elegans, Polysiphonia lanosa, Spermothamnion repens, Ceramium flabelligerum, Ceramium pedicallatum, Antithamnion plumula, Callithamnion tetricum, Brongniartella byssoides,* and *Nemalion multifidum.* With the exception of *Nemalion*, for which carposporophytes were used, tetrasporophytes were collected, branches bearing tetrasporangia (carposporangia for *Nemalion*) were then placed on glass slides in sterile seawater and left overnight. Branches were subsequently removed, and the released spores left undisturbed for two days to ensure attachment. This usually resulted in 150–300 sporelings per slide. Slides were immersed for various times in different toxicant solutions, washed and transferred to culture dishes. Growth was most often scored within seven days. The effects of heavy metals and a variety of other toxic agents on growth and mortality were studied. Comparable procedures for the use of both gametophytes and tetrasporophytes of *Champia parvula* have been developed by Steele and Thursby (1980), and growth has been assayed against different oil types and concentrations.

Stewart (1977) investigated the effect of lead exposure on the growth of four red algae (*Platythamnion pectinatum*, *Platysiphonia decumbens*, *Plenosporium squarrulosum* and *Tiffaniella snyderae*) in which the plants consist of filaments arising from apical cell divisions only, i.e. the daughter cells produced by apical divisions do not divide again in the same plane and therefore do not contribute additional cells which would extend the filament length. This enabled a periodic measurement of not only the number of cells composing the filament, but also the degree of elongation of the individual cells. Cultures were established from apical fragments, various concentrations of lead added to the medium, and every cell of the filaments measured at intervals for up to ten weeks.

4. Ulva growth studies There is a correlation between sewage pollution in coastal waters and abundant *Ulva* growth (Cotton, 1911; Sawyer, 1965; Subbaramaiah and Parekh, 1966; Fogg, 1969; Burrows, 1971; Russell, 1973; Fletcher, 1974). This observation has led to both *in vitro* and *in situ* studies using the growth of discs cut from the thalli of *Ulva* as an indication of the degree of eutrophication in coastal waters (Waite and Gregory, 1969; Cochrane *et al.*, 1970; Burrows, 1971; Rhyne, 1973; Levine and Wilce, 1980; Parker, 1981). A major advantage of this technique is that numerous discs can be obtained from one plant, and genetically uniform test material is useful in circumventing genetic sources of variation in biodetection studies. Laboratory tests involve growing the discs in seawater collected from different sites. Muds from different locations can also be compared by making soil extract supplements from sediments collected at the test sites (Burrows, 1971).

Several factors have been shown to reduce the reliability of the disc approach for *in situ* deployments (Levine and Wilce, 1980). Differentiation and discharge of reproductive cells is a major factor affecting disc growth capacity. In the absence of reproductive cell discharge, disc growth potential varies according to site of disc origin within the thallus. In addition, physical and biological factors result in significant disc disruption during deployment, necessitating the use of replicates, yet position effects between replicates within the deployment units can exist.

As an alternative approach, procedures for *in situ* deployment of genotypically identical *Ulva* plantlets have been developed (Levine and Wilce, 1980). Instead of measuring the growth of *Ulva* discs, the unit of measurement relates to germling size. Male gametophytes of *Ulva* were induced to discharge gametes under controlled conditions. Thousands of genetically identical germlings (derived from one parental plant) were obtained attached to wooden applicator sticks (15 cm long × 2 mm diameter). These germlings were the result of parthenogenetic development of the gametes, i.e. development without syngamy. Germling containing sticks have been deployed *in situ* between the turns of a polypropylene line moored with cinder blocks and kept vertical in the water column with a surface float (Fig. 2). A weight attached to the line below the float enabled germlings to be maintained at relatively constant distances from the surface over the tidal cycle. This is a modification of the deployment unit developed by Ramus *et al.* (1976). In

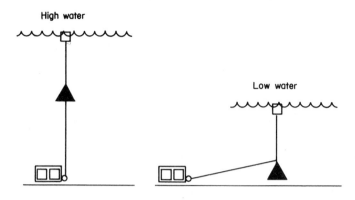

Fig. 2

this way growth and development of *Ulva* were studied throughout the water column at selected sites.

Prior to deployment, the plants were bathed for 30 minutes in saturated solutions of Calcofluor White ST (in sterilized seawater). Calcofluor White (available from American Cyanamid Co.) is a non-toxic fluorescent brightener which is selectively absorbed by cell walls. It is stable, intensely fluorescent within the pH range of 5·0–8·0, and is transported to growing regions within the plant (Cole, 1964). Calcofluor White treatment permits positive identification of plants initially deployed. Other plants which colonize applicator sticks during the deployment period are distinguished by their lack of fluorescence when viewed under ultraviolet irradiation. These procedures permit an accurate knowledge of the plant deployment interval.

B. *Photosynthesis and Respiration Rates*

Photosynthetic and respiration rates are also used as measures of seaweed productivity. The effects of chronic low level oil pollution on primary productivity in the red alga *Porphyra umbilicalis* were investigated by Schramm (1972). This species consists of a discoidal holdfast which expands into a blade one cell layer in thickness. *Porphyra* was collected, acclimatized to laboratory conditions, and the maximum rate of CO_2 uptake was determined while emersed using an infrared absorption technique. The plant was then coated with oil, and the measurement repeated. Three types of crude oil were assayed at oil coating thicknesses ranging from 0·0001–0·1 mm. To coat the seaweed, specimens were attached to a thin wire and immersed in a wide-mouthed container filled with seawater. Subsequently, a pre-determined amount of oil was applied to the water surface, and the alga was slowly pulled up through the layer of oil. This procedure produced uniform coatings on both sides of the plant, corresponding to the thickness of the oil layer at the surface, although at the 0·1 mm level the oil tended to coalesce into droplets on the surface of the seaweed. Alternatively, by allowing the water level to sink below the level of the suspended alga, the coating could be restricted to one side of the plant, a situation which may simulate field conditions more closely.

Using respiration and photosynthetic rate measurements based on the evolution and/or consumption of oxygen, Hellenbrand (1979) combined both *in vivo* and *in situ* procedures for assessing the effects of a pulp mill effluent on productivity of *Fucus vesiculosus*, *Ascophyllum nodosum* and *Chondrus crispus*. Laboratory studies consisted of short-term (up to 14 days) and long-term (180 days) experiments in which plants were maintained in aquaria under controlled conditions and exposed to various concentrations of toxicants. Field studies involved transplanting large stones to which the seaweeds had attached from a control site to sites at various distances from the effluent outfall. Plants were collected from experimental and control locations at monthly intervals, transported in cold aerated containers to the laboratory, and their photosynthetic and respiratory rates measured. Similar procedures were used by Zavodnik (1977), Kusk (1980) and Kindig and Littler (1980).

Using a different approach, Boney (1978) made experimentally based predictions on the detrimental effects of an iron-ore-unloading terminal on productivity of indigenous seaweeds. Plants representing different morphological types were exposed to various ore dust–seawater suspensions, shaken at a constant speed for a predetermined interval, and measured for retention of ore dust and transmission of light through the coated thalli. The degree to which particulate matter adhered to the algal surfaces could be inferred to affect photosynthetic rates by reduction of light received.

Methods exist for the *in situ* estimation of seaweed primary productivity using incorporation of the radioisotope ^{14}C. For species associated with soft bottoms, the technique of covering the alga with transparent and darkened plexiglass cylinders into which $^{14}CO_2$ is injected has been developed (Pomeroy, 1961; Wetzel, 1964). The alternative approach of Drew (1973) involves cutting standardized pieces of tissue from the alga and incubating them within containers in the presence of the ^{14}C isotope. Both techniques are carried out under water, minimizing complications arising from artificial manipulations.

Hsiao *et al.* (1978) used ^{14}C uptake methodologies to assess the effects of oil on primary productivity in two seaweeds. Field collected *Phyllophora truncata* and *Laminaria saccharina* were cleaned, cut into standardized pieces and incubated in light and dark bottles supplied with $^{14}CO_2$ for four to eight hours. Various seawater–oil mixtures were assayed, and the type of oil, method of emulsion preparation, environmental conditions, and species tested were all found to be significant factors affecting photosynthesis. In a similar approach, Hopkins and Kain (1971) took tissue discs from *Laminaria hyperborea* sporophytes and measured their rates of respiration in toxicant solutions over periods of 24 h.

C. Biomass

Biomass estimations are yet another way to use seaweed productivity to characterize environmental conditions. Gonor and Kemp (1978) present a complete account of sampling considerations and quantitative procedures for marine environments. In subtidal localities, the area of interest is typically sampled with standardized quadrats. The sampling strategy depends upon the nature of the site to be characterized. With extremely uniform areas, completely randomized designs can be used. In the absence of such idealized conditions, a stratified random sample is recommended, where random sampling is accomplished within patches of similar nature (Bellamy *et al.*, 1973). The algae are removed, bagged, sorted, dried, and weighed. One useful innovation employs the suction (Venturi effect) created by the rush of air produced by a scuba tank (Fig. 3). One diver scrapes loose all plants within the confines of the quadrat, while a second diver controls the suction created by the degree to which the air valve is opened (Wilce *et al.*, 1978).

D. Reproduction

A number of studies have investigated the effects of pollutants on the reproductive processes of seaweeds. Meiosis is a particularly sensitive phase in the life cycle of most organisms. Steele and Thursby (1980) have found the production of viable tetraspores (as scored by tetraspore germination) in *Champia parvula* tetrasporophytes to be negatively correlated with increasing oil concentrations. Inhibitory effects of hydrocarbons on gamete fusion were also investigated by culturing male and female gametophytes together and scoring for the presence of cystocarps, which only arise as a result of fertilization.

Hopkins and Kain (1971) assayed for attainment of sexual fusion in *Laminaria hyperborea*. Zoospores were cultured in the presence of various toxicants for up to one month. During this interval they germinated into male and female gametophytes and fertilization was scored by the development of sporophytes. Both the time taken for sporophytes to appear and their subsequent growth were measured.

Some of the most subtle biological phenomena involve the tropic responses and motility of reproductive cells. The use of these processes holds promise for the rapid and relatively simple screening of compounds. The study by Pybus (1973) on the effects of detergents on *Laminaria saccharina* zoospore motility has been one of the few attempts of this sort to date.

Steele (1977) has investigated the effects of petroleum on the sexual processes of *Fucus edentatus*. Fertile receptacles were soaked in petroleum–seawater solutions for five minutes, incubated overnight in moist chambers (Petri dishes containing moistened filter paper), and the following morning returned to the petroleum–seawater solutions. Both oospheres and spermatozoids were released, but successful fertilization failed to occur at the toxicant concentrations tested, as indicated by the absence of germinating zygotes. Germination (and therefore fertilization) occurred when this protocol was altered and the receptacles were

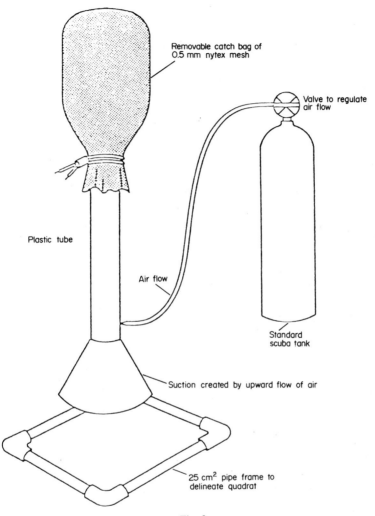

Removable catch bag of
0.5 mm nytex mesh

Valve to regulate
air flow

Plastic tube

Air flow

Standard
scuba tank

Suction created by upward flow of air

25 cm^2 pipe frame to
delineate quadrat

Fig. 3

soaked for five hours in toxicant solutions but later exposed only to sterile seawater. The resulting juvenile plants, however, had reduced growth rates.

The effects of iron ore dust on the sexual process in *Fucus serratus* were studied by Boney (1980). This investigation included the degree of protection afforded by conceptacles, effects on newly released oospheres, and how spermatozoid attraction and fertilization were affected. All portions of seaweed reproductive biology are therefore subject to scrutiny pertaining to disruptive influences attributable to pollutants.

IV. Seaweeds as Bioaccumulators

Algae can be used as continuous sampling monitors for pollutants in coastal waters. To date, phytoplankton have received the bulk of the attention in this regard. Benthic seaweeds have the advantage of algal accumulation properties coupled with a sessile nature and a macrophytic size, enabling a relatively easy collection of large quantities of tissue. As attached plants they reflect environmental conditions at one site over time, a paramount criterion for a successful monitoring tool. Seaweeds have been shown to accumulate heavy metals (Bryan, 1969; Imbamba, 1972; Preston *et al.*, 1972; Yamamoto, 1972; Bryan and Hummerstone, 1973, 1977; Stenner and Nickless, 1974; Young, 1975; Morris and Bale, 1975; Mandelli, 1976; Saenko *et al.*, 1976; Seeliger and Edwards, 1977; Phillips, 1977, 1979; Agadi *et al.*, 1978; Munda, 1978; Sivalingan, 1978, 1980; Shiber and Washburn, 1978; Shiber and Shatila, 1979; Shiber, 1980), hydrocarbons (Youngblood *et al.*, 1971; Youngblood and Blumer, 1973; O'Brien and Dixon, 1976; Lytle *et al.*, 1979; Shaw and Wiggs, 1979), pesticides (Woodwell *et al.*, 1967; Sikka *et al.*, 1976; Amico *et al.*, 1979), PCBs (Parker and Wilson, 1975; Amico *et al.*, 1979; Levine and Wilce, 1980), radionuclides (Alfirmov and Lesiovskii, 1959; Gutknecht, 1961, 1963; Hampson, 1967; Parekh *et al.*, 1968, 1969; Heft *et al.*, 1973; Suzuki *et al.*, 1975; Hamilton and Clifton, 1980), and numerous other compounds from the waters in which they have grown.

Different compounds are accumulated and released to different degrees. Clearly, the chemical properties of a pollutant (Marchyalenene, 1978) and its concentration within the environment are of primary importance, but there are many modifying circumstances. The vertical position of organisms on the shore (or in the water column) affects the degree of pollutant contamination, especially in estuaries (Nickless *et al.*, 1972; Bryan and Hummerstone, 1973; Fuge and James, 1973, 1974). In these environments, river flow frequently dominates tidal flow. River contaminants therefore remain in estuarine surface layers, barring strong mixing forces (Gloyna and Ledbetter, 1969). Some pollutants are primarily restricted to surface layers; other sink relatively rapidly and become associated with the sediments which can subsequently act as a source for these compounds (Blinks, 1952; Duursma and Bosch, 1970; Parry and Hayward, 1973; Hayne *et al.*, 1974; Gardiner, 1974). Inattention to these sampling considerations invalidate many reported conclusions concerning the relative abundances of biota-accumulated pollutants in disparate environments (Phillips, 1977).

While the process of adsorption appears to be the principal mode for accumulation of compounds by algae, secondary processes are also involved. Studies with radioisotopes have shown that uptake frequently involves two stages: (1) an initial passive accumulation involving a rapidly equilibrating adsorption on to the exterior surface, followed by (2) a slower uptake which is dependent upon metabolic processes (Södengren, 1968; Feldt, 1971; Davies, 1973; Parry and Hayward, 1973; Nakahara *et al.*, 1975; Nakamura *et al.*, 1979). The first phase is largely independent of metabolism regulating factors such as light and temperature; the second phase is subject to variation as a function of these parameters.

Chemical uptake is therefore affected by metabolic processes which are specific for the species under consideration, and are influenced by physical parameters within the environment. Metabolic processes within the plant result in active accumulation of some compounds and the degradation or exclusion of others. Since metabolism is affected by such physical factors as temperature, light availability, salinity and degree of desiccation, the influence of different localities and seasons can be significant. The environmental regime at any particular site therefore becomes critical when interpreting bioaccumulation data.

Structural considerations also exist. In many seaweeds, virtually every cell within the organism is in direct contact with the water. Extracts made from the entire plant are therefore more meaningful than would be the case if only localized segments functioned as the primary sites of accumulation. In the more differentiated seaweeds, there is a differential uptake and/or accumulation by different portions of the plant (Bryan, 1969, 1971; Fuge and James, 1973, 1974; Haug et al., 1974; Bohn, 1979). This pattern is particularly evident in radionuclide uptake studies which have demonstrated that most of the radioactivity accumulated is confined to the outermost tissues (Wong et al., 1972). Seaweeds with well-defined, localized meristematic regions have position effects not only between different tissues but also between different regions of the plant (Bryan, 1969; Bryan and Hummerstone, 1973). Generally the older plant parts are found to have accumulated higher concentrations of pollutants.

The perennial seaweed *Ascophyllum nodosum* has a construction which permits a certain degree of historical tracing of water quality conditions. At initiation of the growing season, vesicles form at the apical end of the shoots. Subsequently, the apex adds to the shoot length for the duration of the growth season. The pattern is repeated from year to year, giving rise to alternating vesicles and internodal lengths of the shoot (Fig. 4). This makes possible a subdivision of the alga into year classes which can be analysed separately.

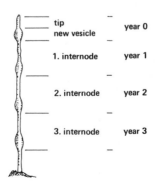

tip	—	year 0
new vesicle		
1. internode	—	year 1
2. internode	—	year 2
3. internode	—	year 3

Fig. 4   Classification of *A. nodosum* age groups (from Myklestad *et al.*, 1978).

There are, however, complications involved in the interpretation of data obtained in this way. Both indigenous and transplanted plants have been analysed (Haug *et al.*, 1974; Skipnes *et al.*, 1975; Foster, 1976; Melhuus *et al.*, 1978; Myklestad *et al.*, 1978, 1979; Klumpp, 1980; Eide *et al.*, 1980). In transplantation studies, the substratum on to which *Ascophyllum* has anchored is undercut and reattached at the experimental location with rapid hardening cement. Myklestad *et al.* (1978) found that when transplanted from a site with high zinc concentrations to one with low zinc concentrations, the newly grown tip segments are characterized by the same low zinc content as the local plants, indicating negligible transfer from older regions. The older regions did, however, experience a slight decrease in zinc content which may be attributable to dilution by new growth and/or a partial release of zinc previously accumulated. Seip (1979) has developed an *Ascophyllum* growth model which has predicted concentration factors for zinc that are in agreement with published results. Other elements can behave differently (Myklestad *et al.*, 1979), and seasonal effects exist (Eide *et al.*, 1980). Each compound of interest must therefore be characterized with respect to its rate of accumulation and the degree to which it is retained under different conditions. Only after all such considerations are taken into account can a meaningful appraisal of a polluted condition be made.

V. Coupling Tissue Extracts with Mutagenicity Assays

It is generally accepted that the increased incidence of cancer today is primarily attributable to environmental agents (Alcantara and Speckman, 1976; Doll, 1977; Barna-Lloyd, 1978; Cairns, 1978; Maugh, 1979). Numerous methods exist to test for the cancer-inducing capacity of man-made chemicals introduced into the environment. One successful line of research involves testing for the ability of these compounds to induce mutations in microbial DNA, since there is a correlation between those chemicals which are carcinogens and those which are mutagens (Ames *et al.*, 1973; McCann *et al.*, 1975). These methods have been applied to water samples in an attempt to pinpoint areas of concern. Owing to dilution considerations, the concentration of large volumes of water is required for detection of most mutagens in aqueous environments (Johnston and Hopke, 1980). This procedural bottleneck can be circumvented by using extracts derived from bioaccumulator species.

Barnes (1980), working with seaweeds from the coast of southern Wales, detected mutagens (as of yet uncharacterized) within tissue extracts made from *Porphyra umbilicalis*, *Fucus vesiculosus*, and *Enteromorpha* spp. using the Ames assay for mutagenicity (Ames *et al.*, 1975). Suitable controls ensured that the mutagens were not a product of the extraction process. Mutagenic activity was lowest in *Fucus*, which may be attributable to its parenchymatous nature. Both *Porphyra* and *Enteromorpha* consist of thalli that are one to two cell layers in thickness, resulting in a large surface area to volume ratio, a property which enhances their bioaccumulation efficiency.

The question of whether mutagens are produced by the seaweeds (endogenously produced) or accumulated by them (exogenously derived) remains unresolved. Five endogenously derived halogenated hydrocarbons which are mutagenic have been isolated from species of the red alga *Plocamium* (Leary *et al.*, 1979). The presence of halogenated compounds is a common occurrence in red algae (Fenical, 1975), a factor which may preclude or reduce their value as mutagen-monitoring bioaccumulators. The data of Barnes tend to support this supposition. Extracts made from the tissues of *Fucus* and *Enteromorpha* displayed a statistically significant correlation between increasing distance of the site sampled from the industrial centre of Swansea, Wales (the presumed site of contaminant input), and decreasing mutagenicity activity. Similarily prepared tissue extracts from the red alga *Porphyra* failed to show this distance relationship.

Preliminary results employing the *Escherichia coli* Pol A assay (Slater *et al.*, 1971; Rosenkranz *et al.*, 1976), in conjunction with extracts prepared for gas chromatography analysis, demonstrated both the presence and the absence of mutagens in extracts derived from different *Ulva* populations situated along the Massachusetts coast (Levine, unpubl. data). These results suggest an exogenous source of mutagens accumulated by *Ulva*. More work is required, however, before the possibility of procedural artefacts or endogenous mutagen production by *Ulva* can be ruled out.

Of primary importance for coastal monitoring is the character of the mutagenic agents. Barnes (1980) detected no significant accumulation of radioactive material in his seaweeds, ruling out radionuclides as the causative agents. Heat inactivation studies indicated that the mutagens were organic in nature. One possibility are the polynuclear aromatic hydrocarbons. These compounds are present in all marine waters as a result of both natural production and man-made inputs (Seuss, 1972); some, such as 3,4-benzpyrene, are potent carcinogens (Falk *et al.*, 1964).

Alternative mutagen candidates are the halogenated compounds present in coastal waters. There is growing documentation that the chlorination process, widely used for the disinfection of drinking water and the treatment of sewage, produces halogenated compounds, some of which are mutagens (Rook, 1974; Bellar *et al.*, 1974; Glatz *et al.*, 1978; Cheh *et al.*, 1980; Marnoka and Yamanaka, 1980). Gas chromatograms of tissue extracts prepared from Massachusetts *Ulva* populations show numerous peaks representing currently unidentified organohalide compounds (Levine and Wilce, 1980). Goldberg *et al.* (1978), reporting on results from the national Mussel Watch Programme, observed a similar spectrum of compounds. It is possible that these chemicals constitute ubiquitous pollutants, some of which may be mutagenic.

Bioaccumulator species can concentrate pollutants orders of magnitude above ambient levels. Their tissues provide samples which represent an integration of the environmental regime to which the organism has been exposed. In view of these properties, the technique of coupling bioaccumulator tissue extracts with microbial mutagenicity assays clearly has potential as a rapid screening procedure for many categories of hazardous chemicals present in coastal waters.

VI. Future Research

The differential susceptibility to pollutants between species of algae is not surprising (Davavin et al., 1975; Kusk, 1980). Perhaps more significant are differences between populations of the same species. In work on phytoplankton, coastal strains have demonstrated greater resistances to pollutants than their open water counterparts. Fisher et al. (1973) reported that growth rates of three phytoplankton species isolated from the Sargasso Sea were more inhibited by PCBs than estuarine strains of the same species. This greater susceptibility of open ocean isolates was also shown relative to compounds non-existent within the oceans (Fisher, 1977).

Estuarine organisms are subjected to an extremely variable environment. It may be that mechanisms which evolved to enable algae to cope with fluctuations in physical factors within the environment also impart some extra degree of resistance to pollutants. For example, adaptation to heterogeneous environments may have involved the development of a highly protective cell membrane, or the evolution of significant metabolic alterations (Michaels, 1979).

Different populations of seaweed species have also been shown to display varying tolerances to pollutants (Russell and Morris, 1970; Hall et al., 1979; Kindig and Littler, 1980; Wilkinson et al., 1980; Hall, 1981), although they may or may not be properly termed "strains" (Kindig and Littler, 1980). Selection for resistance to man-made perturbations (pollutants) has implications for the use of algae as bioindicators (Burrows and Pybus, 1971). Presumably, strains could vary in their response depending upon the environments from which they were isolated.

Use of genetically uniform plants is therefore critical to unravel those factors involved in response variation. Procedures comparable to those developed by Levine and Wilce (1980) for Ulva, and Lüning (1979) for Laminaria, should be applied to other seaweeds. Only in this way can environmental influences be completely separated from genetic effects.

Bioaccumulation studies can also benefit by eliminating or partially controlling genotypic variation. Gas chromatograms of seaweed tissue extracts reveal a spectrum of compounds which represent some combination of natural constituents and accumulated compounds. Definition of the natural constituent pool would permit its subtraction from the total complement of chemicals revealed, leaving only those compounds which characterize the local water mass. Generalized peak patterns could be established against which patterns obtained by analysing field samples would be compared. Anomalous peaks would merit further investigation, and could be characterized with respect to their position relative to known peaks. Eventually a library of non-algal compounds could be established which would enable rapid identification of substances previously characterized.

More work is needed to establish the suitability of coupling seaweed tissue extracts to mutagenicity assays. The question of the endogenous/exogenous origin of mutagens requires additional investigation. The deployment of genotypically uniform plants would also be advantageous in this respect. Given the detection of mutagens from cloned plants deployed at one site, and the absence of mutagens from their counterparts deployed at another site, the conclusion could be drawn that the mutagens were exogenously derived. This does not, however, eliminate the possibility that different portions of the algal genome are being turned on under different environmental regimes.

Populations which give the strongest indication of mutagen accumulation would receive the most intensive chemical investigation in an attempt to identify the mutagenic agent(s). Use of a range of mutagenicity assay systems is also needed since different assays screen for different types of mutagens.

VII. Conclusions

Environmental monitoring involves: (1) laboratory testing for the detrimental effects of compounds on the biota and (2) monitoring for the presence of pollutants in the field. Seaweeds have been used to fulfil both these functions. They are important ecologically as they constitute a significant portion of the food chain base in coastal waters (Ryther, 1963; Riley, 1970; Mann, 1972), and through bioaccumulation they integrate the history of a water mass over prolonged periods of time.

The attached seaweeds can be used to characterize coastal water quality at one site over time. They take nutrients and accompanying chemicals directly from the water, are unable to avoid pollutants, and can be propogated asexually. The propensity of seaweeds as bioaccumulators coupled to the sensitivity of microbial mutagenicity assays for detection of mutagens combine to provide a powerful monitoring technique. Procedures have been, and are being developed to reduce as much of the inherent biological variation as possible. In short, seaweeds have much to recommend them as monitors of pollution in coastal waters.

Acknowledgements

My sincere gratitude is extended to Dr C. Yarish, Dr R. T. Wilce, J. Foertch and K. A. Van den Bosch for their many helpful suggestions and critical review of the manuscript.

References

Agadi, V. V., N. B. Bhosle and A. G. Untawale (1978). *Bot. Mar.* **21**, 247.
Alcantara, E. N. and E. W. Speckman (1976). *Amer. J. Clin. Nutr.* **29**, 1035.
Alfirmov, N. N. and E. M. Lesiovskii (1959). *Botan. Zhur.* **44**, 516.

Ames, B. N., W. E. Durston, E. Yamasaki and F. D. Lee (1973). *Proc. Natl. Acad. Sci. U.S.A.* **70**, 2281.

Ames, B. N., J. McCann and E. Yamasaki (1975). *Mut. Res.* **31**, 347.

Amico, V., G. Oriente, M. Piatelli and C. Tringali (1979). *Mar. Poll. Bull.* **10**, 177.

Baker, J. M. (1976). Biological monitoring—principles, methods and difficulties. *In* "Marine Ecology and Oil Pollution" (J. M. Baker, ed.). J. Wiley, New York.

Barna-Lloyd, G. (1978). *Sci.* **202**, 469.

Barnes, W. S. (1980). Assays for Dispersed Mutagens in Marine Environments Using Extracts of Bioconcentrators. Considerations, Problems and Applications. Ph.D. Dissertation, Botany Dept., Univ. of Mass., Amherst, MA.

Bellamy, D. J., P. H. Clarke, D. M. John, D. Jones, A. Whittick and T. Darke (1967). *Nature* **216**, 1170.

Bellamy, D. J., A. Wittick, D. M. John and D. J. Jones (1973). A method for the determination of seaweed production. *In* "A Guide to the Measurement of Marine Primary Productivity Under Some Special Conditions", pp. 27–33. UNESCO, Paris.

Bellar, T. A., J. J. Lichtenberg and R. C. Kroner (1974). *Sci.* **207**, 90.

Blinks, L. R. (1952). *J. Cell Comp. Physiol.* **39**, 11.

Bohn, A. (1979). *Mar. Poll. Bull.* **10**, 325.

Boney, A. D. (1963a). *J. Mar. Biol. Ass. U.K.* **43**, 643.

Boney, A. D. (1978). *Mar. Poll. Bull.* **9**, 175.

Boney, A. D. (1980). *Mar. Poll. Bull.* **11**, 41.

Boney, A. D. and E. D. S. Corner (1959a). *J. Mar. Biol. Ass. U.K.* **38**, 267.

Boney, A. D. and E. D. S. Corner (1962). *J. Mar. Biol. Ass. U.K.* **42**, 579.

Boney, A. D. and E. D. S. Corner (1963b). *J. Mar. Biol. Ass. U.K.* **43**, 319.

Boney, A. D., E. D. S. Corner and B. W. P. Sparrow (1959b). *Schm. Biochem. Pharm.* **2**, 37.

Borowitzka, M. A. (1972). *Aust. J. Mar. Freshwat. Res.* **23**, 73.

Bryan, G. W. (1969). *J. Mar. Biol. Ass. U.K.* **49**, 225.

Bryan, G. W. (1971). *Proc. R. Soc. Lond., Ser. B.* **177**, 389.

Bryan, G. W. and L. G. Hummerstone (1973). *J. Mar. Biol. Ass. U.K.* **53**, 705.

Bryan, G. W. and L. G. Hummerstone (1977). *J. Mar. Biol. Ass. U.K.* **57**, 75.

Burrows, E. M. (1971). *Proc. Royal Soc. Lond., Ser. B.* **177**, 295.

Burrows, E. M. and C. Pybus (1970). *Proc. Challenger Soc.* **4**, 80.

Burrows, E. M. and C. Pybus (1971). *Mar. Poll. Bull.* **2**, 534.

Cairns, J. (1978). "Cancer: Science and Society." W. H. Freeman, San Francisco, pp. 199.

Cheh, A. M., J. Skochdopole, P. Koski and L. Cole (1980). *Sci.* **207**, 90.

Cochrane, J. J., C. J. Gregory and A. L. Aronson (1970). Water Resources Potential of an Urban Estuary (Saugus River, Pines River and Lynn Harbor complex, Mass.). Mass. Water Resources Research Center, Completion Report, Project No. A-028-MASS.

Cole, K. (1964). *Can. J. Bot.* **42**, 1173.

Cotton, A. D. (1911). On the Growth of *Ulva lactuca* in Excessive Quantity, with Special Reference to the *Ulva* Nuisance in Belfast Lough. Royal Comm. on Sewage Disposal, 7th Rept. II App. IV, p. 121, London, HMSO.

Davavin, I. A., O. G. Mironmv and I. M. Tsimbal (1975). *Mar. Poll. Bull.* **6**, 13.

Davies, A. G. (1973). The kinetics of and a preliminary model for the uptake of radioactive zinc by *Phaeodactylum tricornutum* in culture. *In* "Radioactive Contamination of the Marine Environment". Symp. on the interaction of radioactive contaminants with the constituents of the marine environment. Seattle. Int. Atomic Energy Agency, Vienna.

Doll, R. (1977). Introduction. *In* "Origins of Human Cancer, Book A" (H. H. Hiatt, J. D. Watson and J. A. Winston, eds), pp. 1–12. Cold Spring Harbor.

Drew, E. A. (1973). Primary production of large marine algae measured *in situ* using uptake of $^{14}CO_2$. *In* "A Guide to the Measurement of Marine Primary Production Under Some Special Conditions". UNESCO, Paris.

Duursma, E. K. and C. J. Bosch (1970). *Neth. J. Sea Res.* **4**, 395.

Edwards, P. (1972). *Mar. Poll. Bull.* **3**, 55.

Edwards, P. (1975). *Bot. J. Linn. Soc.* **70**, 269.

Eide, I., S. Myklestad and S. Melson (1980). *Environ. Pollut.* **23**, 19.

Falk, H. L., P. Kotin and A. Mehler (1964). *Archs. environ. Hlth* **8**, 721.

Feldt, W. (1971). Introductory Paper. *In* "Marine Radioecology: The Cycling of Artificial Radionuclides through Marine Food Chains". Seminar on Marine Radioecology, Hamburg, Paris.

Fenical, W. (1975). *J. Phycol.* **11**, 245.

Fisher, N. S. (1977). *Am. Nat.* **111**, 871.

Fisher, N. S., L. B. Graham, E. J. Carpenter and C. F. Wurster (1973). *Nature* **241**, 548.

Fletcher, R. L. (1974). *Mar. Poll. Bull.* **8**, 21.

Fogg, G. E. (1968). "Photosynthesis." English University Press, London.

Foster, P. (1976). *Environ. Pollut.* **10**, 45.

Fuge, R. and K. H. James (1973). *Mar. Chem.* **1**, 281.

Fuge, R. and K. H. James (1974). *Mar. Poll. Bull.* **5**, 9.

Gardiner, J. (1974). *Water Res.* **8**, 157.

Glatz, B. A., C. D. Chriswell, M. D. Arguello, H. J. Sevec, J. S. Fritz, S. M. Grimm and M. A. Thomson (1978). *J. Amer. Water Works Assoc.* **70**, 465.

Gloyna, E. F. and J. O. Ledbetter (1969). "Principles of Radiological Health", pp. 213–284. Marcel Dekker, New York.

Goldberg, E. D. *et al.* (1978). *Environ. Conserv.* **5**, 101.

Goldman, J. C. and J. H. Ryther (1976). Waste reclamation in an integrated food chain system. *In* "Biological Control of Water Pollution" (J. Toubier and R. W. Pierson, Jr, eds). Univ. of Penn. Press.

Gonor, J. J. and P. F. Kemp (1978). Procedures for Quantitative Ecological Assessments in Intertidal Environments. EPA-600/3-78-087, pp. 103.

Grenager, B. (1957). *Nytt. Mag. Bot.* **5**, 41.

Gutknecht, J. (1961). *Limnol. Oceanogr.* **6**, 425.

Gutknecht, J. (1963). *Limnol. Oceanogr.* **8**, 31.

Hall, A. (1981). *Bot. Mar.* **24**, 223.

Hall, A., A. H. Fielding and M. Butler (1979). *Mar. Biol.* **54**, 195.

Hamilton, E. I. and R. J. Clifton (1980). *Mar. Ecol. Prog. Ser.* **3**, 267.

Hampson, M. A. (1967). *J. Exp. Bot.* **18**, 17.

Haug, A., S. Melsom and S. Omang (1974). *Environ. Pollut.* **7**, 179.

Hayne, D. W., T. W. Duke and T. J. Sheets (1974). Pesticides in estuaries. *In* "III Coastal Ecosystems of the U.S." (H. T. Odum, B. J. Copeland and E. A. McMahan, eds). Conservation Foundation in cooperation with the National Oceanic and Atmospheric Administration Office of Coastal Environment. Wash., D.C.

Heft, R. E., W. A. Phillips, H. R. Ralston and W. A. Steele (1973). Radionuclide transport studies in the Humboldt Bay marine environment. *In* "Radioactive Contamination of the Marine Environment". Symp. on the interaction of radioactive contaminants with the constituents of the marine environment. Seattle, Int. Atomic Energy Agency, Vienna.

Hellenbrand, K. (1979). *Proc. Int. Seaweed Symp.* **9**, 161.

Hopkins, R. and J. M. Kain (Jones) (1971). *Mar. Poll. Bull.* **2**, 75.

Hsiao, S. I. C., D. W. Kittle and M. G. Foy (1978). *Environ. Pollut.* **15**, 209.

Imbamba, S. K. (1972). *Bot. Mar.* **15**, 113.

Johnston,·J. B. and P. K. Hopke (1980). *Environ. Mut.* **2**, 419.

Kain, J. M. (1976). *J. Mar. Biol. Ass. U.K.* **56**, 603.

Kindig, A. C. and M. M. Littler (1980). *Mar. Environ. Res.* **3**, 81.

Klavestad, N. (1978). *Bot. Mar.* **21**, 71.

Klumpp, D. W. (1980). *Mar. Biol.* **58**, 257.

Kusk, K. O. (1980). *Bot. Mar.* **23**, 587.

Leary, J. V., R. Kfir, J. J. Sims and D. W. Fulbright (1979). *Mut. Res.* **68**, 301.

Levine, H. G. and R. T. Wilce (1980). *Ulva lactuca as a Bioindicator of Coastal Water Quality.* Mass. Water Resources Research Center. Completion Report, Project No. A-112-Mass, pp. 77.

Littler, M. M. and S. N. Murray (1975). *Mar. Biol.* **30**, 277.

Littler, M. M. and S. N. Murray (1977). Influence of Domestic Wastes on the Structure and Energetics of Intertidal Communities near Wilson Cove, San Clemente Island. California Water Resources Center, Contrib. No. 164, pp. 88.

Lüning, K. (1979). *Mar. Ecol. Prog. Ser.* **1**, 195.

Lytle, J. S., T. F. Lytle, J. N. Gearing and P. J. Gearing (1979). *Mar. Biol.* **51**, 279.

Mandelli, E. (1976). Monitoring of trace elements other than radionuclides. *In* "Manual of Methods in Aquatic Environment Research" (J. E. Portmann, ed.), pp. 27–37. FAO Fish. Tech. Pap. No. 150.

Mann, K. H. (1972). *Mem. Inst. Ital. Idrobiol.* **29**, 353.

Mann, K. H. (1973). Method for determining growth production of *Laminaria* and *Agarum*. *In* "A Guide to the Measurement of Marine Primary Production Under Special Conditions". UNESCO, Paris.

Marchyalenene, E. D. P. (1978). *Soviet J. Ecol.* **9**, 163.

Markham, J. W., K. Lüning and K.-R. Sperling (1979). *Proc. Int. Seaweed Symp.* **9**, 153.

Marnoka, S. and S. Yamanaka (1980). *Mut. Res.* **79**, 381.

Mathieson, A. C. and R. A. Fralick (1973). *Rhodora* **75**, 52.

Maugh, T. H. (1979). *Sci.* **205**, 1363.

McCann, J., E. Choi, E. Yamasaki and B. N. Ames (1975). *Proc. Natl. Acad. Sci. U.S.A.* **72**, 5135.

Melhuus, A., K. L. Seip and H. M. Seip (1978). *Environ. Pollut.* **15**, 101.

Michaels, R. A. (1979). *Amer. Nat.* **113**, 942.

Morris, A. W. and A. J. Bale (1975). *Estuar. Coastal Mar. Sci.* **3**, 153.

Munda, I. M. (1978). *Bot. Mar.* **21**, 261.

Murray, S. N. and M. M. Littler (1978). *J. Phycol.* **14**, 506.

Myklestad, S., I. Eide and S. Melsom (1978). *Environ. Pollut.* **16**, 277.

Myklestad, S., I. Eide and S. Melsom (1979). *Proc. Int. Seaweed Symp.* **9**, 143.

Nakahara, M., T. Koyanagi and M. Saiki (1975). Concentration of radioactive cobalt by seaweeds in the food chain. *In* "Impacts of Nuclear Releases into the Aquatic Environment", Proc. internat. symp. on radiological impacts of releases from nuclear facilities into aquatic environments, pp. 301. Otaniemi, Int. Atomic Energy Agency, Vienna.

Nakamura, R., M. Nakahara, T. Ishii, T. Ueda and C. Shimizu (1979). *Bull. Jap. Soc. Fish.* **45**, 757.

Nickless, G., R. Stenner and N. Terrile (1972). *Mar. Poll. Bull.* 3, 188.

North, W. J. (1977). Algal nutrition in the sea—management possibilities. *In* "The Marine Plant Biomass of the Pacific Northwest Coast" (R. W. Krauss, ed.), pp. 215–230. Oregon State Univ. Press.

O'Brien, P. Y. and P. S. Dixon (1976). *Brit. Phycol. J.* 11, 115.

Oglesby, R. T. (1967). Biological and physiological basis of indicator organisms and communities. *In* "Pollution and Marine Ecology" (T. A. Olsen and F. J. Burgess, eds), pp. 267–269. Intersci. Pub., London.

Parekh, J. M., S. T. Talreja and Y. A. Doshi (1968). *Cur. Sci.* 37, 700.

Parekh, J. M., S. T. Talreja, B. J. Bhalala and Y. A. Doshi (1969). *Cur. Sci.* 38, 308.

Parke, M. (1948). *J. Mar. Biol. Ass. U.K.* 27, 651.

Parker, H. (1981). *Mar. Biol.* 63, 309.

Parker, J. G. and F. Wilson (1975). *Mar. Poll. Bull.* 6, 3.

Parry, G. D. R. and J. Hayward (1973). *J. Mar. Biol. Ass. U.K.* 53, 915.

Phillips, D. J. H. (1977). *Environ. Pollut.* 13, 281.

Phillips, D. J. H. (1979). *Environ. Pollut.* 18, 31.

Pomeroy, L. R. (1961). Isotopic and other techniques for measuring benthic primary production. *In* "Proc. Conf. on Primary Productivity Measurement, Marine and Freshwater" (M. S. Doty, ed.), p. 97. Univ. Hawaii, AEC Div. Tech. Inform. (TID-7633).

Preston, A., D. F. Jefferies, J. W. R. Dutton, B. R. Harvey and A. K. Steele (1972). *Environ. Pollut.* 3, 69.

Pybus, C. (1973). *Mar. Poll. Bull.* 4, 73.

Ramus, J., S. I. Beale, D. Mauzerall and K. L. Howard (1976). *Mar. Biol.* 37, 223.

Rhyne, C. F. (1973). Field and experimental studies on the systematics and ecology of *Ulva curvata* and *Ulva rotundata*. UNC-SG-73-09, pp. 124.

Riley, G. A. (1970). *Adv. Mar. Biol.* 8, 1.

Rook, J. J. (1974). *Water Treat. Examin.* 23, 234.

Rosenkranz, H. S., B. Gutter and W. T. Speck (1976). *Mut. Res.* 41, 61.

Russell, G. (1973). *J. Ecol.* 61, 525.

Russell, G. and O. P. Morris (1970). *Nature* 228, 288.

Ryther, J. H. (1963). Geographic variations in productivity. *In* "The Sea" (M. N. Hill, ed.), pp. 347–380. J. Wiley, New York.

Saenko, G. N., M. D. Koryakova, V. F. Makienko and I. G. Dobrosmyslova (1976). *Mar. Biol.* 34, 169.

Sawyer, C. N. (1965). *J. Wat. Poll. Control Fed.* 37, 1122.

Schneider, C. W. (1981). *J. Therm. Biol.* 6, 1.

Schramm, W. (1972). *Proc. Int. Seaweed Symp.* 7, 309.

Seeliger, U. and P. Edwards (1977). *Mar. Poll. Bull.* 8, 16.

Seip, K. L. (1979). *Ecol. Modelling* 6, 183.

Seuss, M. J. (1972). Polynuclear aromatic hydrocarbon pollution of the marine environment. *In* "Marine Pollution and Sea Life" (M. Ruivo, ed.), p. 568. Unipub. Inc., New York.

Shaw, D. G. and J. N. Wiggs (1979). *Phytochem.* 18, 2025.

Shiber, J. G. (1980). *Hydrobiologia* 69, 147.

Shiber, J. G. and T. Shatila (1979). *Hydrobiologia* 63, 105.

Shiber, J. G. and E. Washburn (1978). *Hydrobiologia* 61, 187.

Sikka, H. C., G. L. Butler and C. P. Rice (1976). Effects, Uptake, and Metabolism of Methoxychlor, Mirex, and 2,4-D in Seaweeds. EPA 600/3-76-048, pp. 39.

Sivalingan, P. M. (1978). *Bot. Mar.* 21, 327.

Sivalingan, P. M. (1980). *Mar. Poll. Bull.* **11**, 106.

Skipnes, O., T. Roald and A. Haug (1975). *Physiol. Plant.* **34**, 314.

Slater, E. E., M. D. Anderson and H. S. Rosenkranz (1971). *Can. Res.* **31**, 970.

Södengren, A. (1968). *Oikos* **19**, 126.

Steele, R. L. (1977). Effects of certain petroleum products on reproduction and growth of zygotes and juvenile stages of the alga *Fucus edentatus* de la Pyl (Phaeophyceae: Fucales). *In* "Fate and Effects of Petroleum Hydrocarbons in Marine Organisms and Ecosystems" (D. S. Wolfe, ed.), pp. 138. Pergamon, New York.

Steele, R. L. and M. D. Hanisak (1979). *Proc. Int. Seaweed Symp.* **9**, 181.

Steele, R. L. and G. B. Thursby (1980). Development of a bioassay using the life cycle of *Champia parvula* (Rhodophyta). International Phycological Society Meeting, Aug 19–21, Glasgow, Scotland.

Stein, J. E. and J. G. Denison (1967). Limitations of indicator organisms. *In* "Pollution and Marine Ecology" (T. A. Olsen and F. J. Burgess, eds), pp. 323–335. Intersci. Pub., London.

Stenner, R. D. and G. Nickless (1974). *Water, Air, Soil Pollut.* **3**, 279.

Stewart, J. G. (1977). *Phycologia* **16**, 31.

Strömgren, T. (1975). *Limnol. Oceanogr.* **20**, 845.

Strömgren, T. (1977a). *J. Exp. Mar. Biol. Ecol.* **29**, 181.

Strömgren, T. (1977b). *Oikos* **29**, 245.

Strömgren, T. (1979a). *J. Exp. Mar. Biol. Ecol.* **40**, 95.

Strömgren, T. (1979b). *J. Exp. Mar. Biol. Ecol.* **37**, 153.

Strömgren, T. (1980a). *J. Exp. Mar. Biol. Ecol.* **43**, 107.

Strömgren, T. (1980b). *Mar. Environ. Res.* **3**, 5.

Subbaramaiah, K. and R. G. Parekh (1966). *Sci. Cult.* **32**, 370.

Sundene, O. (1964). *Nytt. Mag. Bot.* **11**, 83.

Suzuki, H., T. Koyanagi and M. Saiki (1975). Studies on rare earth elements in seawater and uptake by marine organisms. *In* "Impacts of Nuclear Releases into the Aquatic Environment", Proc. internat. symp. on radiological impacts of releases from nuclear facilities into aquatic environments, pp. 77. Otaniemi, Int. Atomic Energy Agency, Vienna.

Thom, R. M. (1980). *J. Phycol.* **16**, 102.

Waite, T. and C. Gregory (1969). *Phycologia* **18**, 65.

Wetzel, R. G. (1964). *Verh. Internat. Verein. Limnol.* **15**, 426.

Widdowson, T. B. (1971). *Bull. S. Calif. Sci.* **70**, 2.

Wilce, R. T., J. Foertch, W. Grocki, J. Kilar, H. Levine and J. Wilce (1978). Flora: Marine Algal Studies, pp. 307–672; 864–906. Vol. 3 of Marine Ecological Studies Related to Operation of the Pilgrim Nuclear Power Station, Boston, MA.

Wilkinson, M., L. MacLeod and I. Fuller (1980). *Bot. Mar.* **23**, 475.

Wilkinson, M. and I. Tittley (1979). *Bot. Mar.* **22**, 249.

Wong, K. M., V. F. Hodge and T. R. Folsom (1972). *Nature* **237**, 460.

Woodwell, G. M., C. F. Wurster, Jr and P. A. Isaacson (1967). *Sci.* **156**, 821.

Yamamoto, T. (1972). *Records of Oceanogr. Works in Japan* **11**, 65.

Young, M. L. (1975). *J. Mar. Biol. Ass. U.K.* **55**, 583.

Youngblood, W. W. and M. Blumer (1973). *Mar. Biol.* **21**, 163.

Youngblood, W. W., M. Blumer, R. L. Guillard and F. Fiore (1971). *Mar. Biol.* **8**, 190.

Zavodnik, N. (1977). *Bot. Mar.* **20**, 167.

Part Three

Terrestrial Ecosystems

7 The Use of Algae as Indicators of Soil Fertility

A. E. Pipe

Department of Plant Sciences
University of Western Ontario
London, Ontario, Canada

L. E. Shubert

Department of Biology
University of North Dakota
Grand Forks, North Dakota, USA

I. Introduction

It has long been recognized that the algae are sensitive indicators of water quality (Patrick, 1972) and that changes in their populations may provide as much information about water chemistry as the chemical determinations themselves. A large amount of research has been conducted on this subject, culminating in the development of standard algal assay procedures for the analysis of nutrients, pesticides, etc. in water (Rand *et al.*, 1976). In contrast, the interaction between algae and soil conditions has received much less attention. This may be a result of several factors.

Notably, there is a general lack of awareness of the presence of algae in the soil. This was acknowledged by Fogg *et al.* (1973), who remarked that the ecology of soil algae has been much less studied than that of the heterotrophic soil microorganisms, creating the impression that they are an unimportant part of the soil microflora. Smith (1978) concluded that "the algae are the most experimentally neglected organisms in the soil". One reason for the paucity of data on soil algae is the difficulty of examining them. Their study is much more demanding than that of their aquatic counterparts. For instance, unless the soil algae are extremely numerous, a culturing technique is necessary before they can be satisfactorily observed and identified. Extreme caution must then be exercised in the interpretation of species lists produced from the observation of such cultures, since all organisms identified may not have been growing actively in the soil. Many algae

ALGAE AS ECOLOGICAL INDICATORS
ISBN 0 12 640620 0

exist in the soil in the form of resting stages. These include the ornamented or thick-walled spores of many green algae and the akinetes produced by blue–green algae (Trainor, 1978). Other algae, while not producing spores as such, may exhibit features conferring upon them the advantages of true resting stages. For example, green and blue–green algae may possess thick mucilaginous investments, enabling them to withstand extreme desiccation. This is particularly well documented for the blue–green algae (Durrell and Shields, 1961; Shields and Durrell, 1964). Similarly, soil diatoms are able to survive adverse conditions by the storage of oil, build-up of inner plates, and reduction in cell size (Patrick, 1977). It is therefore obvious that the presence of some types of algae in soil cultures may represent the germination of spores, or the proliferation of algae normally surviving at low population levels by virtue of modifications in vegetative form and function. A further difficulty in the examination of soil algae is their identification. This is particularly true of the green algae, many of which appear as flagellated zoospores, non-motile unicells, or aggregations of cells, depending on the stage of their development at the time of examination and on the cultural conditions employed. Consequently, their accurate identification relies on their isolation in unialgal culture and on a close observation of their subsequent developmental stages.

Even if methods of studying the algae in the soil were as simple and well researched as those of studying aquatic forms, their interaction with soil conditions would probably still be less well understood than that of aquatic algae with water quality: deterioration of water quality is visually more noticeable than that of soil quality and is therefore more likely to provoke public concern and to be the more closely studied. A similar surveillance of soil quality, particularly as this relates to agriculture, is becoming increasingly necessary (Kimball, 1972). This is supported by Rust *et al.* (1972), who recommended the development of a mathematical "soil quality index" relating to the environmental impact of continuing or sustained use of chemical amendments in crop production or certain other land uses. They listed concerns to include the use of soluble fertilizers (particularly nitrogen and phosphorus), applications of pesticides, and distribution of industrial and domestic wastes that may contain heavy metals. The authors stressed the complexity of such an undertaking, which would require the measurement of a large number of parameters (physical, chemical and microbiological) pertaining to soil conditions. An alternative to such an approach would be the use of the soil organisms themselves to indicate the quality of their habitat. Many of the soil microorganisms are potentially capable of indicating soil conditions (Shubert, 1980). However, the algae (and particularly the Chlorophyceae) are very similar to the higher plants in terms of their cell structure, metabolism and nutrient requirements, so they would obviously be appropriate indicators of soil fertility (Cullimore, 1964, 1965a; Tchan, 1959). Algae are also sensitive indicators of pesticide residues in the soil (McCann and Cullimore, 1979), and their use can provide information on the accumulation of such chemicals and their possible detrimental effects on crop plants. A more detailed account of the relationship between algal growth and soil conditions is presented in the following section.

II. Algal Growth as Affected by Soil Conditions

In the aquatic environment, algal growth is affected by a variety of physical and chemical factors, including temperature, light intensity, pH and nutrient concentration (Patrick, 1972). Any change in these conditions or any introduction of toxicants into the system is accompanied by an alteration in the composition of the algal flora. Similar responses may be expected to be exhibited by the soil algae. The factors influencing the growth of soil algae have been reviewed by a number of authors (Dommergues and Mangenot, 1970; Lund, 1962; Starks *et al.*, 1981). The most important of these factors include physical parameters, such as moisture, temperature, light and soil texture, and the chemical condition of the soil. The physical characteristics are governed by the relatively stable conditions of climate and/or natural soil structure and will not be discussed here. The chemical quality of the soil is determined basically by its composition and biota but may also be influenced by agricultural practices. Fertilizers added to the soil change the balance of mineral nutrients, and pesticide use may result in an accumulation of toxic residues. The composition of the soil microflora may also be altered; this may be undesirable considering that microorganisms have an important role in soil fertility (Cullimore, 1971). Soil chemistry is therefore susceptible to change and requires careful monitoring. It is by the indication of such changes in soil chemistry that the algae could provide information which would aid in the maintenance of soil fertility. Some important aspects of soil chemistry governing soil algal growth are discussed below.

A. Mineral Nutrients

The nutritional requirements of the algae for the major elements (calcium, magnesium, potassium, carbon, nitrogen, phosphorus, sulphur) are the same as those of higher plants (O'Kelley, 1974). Certain trace elements (iron, manganese, silicon, zinc, copper, cobalt, molybdenum are also essential (Round, 1965). The availability of these nutrients has an important influence on both algal species diversity and biomass (assuming that moisture, temperature and light are not limiting).

In studies concerned with the effect of soil nutrients on algae, some elements have been suggested as either limiting or inhibitory. Carson and Brown (1978) reported a high positive correlation of generic diversity and abundance with levels of phosphorus, potassium, calcium, magnesium, manganese, NH_4^+-N and NO_3^--N in soils on a dated Hawaiian volcanic substrate over time. Although levels of PO_4-P in soils exceeded published minimum requirements, King and Ward (1977) showed that phosphorus was the major growth-limiting nutrient in soil. Shubert and Starks (1980) related soil properties with the algae present in surface mine soils by cluster analysis and linear regression. High sodium and low nitrogen levels were suggested as the major factors influencing the algal communities. Manganese was one of the major limiting factors included in a conceptual model developed by Hunt *et al.* (1979) to predict algal abundance in two successional fields and a climax forest. Starks and Shubert (1979) reported a significant negative correlation between manganese levels and algal abundance, whereas sodium correlated positively with abundance on a revegetation test plot in western North Dakota. In a related study, Shubert and Starks (1979) demonstrated positive correlations of algal abundance during succession with levels of nitrogen, PO_4-P, silicon, manganese, aluminium, and lead while sodium, copper, lithium, molybdenum and strontium had negative correlations. Most recently, Starks and Shubert (1982) developed a conceptual model that demonstrated the positive and negative influence of soil chemical properties on algal abundance associated with soils disturbed by mining. Positive nutrient factors were r-Na and e-Fe and negative nutrient factors were w-K, w-Ca, w-Na; w-Mg; r-Mg; r-Li, r-Mn, rSr; E-Cd; EMn; and E.C. These data illustrated the complexity of the system and demonstrated the need for specific models.

B. pH

Nearly neutral or slightly alkaline soils support the most varied algal flora (Dommergues and Mangenot, 1970; Lund, 1962). Although members of the Chlorophyceae may be found in soils of a wide pH range (Dommergues and Mangenot, 1970; Lund, 1945, 1962), the Bacillariophyceae and the Cyanophyceae are less well represented in acid soils (Shubert, 1980). Growth of blue–greens is extremely rare below a pH of 5 (Granhall, 1970; Jurgensen and Davey, 1968). This probably accounts for the abundance of blue–greens in arid soils (Anantani and Marathe, 1972; Cameron, 1964), which are all alkaline. It was postulated by Fogg *et al.* (1973) that the inability of the Cyanophyceae to tolerate acidic soils may be due to a decreased availability of molybdenum, essential for nitrogenase and nitrate reductase activity. There appears to be an absence of data concerning the possible reasons for the avoidance of acid soils by the diatoms. Peterson (1935) considered that the paucity of the diatom flora in such soils was not entirely due to acidity, since he observed large numbers of *Pinnularia* and *Eunotia* in certain acid high-moors. Lund (1945) argued that these diatoms were aquatic and not true terrestrial forms, and were in fact poorly represented in Peterson's own collections from acid soils.

   Both the nutrient status of the soil and its pH are susceptible to considerable alterations by the use of fertilizers. The quality of the soil and the populations of algae therein may also be affected by the introduction of toxicants into the soil. The most important of these, and the only ones considered here, are the pesticides.

C. Pesticides

Although a limited number of studies have been conducted on the influence of pesticides on soil algae under field conditions, it has been established that the organisms are susceptible to a wide variety of these chemicals, the herbicides producing the most adverse effects (McCann and Cullimore, 1979). The data obtained (by different workers) in experiments to determine the influence of herbicides, fungicides and insecticides on soil algae, under both *in vivo* and *in vitro* conditions were summarized by McCann and Cullimore (1979) (Table 1). The chemical names of all pesticides mentioned in this table (and elsewhere in the chapter), together with an indication of their usage (as herbicide, fungicide or insecticide) are given in Table 2.

Table 1   Toxicity of pesticides to soil algae (from McCann and Cullimore, 1979)

| | Toxicity level[a] | | | | | | | |
| | *In vivo* experiments | | | | *In vitro* experiments | | | |
Pesticide	1	2	3	4	1	2	3	4
A-18-17				x				
Amitrol-T[b]							x	x
Asulam							x	
Atratone[b]							x	
Atrazine[b]			x				x	x
Barban[b]					x	x	x	
BHC							x	x
Bromacil						x		
Bromoxynil								x
Ceresan								x
Chloranil[b]					x			
2,4-D[b]				x	x		x	
Dalapon[b]					x		x	
2,4-DB					x			
DDT								x
Dicamba[b]					x		x	
Dichlobenil[b]					x		x	
Dinoseb							x	
Diphenamid					x			
Diquat[b]							x	x
Diuron		x				x		
EPTC[b]					x	x		
Fenuron						x		
Fluometuron						x		

Table 1 (continued)

Pesticide	In vivo experiments				In vitro experiments			
	1	2	3	4	1	2	3	4
Ioxynil								X
Linuron		X				X		
Malathion[b]					X			
Mancozeb		X						
M and B 8882								X
MCPA[b]					X	X		
Methabenzthiazuron	X							
Methoxychlor								X
Metobromuron[b]						X		X
Metribuzin			X			X		
Mirex[b]								X
Monuron		X				X		
Neburon						X		
Paraquat[b]							X	X
Pentachlorophenol			X			X		
Picloram	X					X		
Prometryne			X					
Propanil						X		
Propazine						X		
Siduron[b]	X							X
Simazine[b]				X			X	X
Sodium pentachlorophenate			X			X		
Solan					X			
2,4,5-T[b]	X							X
TCA	X	X				X		
Trifluralin						X		
Vapam					X			

For chemical names of pesticides in this table, see Table 2.

[a] Toxicity level defined as follows:
  1: No toxicity observed
  2: Toxic to some algae at low concentrations (<0·1 p.p.m.)
  3: Toxic to some algae at field application levels (0·1 to 5 p.p.m.)
  4: Toxic to most or all algae at high concentrations (>5 p.p.m.)
[b] Wide range of results obtained by different workers with this pesticide.

Table 2 Chemical names of pesticides[a] mentioned in text

Name used in text	Chemical name
A-18-17	33·3% simazine + 16·7% prometryne
Ametryne	4-ethylamino-6-isopropylamino-2-methylthio-1,3,5-triazine
Amitrol-T	3-amino-1,2,4-triazole-ammonium thiocyanate mixture
Asulam	methyl N-(4-aminobenzenesulphonyl) carbamate
Atratone	2-methoxy-4-ethylamino-6-isopropylamino-s-triazine
Atrazine	2-chloro-4-ethylamino-6-isopropylamino-s-triazine
Barban	4-chloro-2-butynyl N-(3-chlorophenyl) carbamate
BHC[c]	benzenehexachloride
Bromacil	5-bromo-6-methyl-3-s-butyl uracil
Bromoxynil	3,5-dibromo-4-hydroxy-benzonitrile
Ceresan[c]	N-(ethylmercury)-p-toluenesulphonilide
Chloranil[b]	2,3,5,6-tetrachloro-1,4,-benzoquinone
2,4-D	2,4-dichlorophenoxyacetic acid
Dalapon	2,2-dichloropropionic acid
2,4-DB	4-(2,4-dichlorophenoxy) butyric acid
DDT[c]	2,2-bis(p-chlorophenyl)-1,1,1-trichloroethane
Dicamba	3,6-dichloro-2-methoxy-benzoic acid
Dichlobenil	2,6-dichloro-benzonitrile
Dinoseb	2,4-dinitro-6-s-butyl-phenol
Diphenamid	N,N-dimethyldiphenylacetamide
Diquat	1,1'-ethylene-2,2'-bipyridylium dibromide
Diuron	3-(3,4-dichlorophenyl)-1,1-dimethylurea
EPTC	S-ethyl dipropylthiocarbamate
Fenuron	3-phenyl-1,1-dimethylurea
Fluometuron	3-(α,α,α-trifluoro-4-methyl-phenyl)-1,1-dimethylurea
Ioxynil	4-hydroxy-3,5-diiodobenzonitrile
Linuron	3-(3,4-dichlorophenyl)-1-methoxy-1-methylurea
Malathion[c]	O,O-dimethylphosphorodithioate of diethyl mercaptosuccinate
Mancozeb[b]	zinc ion and manganese ethylene bis-dithiocarbamate
M and B 8882	methyl N-(4-nitrobenzene-sulphonyl) carbamate
MCPA	2-methyl-1-4-chlorophenoxyacetic acid
Methabenzthiazuron	3-(2-benzothiazolyl)-1,3-dimethylurea
Methoxychlor[c]	1,1,1-trichloro-2,2-bis(4-methoxyphenyl) ethane
Metobromuron	3-(4-bromophenyl)-1-methoxy-1-methylurea
Metribuzin	3-methylthio-4-amino-6-tert-butyl-1,2,4-triazine-5(4H)-one
Mirex[c]	dodecachloro-octahydro-1,2,4-metheno-2H-cyclobuta(cd)pentalene
Monuron	3-(4-chlorophenyl)-1,1-dimethylurea
Neburon	1-butyl-3-(3,4-dichlorophenyl)-1-methylurea

Table 2 (continued)

Name used in text	Chemical name
Paraquat	1,1'-dimethyl-4,4'-bipyridylium dichloride
Pentachlorophenol	—
Picloram	4-amino-3,5,6-trichloropicolinic acid
Prometryne	2-methylthio-4,6-bis(isopropylamino)-s-triazine
Propanil	3,4-dichloropropionanilide
Propazine	2-chloro-4,6-bis(isopropylamino)-s-triazine
Siduron	3-phenylurea-1-(2-methylcyclohexyl)
Simazine	2-chloro-4,6-bis(ethylamino)-s-triazine
Sodium pentachlorophenate	—
Solan	N-(3-chloro-4-methylphenyl)-2-methyl-pentanamide
2,4,5-T	2,4,5-trichlorophenoxyacetic acid
TCA	trichloroacetic acid
Trifluralin	2-6-dinitro-N,N-dipropyl-4-trifluoromethylaniline
Vapam[d]	sodium N-methyldithiocarbamate dihydrate

[a] All herbicides unless otherwise indicated:
[b] Fungicides
[c] Insecticides
[d] May be used as fungicide, herbicide or nematocide

Since the chemical status of the soil has such a marked influence on algal growth and distribution, it is reasonable to assume that these organisms may be satisfactorily used to assay for chemical parameters related to soil quality and fertility. The use of such assay procedures is discussed below.

III. Development of Algal Assay Procedures for Monitoring Soil Quality

A. Algal Assays versus Chemical Analyses

In consideration of the use of algal assays, it is first necessary to justify such procedures as being as efficient and reliable as the traditional chemical determinations. The advantages and disadvantages of bioassays and chemical methods have been discussed by several authors (Hance and McKone, 1976; Rand et al., 1976; Wright, 1978), and the most important of the factors mentioned in these accounts are outlined below.

First, chemical determination of soil elements, and particularly of pesticides, is frequently time-consuming and expensive, requiring elaborate extraction procedures and analytical techniques. On the contrary, algal assays generally require a limited amount of equipment and can be conducted rapidly and simply. In addition, the efficiency of the chemical procedures is not always as high as is desirable. Extractability of pesticides from soil samples may decrease if these are stored between collection and analysis. Secondly, chemical procedures generally determine the total concentration in the soil of the material in question. This amount is not necessarily that available for uptake by the plant, and consequently does not provide information concerning its potential toxicity (considering pesticides, etc.) or benefits (considering plant nutrients) to plant growth. These determinations may therefore be less useful than bioassays in evaluating the quality of the soil as it affects agriculture. However, bioassays may fail to distinguish between related chemicals, for example, pesticides and their transformation products (formed by degradation of the parent compound). The transformation products, and not only the parent molecule, may be biologically active (Hill and Arnold, 1978) and hence detectable by bioassay.

There are few reports of direct comparisons between bioassays and chemical methods (Hance and McKone, 1976), particularly for soil systems. Cullimore (1965b) reported that a bioassay for potassium using pure strains of *Chlorella vulgaris* and *Stichococcus bacillaris* produced results which compared favourably with chemical methods. He also concluded (1966c) that a bioassay method using *Hormidium flaccidum* was more convenient for the determination of nitrogen, phosphorus and potassium than the chemical testing procedure, since it allowed simultaneous assessment of all three plant nutrients. Wong and Chung (1979) found that soils determined by chemical analysis to be rich in macronutrients also supported luxuriant growth of *Chlorella pyrenoidosa*. Chiou (1980), investigating the application of bioluminescence assays* to the monitoring of herbicides in the natural (aquatic and terrestrial) environment, compared these methods and a gas liquid chromatography method to detect diuron in water used to flush the herbicide from the surface of treated irrigation channels. The bioassay method generally indicated higher concentrations of diuron than did the chromatography procedure. Chiou (1980) postulated that this discrepancy was caused by detection of breakdown products as well as the parent herbicide by the bioluminescence method.

In considering the employment of a bioassay method in preference to chemical analysis for a particular determination, the use of algae instead of higher plants as assay organisms may be questioned. Although assays using higher plants are often quite sensitive and specific, the length of time taken to obtain a result is a considerable disadvantage. This is particularly true of conventional whole plant assays, root and shoot methods being somewhat faster (Hance and McKone, 1976). Algal assays, in contrast, are more convenient, due to the ease and rapidity with which the organisms can be cultured and their response measured. They also place much less demand on equipment, space and the time of personnel (Wright, 1978).

---

* Developed by Tchan *et al.* (1975). Method involves the detection of herbicidal interference with algal photosynthetic oxygen evolution by the bioluminescence system of a photobacterium sensitive to low levels of oxygen.

B. Selection of an Assay Alga

The selection of an assay alga depends primarily on its sensitivity to the material to be detected, and on its suitability for the procedure chosen, that is, its ability to grow satisfactorily under the experimental conditions imposed upon it.

1. Sensitivity of the alga to the material being assayed   Sensitivity to the material to be detected is essential in the consideration of the potential of an organism for assay purposes: the organism should be sensitive to low concentrations of the material. This is particularly true for the assay of herbicides and other pesticides, which are usually present in the soil at concentrations below 2 p.p.m. (Cullimore, 1975). The exact nature of the response of the alga to the chemical is also important. Cullimore (1975), in a study of the sensitivity of a range of members of the Chlorophyceae to a variety of herbicides, identified seven different response patterns. These were:

(*a*) A gradually increasing sensitivity to the herbicide with increasing concentration;

(*b*) Resistance to low concentrations but sensitivity to high concentrations of the herbicide;

(*c*) Initial sensitivity to the herbicide with eventual resistance and growth after the induction period;

(*d*) Resistance even at high concentrations, followed by sensitivity with a subsequent dying off of the algal cells;

(*e*) Mild stimulation of algal growth over a limited range of low concentrations of the herbicide;

(*f*) More marked stimulation of growth at a higher concentration of the herbicide;

(*g*) Complete resistance to the herbicide.

Clearly, the interaction described in (a) would be most desirable for bioassay of pesticides and other toxicants. A corresponding linear relationship between algal growth and increasing nutrient is likewise desirable for the assay of plant nutrients.

As a result of the study described above, Cullimore (1975) recommended the consideration of a number of potentially suitable algal assay organisms for a specific herbicide or group of herbicides. *Hormidium barlowi* was identified as the most suitable alga for a broad spectrum of herbicides, but its poor sensitivity rendered it unsatisfactory for direct soil assay where the highest concentration commonly reported following field applications is 2 p.p.m. A strain of *Haematococcus lacustris*, on the contrary, was found to be highly sensitive to low concentrations (0·05 p.p.m. in some cases) of a wide spectrum of herbicides, particularly the ureas and the bipyridyls.

In the evaluation of data concerning the sensitivity of algal assay organisms to different materials in studies such as that described above, it is essential to consider the experimental procedure used. Of particular importance is the type of algal growth medium. Alternative types of media which have been used in various assay procedures include liquid media and soil or other solid media (Wright, 1978). The medium selected may influence the results obtained by affecting the availability of the material to be detected to the assay organism. For example, many pesticides become adsorbed to organic matter (Grover, 1975) or clay particles (Adams, 1973) in the soil, and hence may be rendered unavailable for algal uptake (Hance and McKone, 1976). The sensitivity of algal assay organisms used to detect such pesticides might therefore appear to be greater in liquid than in soil media.

2. *Suitability of the alga for the assay procedure* Assuming that an alga of satisfactory sensitivity has been selected, it is necessary to evaluate the organism on the basis of its suitability for the method of assay to be used. For example, *Hormidium barlowi*, suggested as a possible assay organism by Cullimore (1975) on the strength of its broad-spectrum (although rather low) sensitivity to the range of herbicides used, is a filamentous form which does not disperse homogeneously in liquid medium. Consequently, growth on the filter paper discs used by Cullimore (1975), impregnated with a suspension of the alga, may be rather patchy. A unicellular alga, such as *Haematococcus lacustris* or *Chlorella* spp. would be more satisfactory for most types of assay.

The algae used by Cullimore (1975) were all forms which may be encountered in the soil. The use of soil algae for assays of soil chemicals has certain advantages over the use of those isolated from other environments. An alga isolated from soil is more likely to grow luxuriantly in a soil-based medium than an alga isolated from a totally different habitat. A similar argument was used by Shoaf (1978) in support of the use of indigenous aquatic algae as assay organisms in tests involving water samples. Aquatic algae have occasionally been used in assessing the nutrient status of the soil. The planktonic *Selenastrum capricornutum*, one of the assay organisms recommended for the standard algal assay procedure developed for aquatic systems (Rand *et al.*, 1976), was used by King and Ward (1977) to determine the phosphate-phosphorus in water leachates of soils. The usefulness of such a procedure for soil systems has been questioned (Starks *et al.*, 1981).

Certain physiological characteristics which may be unique to soil algae should be borne in mind if they are to be used in assays. For instance, soil algae may be able to extract nutrients from soil particles (Starks *et al.*, 1981), contrary to the generally held belief (Hance and McKone, 1976) that chemicals adsorbed are biologically unavailable. Such behaviour has been observed by Bose *et al.* (1971), who found that 17 strains of terrestrial blue–green algae (belonging to the genera *Anabaena*, *Aulosira*, *Nostoc* and *Tolypothrix*) were capable of solubilizing bound (tricalcium) phosphate, making it available for uptake by themselves and other organisms. Venkataraman (1972) observed that extracellular substances of blue–green algae are able to chelate the surrounding medium, leading to greater solubility and availability of soil phosphates. Therefore, a soil alga may exhibit a different response to augmentations of soil nutrients than an aquatic alga.

C. *Available Algal Assay Procedures*

Tchan (1959) was the first to suggest the use of algal assays as an alternative to the traditional chemical analyses for the measurement of nutrients in the soil. Since this work, a variety of different techniques have been developed for the determination of both the nutrient status of the soil and the presence therein of toxicants such as pesticides. These procedures range from the simple monitoring of algal growth in the presence of the chemical in question to elegant techniques such as the bioluminescence method developed by Tchan *et al.* (1975) for the determination of photosynthetic inhibitors. Some of the most important algal assay studies are listed in Table 3.

The procedures most frequently used for the assay of both nutrients and toxicants are those involving the use of filter paper as a vehicle for the algal inoculum and/or the material being assayed. The filter paper is then placed on the soil surface. This technique is simple, reliable and sensitive, and requires little equipment. It does, however, suffer from one major problem common to most bioassays of materials in soils; that of the sterilization of the soil.

The routine methods of soil sterilization are autoclaving, fumigation, (using, for example, chloroform (Parkinson *et al.*, 1971), methyl bromide and 1,2-dichloro-ethane (Crosby, 1976)) and gamma irradiation. Neither autoclaving nor fumigation is entirely desirable, since both risk fundamental changes in soil structure and chemical composition. The high temperatures and pressures of steam sterilization may alter soil surface adsorption characteristics (Furmidge and Osgerby, 1967) and cause some organic matter degradation (Warcup, 1957). Fumigation may affect the chemical reactivity and adsorptive properties of the soil (Hill and Arnold, 1978). Gamma irradiation, although not altering cation exchange capacity, particle size distribution or moisture desorption properties of soil, results (as does autoclaving) in an increased extractability of nitrogen, phosphorus, sulphur and manganese (Eno and Popenoe, 1964). It also promotes greater release of hydrocarbons into the soil and air than does autoclaving (Rovira and Vendrell, 1972), and fails to remove all biological activity (Hill and Arnold, 1978). In the case of sterilization of soils containing pesticides to be assayed, both autoclaving and gamma irradiation may promote the decomposition of the pesticide (Crosby, 1976).

The use of unsterilized soil, however, may also cause problems interfering with the validity of the assay results. These include the interference with the growth of the assay organism by that of the indigenous soil microflora, and the utilization of the material being assayed by organisms other than the assay organism. The former would have the effect of producing erratic and unreliable growth of the assay organism, and the latter of reducing the amount of nutrient or toxicant available to it. Such complications undoubtedly helped to provoke attempts to assay the concentration of nutrient or toxicant in liquid extracts of the soil (as conducted, for example, by Addison and Bardsley, 1968) in favour of direct determinations in the soil itself.

All the procedures listed in Table 3 involve the use of an alga isolated from a terrestrial or aquatic habitat and inoculated back into the culture medium for the

Table 3 Algal assay techniques employed for the determination of nutrients and toxicants in the soil

1. Nutrients

Reference	Procedure	Chemical(s) assayed
Tchan (1959)	Growth of crude soil algal culture compared (by extracted chlorophyll measurement) in media with and without nutrients being assayed	Nitrogen, phosphorus, potassium
Cullimore (1962)	Paper zones, impregnated with nutrient and pure culture of assay alga (*Chlorella vulgaris*, *Stichococcus bacillaris*, *Hormidium flaccidum*, *Vaucheria sessilis*), placed on soil surface. Growth measured visually or with reflectance photometer at 420 nm	Potassium
Tchan *et al.* (1963)	As for Tchan, 1959	Sulphur
Cullimore (1965a)	Soil saturated with culture of *Hormidium crenulatum*. Filter paper discs, impregnated with nutrient, placed on soil. Growth measured by determination of diameter of zone of stimulation around disc	Calcium, copper, iron, magnesium, manganese, nitrogen, phosphorus
Cullimore (1966a)	As for Cullimore (1965a) but with *Hormidium flaccidum* instead of *H. crenulatum*, and filter paper rosette instead of discs	Phosphate-phosphorus
Cullimore (1966b)	As for Cullimore (1962), using pure cultures of *Hormidium crenulatum* or *Ulothrix* sp. Also, soil and inoculum of either of above two algae added to liquid growth medium; growth measured by determination of extracted chlorophyll	Sulphur
Cullimore (1966c)	As for Cullimore, 1962, using filter paper rosettes and *Hormidium flaccidum*	Nitrogen, phosphorus, potassium

1. Nutrients (continued)

Reference	Procedure	Chemical(s) assayed
King and Ward (1977)	*Selenastrum capricornutum* inoculated into liquid media containing soil leachates. Growth measured spectrophotometrically	Phosphate-phosphorus
Wong and Chung (1979)	*Chlorella pyrenoidosa* cultivated in water extracts of several soils	Total nitrogen, water soluble phosphorus, exchangeable calcium, magnesium, potassium, sodium

2. Toxicants

Reference	Procedure used	Chemical(s) assayed
Jansen *et al.* (1958)	Mixed algal cultures spot inoculated on to surface of smooth inert perlite substrate followed by spray application of herbicides. Effect determined by measurement of diameter of film of growth	Substituted urea and *s*-triazine herbicides
Atkins and Tchan (1967)	Herbicide treated soil in cellophane dialysis tubes placed in cultures of *Chlorella* in liquid media. Growth measured by determination of turbidity and chlorophyll extraction	Atrazine
Addison and Bardsley (1968)	*Chlorella vulgaris* inoculated into extract from herbicide treated soil. Growth determined by measuring transmittance of culture	Diuron
Pillay and Tchan (1969)	Unialgal culture inoculated on to paper disc placed on herbicide treated soil. Growth assessed visually	Phenylurea herbicides
Kruglov and Paromenskaya (1970)	As for Pillay and Tchan (1969) but growth measured by determination of ethanol extracted pigments	Simazine
Helling *et al.* (1971)	*Chlorella sorokiniana* aspirated on to soil-covered plate. Movement of pesticides across plate detected by inhibition of algal growth	11 triazine, 9 phenylurea and 13 miscellaneous herbicides; 2 methyl-carbamate insecticides

2. Toxicants (continued)

Reference	Procedure used	Chemical(s) assayed
Kratky and Warren (1971a)	*Chlorella pyrenoidosa* inoculated into liquid medium containing toxicant. Growth measured by chlorophyll analysis	42 photosynthetic and respiratory inhibitors
Kratky and Warren (1971b)	As for Kratky and Warren (1971a) using liquid media containing soil leachates	9 photosynthetic and 2 respiratory inhibitors
Pillay and Tchan (1972)	As for Pillay and Tchan (1969)	Ametryne, atratone, atrazine, bromacil, 2,4-D, dalapon, diuron, monuron, neburon, simazine
Noll and Bauer (1973)	*Phormidium* inoculated on to filter paper impregnated with herbicide. Effect monitored by extent of migration of trichomes across paper	22 herbicides, including carbamates, bipyridylium derivatives, naphthoquinones, phenoxycarboxylic acids, phenylureas, triazines
Kruglov and Kwiatkowskaja (1975)	*Chlorella vulgaris* impregnated on to Seitz filter pads placed on herbicide treated soil. Growth measured by determination of ethanol extracted pigments	Phenylurea herbicides
Tchan *et al.* (1975)	Bioluminescence technique. Influence of photosynthetic inhibitor on oxygen production by *Chlorella NMI* estimated by concomitant reduction in luminescence of photobacterium	Atrazine, diuron, monuron, neburon
Wright (1975)	Six-mm diameter paper discs impregnated with herbicide placed on agar inoculated with *Chlorella pyrenoidosa*. Effect determined by diameter of zone of inhibition of growth	Phenylcarbamate, phenylurea and triazine herbicides
Chiou (1980)	*Dunaliella tertiolecta* used as assay organism in bioluminescence method of Tchan *et al.* (1975)	Diuron

228   *A. E. Pipe and L. E. Shubert*

assay. Another approach with potential for assays in soil is the use of the indigenous soil algal flora *in situ*. This was used by Pipe and Cullimore (1980) in the study of the influence of the phenylurea herbicide, diuron, on soil algae. The procedure consists of a modification of the Rossi–Cholodny buried slide technique (Cholodny, 1930; Rossi, 1928). In this study and related work (Pipe, 1982), the soil diatom *Hantzschia* was found to be a useful indicator of the presence and concentration of the phenylurea herbicides in soils. The alga is ubiquitous in soils (Bristol–Roach, 1927; Round, 1965), and as such has potential as a standard assay organism. Being unicellular, it is also easier to count on the slide surfaces than are the filamentous or colonial soil algae. Figure 1 shows a group of the diatoms colonizing a slide which was incubated in a sandy loam soil for 30 days. Details of the implanted slide

Fig. 1 *Hantzschia* frustules colonizing the surface of a microscope slide which was incubated in the soil for 30 days.

procedure are given in Table 4. Although this method has so far been tested extensively only for the phenylurea herbicides, it is potentially applicable to the determination of other toxicants and of soil nutrients. Major advantages which it offers include the provision of *in situ* data, the use of unsterilized soil, and its requirement for an absolute minimum of equipment and labour.

Table 4   Procedure for the assay of soil chemicals using the implanted slide technique

1. *Standards*
   (a) Treat well-mixed, freshly collected soil with chemical to be assayed. Re-mix soil thoroughly
   (b) Add equal weights of soil to paper cups (bases perforated with small holes). Use at least three cups per treatment level
   (c) Insert a pair of clean microscope slides (back-to-back) into soil in each cup until top 15 mm only remain above soil surface
   (d) Moisten soil to 50% water holding capacity with distilled water or Bold's basal medium (Nichols and Bold, 1965), depending on type of assay
   (e) Cover with plastic and incubate at 20°C under continuous illumination for 30 days. Check moisture level (by weighing) at weekly intervals; supplement if necessary by absorption of distilled water through holes in base of cup
   (f) Test for algal colonization of slides in untreated soil: mark soil surface on slides with wax pencil; cut cup on opposite sides; peel paper away from soil; gently ease soil away from slides; stand slides on end until dry; remove large soil particles by tapping slides against hard surface; spray with molten 2% water agar; air dry until agar is a thin film; make mark with wax pencil 10 mm below soil surface level already marked; add a few drops of distilled water and a cover slip; examine microscopically, counting number of cells of alga colonizing slides the most evenly and consistently in marked 10 mm zone
   (g) Assuming adequate colonization of slides in untreated soils, remove slides from treated soils and examine. Extend incubation for a further 15 days if colonization is inadequate

2. *Unknowns*
   (a) Mix thoroughly soil to be tested
   (b) Add to paper cups and treat as in sections 1b to 1e
   (c) Remove slides from soil and examine as in 1f
   (d) Compare colonization with that of slides from soils used as standards

3. *General remark*
   Slides may be stored indefinitely by removing cover slips and allowing agar to redehydrate. Agar therefore serves dual purpose of sealing algae into position on slide and preserving them in a form in which they can be identified years later.

IV. Conclusions and Recommendations for Future Research

Much more data are available on nutrient/pesticide–algal interactions in aquatic compared to terrestrial ecosystems. Soil algal research must proceed beyond the descriptive flora determinations before advances can be made on the use of algae as bioindicators. Evidence has been cited in this chapter which clearly illustrates the potential use of algae to demonstrate changes in soil fertility. Allen (1977) eloquently emphasized that algae represent a microscale for ecological investigations.

Refinement of the methods for the qualitative and quantitative examination of algae is essential. Ironically, the procedures available today are relatively unchanged from those developed by the pioneers of soil phycology at the beginning of the century. More reliable and reproducible techniques will permit the more accurate determination of the algal response to specific soil nutrients and toxicants. The most meaningful data would be obtained from *in situ* experiments, such as the implanted slide technique (Table 4). However, some *in vitro* experiments under controlled environment conditions, such as the microcosm technique, can provide valuable physiological data (Van Voris *et al.*, 1980).

For an assay procedure to be accepted as a standard, it must be simple, reliable and inexpensive. Consequently, methods that involve either filter paper placed on the surface of the soil or implanted microscope slides are preferable to more elaborate procedures. The assay organism should be preferably isolated from the soil and possess the following characteristics as outlined by Shoaf (1978):

(*a*) The ability to tolerate a wide range of environmental conditions;
(*b*) A growth potential correlating closely with that of algae indigenous to the sample being examined;
(*c*) A growth potential varying widely with different environments;
(*d*) Qualities permitting accurate and rapid monitoring of its growth.

The soil diatom *Hantzschia* was suggested as a potentially useful standard soil test organism. As it is ubiquitous in soils (Bristol-Roach, 1927), it would presumably satisfy requirements (a) and (b). The alga is unicellular and therefore lends itself to accurate growth measurements (d) more readily than would a colonial or filamentous form. For example, it forms the homogeneous suspensions necessary if filter paper is to be inoculated evenly or pigments extracted efficiently. It forms distinct colonies on agar. Because of its comparatively large size (up to 100 µm in length) it is readily observable among soil particles on slides which have been incubated in the soil (Fig. 1). In order to satisfy requirement (c), experiments would need to be conducted on the growth potential of *Hantzschia* under a variety of different conditions.

There is a great need to develop new and reliable methods that involve algae as indicators of soil fertility. In particular, methods that measure low levels of nutrients or toxicants would increase the sensitivity. Radioisotopic methods would be extremely useful.

An increased understanding of the interactions between soil algae and soil conditions will aid in assessing the effects of human activities (agriculture, waste disposal, urban development, etc.) on soil fertility. It is apparent that an ever increasing world population will place greater demands on food production, which will have a deteriorating effect on soil fertility. We must be prepared to assess these changes.

References

Adams, R. S. Jr. (1973). *Res. Rev.* 47, 1–54.
Addison, D. A. and C. E. Bardsley (1968). *Weed Sci.* 16, 427–429.
Allen, T. F. H. (1977). *Phycologia* 16, 253–257.
Anantani, Y. S. and K. V. Marathe (1972). *Bombay Univ. J.* 41 (68), 88–93.
Atkins, C. A. and Y. T. Tchan (1967). *Plant Soil* 27 (3), 432–442.
Bose, P., U. S. Nagpal, G. S. Venkataraman and S. K. Goyal (1971). *Curr. Sci.* 40 (7), 165–166.
Bristol-Roach, B. M. (1927). *Proc. 1st Internat. Cong. Soil Sci.* 3, 30–38.
Cameron, R. E. (1964). *Trans. Amer. Microscop. Soc.* 83 (2), 212–218.
Carson, J. L. and R. M. Brown Jr (1978). *J. Phycol.* 14, 171–178.
Chiou, A. C. M. (1980). Contribution to the Development of New Bioluminescence Assay Systems with Special Reference to Herbicides. Ph.D. thesis. Department of Microbiology, University of Sydney, Australia.
Cholodny, N. (1930). *Archiv für Mikrobiologie* 1, 620–652.
Crosby, D. G. (1976). Nonbiological degradation of herbicides in the soil. *In* "Herbicides: Physiology, Biochemistry, Ecology", Volume 2 (L. J. Audus, ed.), pp. 65–97. Academic Press, London and New York.
Cullimore, D. R. (1962). Studies on Soil Algae. Ph.D. thesis. University of Nottingham, England.
Cullimore, D. R. (1964). *Biol. Sol* 1, 10–14.
Cullimore, D. R. (1965a). *Biol. Sol* 3, 19–20.
Cullimore, D. R. (1965b). *Plant Soil* 23 (1), 34–43.
Cullimore, D. R. (1966a). *Nature* 209, 326–327.
Cullimore, D. R. (1966b). *J. Sci. Food Agric.* 17 (7), 323–326.
Cullimore, D. R. (1966c). *J. Sci. Food Agric.* 17 (1), 321–322.
Cullimore, D. R. (1971). *Res. Rev.* 35, 65–80.
Cullimore, D. R. (1975). *Weed Res.* 15, 401–406.
Dommergues, Y. and F. Mangenot (1970). "Écologie Microbienne du Sol." Masson, Paris.
Durrell, L. W. and L. M. Shields (1961). *Trans. Amer. Microscop. Soc.* 80, 73–79.
Eno, C. F. and H. Popenoe (1964). *Proc. Soil Sci. Soc. Amer.* 28, 533–535.
Fogg, G. E., W. D. P. Stewart, P. Fay and A. E. Walsby (1973). "The Blue-Green Algae." Academic Press, London and New York.
Furmidge, C. G. L. and J. M. Osgerby (1967). *J. Sci. Food Agric.* 18, 269–273.
Granhall, U. (1970). *Oikos* 21, 330–332.
Grover, R. (1975). *Can. J. Soil Sci.* 55, 127–135.
Hance, R. J. and C. E. McKone (1976). The determination of herbicides. *In* "Herbicides: Physiology, Biochemistry, Ecology", Volume 2 (L. J. Audus, ed.), pp. 393–445. Academic Press, London and New York.

Helling, C. S., D. D. Kaufman and C. T. Dieter (1971). *Weed Sci.* **19** (6), 685–690.

Hill, I. R. and D. J. Arnold (1978). Transformations of pesticides in the environment—the experimental approach. *In* "Pesticide Microbiology" (I. R. Hill and S. J. L. Wright, eds), pp. 203–246. Academic Press, London and New York.

Hunt, M. E., G. C. Floyd, and B. B. Stout (1979). *Ecology* **60**, 362–375.

Jansen, L. L., W. A. Gentner and J. L. Hilton (1958). *Weeds* **6**, 390–398.

Jurgensen, M. F. and C. B. Davey (1968). *Can. J. Microbiol.* **14**, 1179–1183.

Kimball, T. L. (1972). Why environmental quality indices? *In* "Indicators of Environmental Quality" (W. A. Thomas, ed.), pp. 7–14. Plenum Press, New York.

King, J. M. and C. H. Ward (1977). *Phycologia* **16**, 23–30.

Kratky, B. A. and G. F. Warren (1971a). *Weed Res.* **11**, 257–262.

Kratky, B. A. and G. F. Warren (1971b). *Weed Sci.* **19** (6), 658–661.

Kruglov, J. W. and L. B. Kwiatkowskaja (1975). *Roczniki Glebozn* **26**, 145–148.

Kruglov, Yu. V. and L. N. Paromenskaya (1970). *Mikrobiologiya* **39**, 139–142.

Lund, J. W. G. (1945). *New Phytologist* **44**, 196–219.

Lund, J. W. G. (1962). Soil Algae. *In* "Physiology and Biochemistry of Algae" (R. A. Lewin, ed.). Academic Press, London and New York.

McCann, A. E. and D. R. Cullimore (1979). *Res. Rev.* **72**, 1–31.

Nichols, H. W. and H. C. Bold (1965). *J. Phycol.* **1**, 34–38.

Noll, M. and U. Bauer (1973). *Zentbl. Bakt. ParasitKde Abt. I* **157**, 178–183.

O'Kelley, J. C. (1974). Inorganic nutrients. *In* "Algal Physiology and Biochemistry" (W. D. P. Stewart, ed.), pp. 610–635. University of California Press, Berkeley.

Parkinson, D., T. R. G. Gray and S. T. Williams (1971). "Methods for Studying the Ecology of Soil Micro-organisms", IBP Handbook No. 19. Blackwell, Oxford.

Patrick, R. (1972). Aquatic communities as indices of pollution. *In* "Indicators of Environmental Quality" (W. A. Thomas, ed.), pp. 93–99. Plenum Press, New York.

Patrick, R. (1977). Ecology of freshwater diatoms—diatom communities. *In* "The Biology of Diatoms" (D. Werner, ed.), pp. 284–332. Botanical Monographs, Volume 13. University of California Press, Berkeley.

Peterson, J. B. (1935). *Dansk. botanisk Arkhiv* **8**, 1–80.

Pillay, A. R. and Y. T. Tchan (1969). *Soil Biol.* **10**, 23–25.

Pillay, A. R. and Y. T. Tchan (1972). *Plant Soil* **36**, 571–594.

Pipe, A. E. (1982). The Influence of the Phenylurea Herbicides on the Algae of some Saskatchewan Soils. Ph.D. thesis. University of Regina, Saskatchewan, Canada.

Pipe, A. E. and D. R. Cullimore (1980). *Bull. Environ. Contam. Toxicol.* **24**, 306–312.

Rand, M. C., A. E. Greenberg and M. J. Taras (eds) (1976). "Standard Methods for the Examination of Water and Wastewater", (14th edit). Amer. Public Health Assoc., 1193 pp.

Rossi, G. (1928). Festschrift Jul. Stoklasa, Berlin.

Round, F. E. (1965). "The Biology of the Algae." Edward Arnold, London.

Rovira, A. D. and M. Vendrell (1972). *Soil Biol. Biochem.* **4**, 63–69.

Rust, R. H., R. S. Adams Jr. and W. P. Martin (1972). Developing a soil quality index. *In* "Indicators of Environmental Quality" (W. A. Thomas, ed.). Plenum Press, New York.

Shields, L. M. and L. W. Durrell (1964). *Bot. Rev.* **30**, 92–128.

Shoaf, W. T. (1978). *Hydrol. Sci. Bull.* **23**, 439–444.

Shubert, L. E. (1980). Soil algae. *In* "Vegetation-Environment Relationships of Woodland

and Shrub Communities and Soil Algae in Western North Dakota" (M. K. Wali, K. T. Killinbeck, R. H. Bares and L. E. Shubert, eds). Final Report, North Dakota Regional Environmental Assessment Program, Project 7-01-1.
Shubert, L. E. and T. L. Starks (1980). *Brit. Phycol. J.* 15, 417–428.
Shubert, L. E. and T. L. Starks (1979). *Bot. Rev.* 30, 92–128.
Smith, D. W. (1978). Water relations of microorganisms in nature. *In* "Microbial Life in Extreme Environments" (D. J. Kushner, ed.). Academic Press, London and New York.
Starks, T. L. and L. E. Shubert (1979). Algal colonization on a reclaimed surface mined area in western North Dakota. *In* "Ecology and Coal Resource Development" (M. K. Wali, ed.), pp. 652–660. Pergamon Press, New York.
Starks, T. L. and L. E. Shubert (1982). *J. Phycol.* 18, 99–107.
Starks, T. L., L. E. Shubert and F. R. Trainor (1981). *Phycologia* 20, 65–80.
Tchan, Y. T. (1959). *Plant Soil* 10, 220–232.
Tchan, Y. T., L. N. Balaam and F. Draetta (1963). *Plant Soil* 19, 233–240.
Tchan, Y. T., J. E. Roseby and G. R. Funnell (1975). *Soil Biol. Biochem.* 7, 39–44.
Trainor, F. R. (1978). "Introductory Phycology." J. Wiley, New York.
Van Voris, P., R. V. O'Neill and H. H. Sugart Jr (1980). *Ecology* 61, 1352–1360.
Venkataraman, G. S. (1972). "Algal Biofertilizers and Rice Cultivation." Today and Tomorrow's Printers and Publishers, New Delhi.
Warcup, J. H. (1957). *Soil Fertil.* 20, 1–5.
Wong, M. H. and K. Y. Chung (1979). Cultivation of unicellular green algae as an assessment of soil fertility. *In* "Environmental Pollution", pp. 83–91. 0013-9327/79/0020-0083. Applied Science, England.
Wright, S. J. L. (1975). *Bull. Environ. Contam. Toxicol.* 14, 65–70.
Wright, S. J. L. (1978). Interactions of pesticides with micro-algae. *In* "Pesticide Microbiology" (I. R. Hill and S. J. L. Wright, eds), pp. 535–602. Academic Press, London and New York.

Part Four

Toxic Substances, Heavy Metals and Vitamins

8 The Effect of Environmental Contaminants on Aquatic Algae

T. P. Boyle

Columbia National Fisheries Research Laboratory
US Fish and Wildlife Service
Columbia, Missouri, USA

I. Introduction

The presence of chemical contaminants has become a pervasive threat to many natural aquatic ecosystems. Environmental contaminants can have toxic effects on many different types of organisms and affect biological processes at the cellular, population, community, and ecosystem levels of organization. For purposes of this chapter, environmental contaminants are defined as man-generated chemicals that have the potential to damage or injure organisms that comprise the aquatic ecosystem. This definition implies that contaminants are toxic and thus excludes chemicals, such as inorganic nutrients, which are sometimes classed as pollutants.

Environmental contaminants originate from a variety of sources that are usually classed as point source or non-point source pollutants. Table 1 lists a general outline

Table 1   Human activities as sources of aquatic contaminants.

SILVICULTURE	INDUSTRY
Insecticides	Manufacturing chemicals (many types)
Herbicides	Mining chemicals
AGRICULTURE	ENERGY DEVELOPMENT
Insecticides	Mining and drilling chemicals
Herbicides	Processing or refining chemicals
Fungicides	Waste chemicals
Drying agents	
	URBAN
	Sewage
	Storm runoff

ALGAE AS ECOLOGICAL INDICATORS
ISBN 0 12 640620 0

of categories of human activities that are the most frequent contributors of environmental contaminants. In a historical sense, pesticides, primarily the organochlorine insecticides, have been the most obvious contaminants in aquatic systems; however, toxic chemicals from many other sources are also recognized as a threat to natural resources. For example, Keith and Telliard (1979) have listed 21 industrial categories from which they have selected 129 high priority pollutants. These chemicals were classified into four different types of organic chemicals and heavy metals.

The problem of the impact of contaminants on aquatic ecosystems is complex. There is a definite continuing need for assessment and mitigation of this impact, and all legal, social, economic, and biological aspects should be considered.

II. Environmental Contaminants and Aquatic Algae

The direct, short-term effects of environmental contaminants on aquatic algae have been the subject of a myriad of different types of laboratory studies. However, the secondary or indirect effects of contaminants on algae, that is the change in the structure and function of algal communities resulting from toxic effects on other organisms in the aquatic community, have received less attention. The role of algae in the disappearance and degradation of environmental contaminants has only recently been addressed.

A. Direct Effects of Contaminants on Algae

The effects of toxic chemicals have been measured on a wide variety of algal species by using different cultural methods and a number of different biological responses. Algae commonly tested range from common, to rare and exotic freshwater, and marine species of various divisions. Cultural methods vary in composition of solutes in the nutrient medium; in nutrient concentration; in temperature; and in intensity, periodicity, and quality of light. Various measurements of algal response to chemical contaminants include: (1) photosynthetic uptake of radiolabelled carbon dioxide, an indication of the functioning of Photosystem I associated with chlorophyll *a*; (2) evolution of oxygen, a measure of rate of the Hill reaction in Photosystem II associated with chlorophyll *b*; (3) measures of relative population growth in time based on changes in cell numbers, amount of chlorophyll *a* extracted *in vitro*, turbidity of a cell suspension, and change in the dry weight of the culture; and (4) measures of critical physiological and biochemical rates, such as synthesis of lipids, protein, and nucleic acids, as well as the uptake of organic and inorganic nutrients from nutrient media.

The response of a particular species of algae to a potentially toxic contaminant is very dependent on cultural conditions. Many physical and chemical factors such as temperature, water hardness, pH, and nutrient conditions may affect the toxicity of a chemical.

To best understand the potential direct effects of environmental contaminants on aquatic algae it is important to first consider examples of these effects on processes at the cellular level of organization.

1. Photosynthesis   Most examples of inhibition of photosynthesis are from the more frequently tested chemicals classed as pesticides and common industrial contaminants. Walsh (1972) listed a variety of commercial herbicides that inhibit photosynthesis. Two triazine herbicides, atrazine and simizine, inhibit photosynthesis in aquatic algae by blocking electron transport in the Hill reaction in Photosystem II (Butler *et al.*, 1975b; Hawxby *et al.*, 1977).

Organochlorine insecticides such as toxaphene, endrin, dieldrin, as well as DDT are quite refractory and persistent in marine and freshwater systems and act as photosynthetic inhibitors (Wurster, 1968; Batterton *et al.*, 1971; Bowes, 1972; MacFarlane *et al.*, 1972). However, Fisher (1975) suggested that DDT affects photosynthesis in a secondary manner by first reducing the overall algal growth rate, indicating that some other biochemical process was more directly affected.

Several organophosphorus insecticides such as Dursban, Baytex, and Abate also reduced photosynthetic activity in aquatic algae (Derby and Rucker, 1970; Brown *et al.*, 1976) by inhibiting the Hill reaction. These compounds were more toxic than the organochlorines tested in reducing photosynthesis; however, they degrade under natural conditions and do not represent a chronic contaminant threat to algae unless they are continuously added to the environment.

Several ubiquitous types of industrial contaminants have also been found to block photosynthesis. Copper, cadmium, mercury, lead, zinc, and possibly some other bivalent heavy metals reduce photosynthesis by causing structural damage to chloroplasts (Fitzgerald and Faust, 1963; Blinn *et al.*, 1977; Wong *et al.*, 1979; Hollibaugh *et al.*, 1980). Halogens such as chlorine and bromine chloride reduce photosynthesis by reducing the amount of chlorophyll *a* and thus affect Photosystem I (Brungs, 1976; Gentile *et al.*, 1976; Brooks and Liptak, 1979; Liden *et al.*, 1980). Polychlorinated biphenyls (PCBs), similar in chemical structure to DDT, have become a worldwide contaminant and are found in a wide variety of aquatic organisms of different trophic levels. PCBs also directly inhibit photosynthesis in a number of aquatic algal species (Harding, 1976). PCBs have been found in relatively high concentrations in a number of fish species and have been implicated in toxic responses. However, Fisher (1975) estimated that concentrations of PCB contamination found in the environment may cause shifts in predominant algal species but not in the total primary production.

The pumping, refining, transportation, and use of petroleum products are widespread activities that frequently affect aquatic ecosystems. The water-soluble fractions of both crude and processed petroleum comprise a plethora of organic compounds. Dennington *et al.* (1975) found that several refined petroleum products affected photosynthesis in sensitive algal species such as diatoms, but not in the less sensitive euglenoids.

2. Production of nucleic acids Deoxyribonucleic acid (DNA) is the primary genetic material that transmits the genetic code from one generation of cells to another. Ribonucleic acid (RNA) serves as the link between DNA and protein synthesis. Several known pollutants and potential contaminants are known to affect the synthesis and cellular content of nucleic acids. Divavin *et al.* (1978) found that phenols and water extractable portions of petroleum reduced the concentration of both RNA and DNA in algae of several different divisions. Biggs *et al.* (1979) showed that the cell size of several species of algae was reduced when they were exposed to several organochlorines. Powers *et al.* (1977) found not only that the organochlorine insecticide dieldrin was toxic and caused cell rupture, but that surviving cells removed from the toxicant stress yielded at least four generations of significantly smaller cells. This continued reduction in cell size indicates a probable genetic effect involving the synthesis and replication of nucleic acids.

3. Biosynthesis Other results of nucleic acid reduction may appear as a reduction in protein synthesis or reduction in other types of biosynthesis that are under the influence of proteinaceous enzymes. Morgan and Lackey (1958) found that copper ions interfered with the synthesis of protein. Furthermore, a wide variety of herbicides have been shown to reduce the rate of lipid biosynthesis in algal cultures (Sumida and Ueda, 1975). Lipids are an important structural element in the cell membrane and in the membranes of various cell organelles and also control the movement of substances into the cell as well as the movement and metabolism of materials within the cell.

4. Nutrient uptake The uptake of inorganic nutrients, especially forms of nitrogen and phosphorus, is critical to growth and reproduction of aquatic algae. The production of algae in culture media as well as in natural waters is frequently limited by one of these two elements. The uptake of phosphorus from artificial media was inhibited in three species of microalgae by the herbicides MCPA and MCPB (Kirkwood and Fletcher, 1970). However, exposure to heavy metals such as lead, zinc, copper, and cadmium may stimulate uptake of phosphorus, which is deposited in polyphosphate bodies in the central vacuole of several species (Stoermer *et al.*, 1980). Low concentrations of chlorine and chloramine, common contaminants from power plants and domestic sewage treatment facilities, destroyed or inactivated enzymes in the cell membrane that are responsible for the uptake of nitrate, resulting in reduced ammonia and nitrate uptake by planktonic algae (Toetz *et al.*, 1977).

Changes in the nutrient uptake ability of various species of algae may result in changes in dominant species and could be important in changes in different functional groups such as population shifts from green algae to blue-green algae (Wetzel, 1975).

5. Nitrogen fixation   The fixation of nitrogen by blue-green algae is directly proportional to the number of heterocysts present (Fogg, 1971). Wurtsbaugh and Apperson (1978) measured the effects of several mosquito larvicides on blue-green algae and found that methoprene, propoxur, temephos, and methoxychlor increased the number of heterocysts and the amount of nitrogen fixed by natural assemblages of phytoplankton species. Diflubenzuron, an insect growth regulator reduced the nitrogen fixation rate only slightly. DaSilva *et al.* (1975) found that a number of herbicides reduce nitrogen fixation in several common genera of aquatic blue-green algae. Horne and Goldman (1974) found that trace amounts of copper, above the chelation capacity of a eutrophic lake, could completely inhibit nitrogen fixation thus reducing the lake's nitrogen budget by half and slowing the eutrophication process.

The ecological significance of an effect on the nitrogen-fixing capacity of an ecosystem varies. An increase in the nitrogen content of aquatic systems such as rice fields would be desirable; however, high rates of nitrogen fixation and growth of blue-green algae can be a detriment to water quality.

B. Indirect Effects of Contaminants on Algae

Aquatic algal communities are controlled by many biotic and abiotic environmental factors. One or more of these factors may be directly affected by a chemical contaminant and produce changes indirectly in the structure and function of algal communities. Hurlbert (1975) included an extensive discussion of the nature of secondary effects of pesticides on biotic communities. This section will summarize recent research on this topic.

1. Macrophyte–phytoplankton interaction   There has been much discussion of the role of rooted macrophytes in aquatic ecosystems. One hypothesis is that aquatic macrophytes absorb dissolved inorganic nitrogen and phosphorus, thus competing with planktonic algae for nutrients. Toetz (1971, 1973) showed that the submersed macrophyte *Ceratophyllum* took up both nitrate and ammonia directly from the water. Experiments with artificial ponds in greenhouses indicate that certain levels of nitrogen and phosphorus added to the water stimulate the growth of the rooted submersed macrophytes, *Elodea canadensis*, *Myriophyllum spicatum*, and *Potamogeton crispus* (Mulligan and Baranowski, 1969; Ryal *et al.*, 1972). In outdoor experimental pond ecosystems that had been fertilized with various amounts of nitrogen and phosphorus, the same species of macrophytes were eliminated due to the heavy increases in phytoplankton density (Mulligan *et al.*, 1976). Goulder (1969) attributed the observed inverse relationship between the abundance of phytoplankton and density of submersed macrophytes in two ponds to the shading effect of the macrophyte and competition for available light.

Other investigators studying the nutrition and relationship of aquatic macrophytes to phytoplankton have shown that macrophytes absorb inorganic minerals and plant nutrients from the hydrosoil through the roots and translocate them to the shoots, where various elements and organic compounds are lost to the water (Wetzel and Manny, 1972; McRoy *et al.*, 1972; Bristow and Whitcombe, 1971; DeMarte and Hartman, 1974). Thus, littoral macrophytes are an important factor controlling a major portion of nutrients in certain lentic ecosystems.

Direct impact of contaminants on rooted macrophytes can produce indirect effects on algal populations in several ways. Death of dense stands of macrophytes under certain conditions may potentially release large amounts of inorganic nutrients to aquatic ecosystems (Jewell, 1971; Nichols and Keeny, 1973; Strange, 1976). However, studies of the consequences of herbicides on dense stands of aquatic macrophytes indicate that saltatory increases in forms of dissolved nitrogen and phosphorus available to algae occur only after severe deoxygenation of the water (Walker, 1963; Pokorny *et al.*, 1971). Some investigations have shown that increases in light and inorganic forms of nitrogen and phosphorus, after the killing of large masses of aquatic macrophytes with herbicides, are sometimes responsible for the development of dense growths of planktonic algae (Fish, 1966; Pokorny *et al.*, 1971; Walsh *et al.*, 1971). However, other situations where macrophytes have been killed with herbicides have shown little or no increase in planktonic algae (Way *et al.*, 1971; Simpson and Pimental, 1972; Brooker, 1974). Brooker and Edwards (1975) suggested that severe deoxygenation of the water due to the increased biochemical oxygen demand caused by decaying plants is necessary before a pulse of nutrients is released into the water and result in algal blooms. It is not clear whether the reduction of the oxidation–reduction potential of the benthic water and mud due to deoxygenation allowed the solubilization and release of nutrients from the sediment (Mortimer, 1971), or whether decomposition of plant tissue in anoxic water with low oxidation–reduction potential results in large pulses of nutrients that are usable by planktonic algae. Finally, planktonic or macrophyte algae such as *Chara* may survive or replace angiosperms because they may be more resistant to chemical toxicants (Matthews, 1967).

Prevention of macrophyte growth presents a different set of factors that appear to reduce the amount of planktonic algae. In a set of experiments with replicate ponds where macrophyte growth was prevented by using herbicides or reduced by the addition of grass carp (*Ctenopharyngodon idella* Val.), relative algal growth was compared with that in ponds with heavy growth of rooted submersed macrophytes (Boyle, 1979). The herbicides used were not directly toxic to algae at the concentrations used. The suppression of macrophyte growth resulted in an increased oxidation–reduction potential of the bottom mud due to the ability of the water to circulate freely, uninhibited by dense stands of vegetation. The higher oxidation–reduction potential prevented the solubilization of usable forms of nitrogen and phosphorus and resulted in reduced nutrient concentrations in the water (Boyle, 1980). Reduced nutrient concentrations appeared to be responsible for a several-fold decrease in planktonic algae, as indicated by chlorophyll *a* (Table 2).

Table 2   Indirect effect of herbicides and grass carp in experimental ponds.

	Treatment means[a]			
	Control	Dichlorobenil	Fenac	Grass carp
Oxidation–reduction potential mud (mv)	−390	130	300	−180
Nutrients (mg l^{-1})				
NO_2–NO_3	0·02	0·02	0·02	0·01
NH_3	0·08	0·02	0·02	0·02
Ortho P	0·01	0·01	0·01	0·01
Total P	0·15	0·03	0·05	0·06
Chlorophyll *a* ($\mu g\ l^{-1}$)	23·9	5·5	4·6	21·2

[a] Treatment means calculated from three ponds each measured repeatedly throughout summer (Boyle 1979)

2. Zooplankton–algae interaction   Algae are a major food item supporting aquatic invertebrates. Invertebrate grazing can affect algae in two ways. (1) A moderate amount of grazing can stimulate the growth and production of algae by increasing the rate of cycling of critical nutrients and cropping to keep the algae in the exponential phase of growth. (2) Heavy grazing will reduce algae abundance. Thus, the expected consequences of contaminants on the aquatic invertebrate community could be either an increase or a decrease in algal abundance.

Hurlbert *et al.* (1972), Butcher *et al.* (1977), Krzeczkowska-Wolosyzn (1979) and Papst and Boyer (1980) have shown that one or more parameters of phytoplankton production and abundance greatly increase when invertebrate numbers are reduced by the application of insecticides to small ponds. Besides an increase in phytoplankton biomass and actual production rates, nitrogen fixation by blue-green algae also sometimes increased. This increase in algae was shown to be an indirect effect by Brown *et al.* (1979), who demonstrated that even low concentrations of Dursban, the insecticide used in three of the pond studies, reduced the growth of several species of algae. However, it appears that the indirect effect of reducing the grazing pressure by invertebrates can more than compensate for any direct toxic effect.

C. *Bioconcentration*

The uptake and concentration of environmental contaminants are important in understanding the toxicity of various compounds and their distribution into different biotic and abiotic compartments of aquatic ecosystems. The phenomenon of uptake and concentration of a chemical by organisms is commonly termed bioaccumulation or bioconcentration and is defined as the ratio of the concentration of a chemical in an organism to the concentration in the water (Kenega, 1972). The bioaccumulation ratio ranges from less than one, indicating little uptake, to 10^5 or 10^6 for some organic compounds with high accumulation potential.

The bioaccumulation of environmental contaminants by algae performs three functions of ecological importance. (1) The degree to which a compound bioaccumulates affects the concentrations at the site of action and is an important parameter of the toxic effect on an organism. (2) Bioaccumulation of a toxic contaminant may render it, temporarily at least, unavailable to invertebrates and fish, thus affording some measure of ecological resistance to toxic impact on other organisms. (3) Bioaccumulation of persistent refractory organic chemicals or heavy metals by algae may be an important factor in the physical transport of toxic materials from one place to another and the bioaccumulation through a food chain to consumer organisms in the upper trophic levels.

The bioaccumulation of a compound is the function of a combination of many factors, which can be classified into three categories: (1) The characteristics of different algae in relation to chemical uptake. (2) The effect of environmental factors on availability of the compound to algae. (3) The physical and chemical properties of the contaminant.

Bioaccumulation can also be considered as the net difference in the ratio of biological uptake and elimination and thus may be related to metabolism or the rate of specific biochemical processes. The physical and chemical differences of various species of algae are important factors in determining uptake or bioaccumulation of chemical contaminants. Differences in the surface area to volume ratio among algal species mean that a small algal species would have a larger exposed surface on which to sorb chemical contaminants than would the same weight of a large species of algae. Large colonial forms present less of their immediate surface to the environment.

There are also gross differences in the surface composition of algae which range from the exposed cell membrane in green algae, to mucilaginous sheaths in blue-green algae, to cellulose tests in some colonial forms, and to the outer specialized cellular layers of macrophytic algae. The cellular contents and morphology also vary from relatively structureless procaryotic blue-green algae to highly structured membraneous cells with large vacuoles and starch inclusions in other phyla to the highly differentiated cells of macrophytic algae.

Environmental factors affecting contaminant uptake include pH, temperature, and the total amount of material in the water on which contaminants can sorb. The most important physical–chemical properties of contaminants affecting bioconcentration appear to be the water solubility and the n-octanol water partition coefficient (Kenaga, 1972; Chiou *et al.*, 1977).

Other important factors affecting the bioaccumulation of contaminants are the latent heat of the strength of binding to a nonspecific surface, polarity which would be a predictor of the ability to bind on charged and neutral surfaces, volatility in that it affects the amount of the compound present in the aquatic environment, light and photolysis and how they affect the energy state of a molecule and its degradation rate, and finally the nature of molecular transformation of the compound (Kenaga, 1972).

The uptake of contaminants from water by algal cells may be the result of several processes. The chemical may be metabolically active, act as an essential nutrient or mineral, and be transported across the cell membrane and thus enter into biochemical processes. The bioconcentration potential for organic compounds has been linked to the n-octanol water partition coefficient. This partition coefficient has been related to the aqueous solubility of a wide variety of compounds, including aliphatic and aromatic hydrocarbons, aromatic acids, organochlorine and organophosphate pesticides, and polychlorinated biphenyls. Therefore, there is a functional correlation between the amount of lipid in an organism, the bioconcentration factor, and the aqueous solubility of a chemical (Chiou *et al.*, 1977). Claeys *et al.* (1975) noted that PCB and chlorinated pesticide levels differed among different marine organisms with similar lipid composition. They suggested that this difference in bioconcentration factor may be due to physical location of the organisms in the environment, the level of activity, or the trophic level of the organism. Thus, contaminant levels in aquatic organisms may be initially due to direct uptake of the chemical from the water; however, concentrations of contaminants in their food also is a factor in determining the final contaminant body burden of an organism. The degree to which algae bioconcentrate a persistent chemical may affect the concentration of the chemical in higher organisms. Hansen (1976) also found that actual bioconcentration factors for PCBs in Florida bays were an order of magnitude higher than those found in acute laboratory bioassays, also suggesting that the bioaccumulation route through contaminated food is important.

Lakshminarayana and Bourque (1980) found that the organophosphate forest insecticide, Fenitrothion, was actively taken up by natural communities of planktonic and benthic algae in a Canadian lake after treatment of the watershed. Bioconcentration ratios for Fenitrothion in algae ranged from 122 to 177 and the degree of uptake was dependent on population density, seasonal variations in bloom formations, and growth rates. They further suggested that analysis of Fenitrothion accumulated in algae would be an indicator of contamination of aquatic systems when the concentration was below detectable limits. Moreover, the bioaccumulation of Fenitrothion by algae could contribute to its rapid disappearance from the water and concentration through a food chain to shellfish and other consumer organisms.

The rates and magnitude of bioconcentration of chemicals are governed by different chemical and physical processes that vary among different taxa of algae. Sodergren (1968, 1971) showed that algae accumulate DDT and other chlorinated hydrocarbons. Rice and Sikka (1973) showed that there was more than an order of magnitude difference in the final concentration and the bioaccumulation factor among five species of marine algae exposed to the organochlorine insecticide, dieldrin, for two hours (Table 3). The same species of algae exposed to concentrations up to 1 part per million (1 mg l^{-1}) accumulated over 1 part per

Table 3 Bioaccumulation of dieldrin by five species of algae after 2-h exposure to 1·7 µg l^{-1}.

Species	Algal concentration (µg g^{-1})	Bioconcentration factor
Skeletonema costatum	27·0	15 882
Tetraselmis chuii	14·6	8 588
Isochrysis galbana	14·0	8 238
Olisthodisus luteus	8·3	4 900
Cyclotella nana	8·2	4 810
Amphidinium carteri	1·7	982

Table modified from Rice and Sikka (1973)

thousand (1 mg g^{-1}) dieldrin. The authors statistically tested the correlation between the bioaccumulation factor and the number of cells per unit, an index of the surface to volume ratio, and found it was not important in explaining differences in uptake. Schauberger and Wildman (1977) further investigated the bioconcentration of dieldrin and aldrin, both organochlorine insecticides, in freshwater blue-green algae and found that there was a great difference in uptake among three species. Moreover, they found that the pattern of bioconcentration ratios of dieldrin and aldrin was different among these three species (Table 4). However, these differences in uptake could not be related to differences in toxic effects on the algae. Glooschenko et al. (1979) showed that *Scenedesmus quadricauda* accumulated chlordane at the same rate alive as when heat killed and that bioaccumulation was essentially complete in 24 h. These studies suggest that bioaccumulation in algae is not affected by the rate of cell growth or metabolism, at least for organic compounds.

Table 4   Bioconcentration factors of aldrin and dieldrin for three species of algae exposed to the chemicals at 1 µg l^{-1}.

	Chemical	
Species	Aldrin	Dieldrin
Anabena cylindrica	1 292	201
Anacystis nidulans	991	504
Nostoc muscorum	Trace	18 484

Table modified from Schauberger and Wildman (1977)

In a comparative study of the uptake of the organochlorine, DDT, and the organophosphate, Fenvalerate, by *Daphnia*, snails, fish, and algae in a model ecosystem, Ohkawa *et al.* (1980) showed algae had the highest bioaccumulation ratio for DDT and a substantially high ratio for Fenvalerate (Table 5). These exposures were for only seven days, and the authors suggested a probable path for further bioaccumulation through a food chain would occur from algae to snails and algae to fish.

There is no factor such as the partition coefficient that will explain whether elements are bioconcentrated (Leland *et al.*, 1979). Lead concentration in planktonic algae grown in water from an Ontario snow dump had 10 to 30 times the lead level of algae grown as controls. Four times more lead was desorbed into water from snow dump sediments than from pristine sediments. Laube *et al.* (1979) showed that the heavy metals, cadmium and copper, present in the sediments are bioconcentrated substantially by planktonic algae. They determined that *Anabena*

Table 5   Distribution and comparative uptake of DDT and Fenvalerate in a model laboratory ecosystem.

	DDT		Fenvalerate	
Medium	Concentration (µg/g)	BR[a]	Concentration (µg/g)	BR
Soil	432	—	277	—
Water	0·31	—	0·49	—
Fish	166	1 330	86	122
Soil	856	2 670	320	617
Daphnia	33	2 500	246	683
Algae	590	5 240	190	477

[a] Bioconcentration ratio: data from Ohkawa *et al.* (1980)

and *Ankistrodesmus* contained substantially more binding sites on a dry weight basis than dry sediment and that the algae were able to affect the flux of heavy metals from the sediments through the water. Thus algae may be important in making heavy metals normally sequestered in the sediment, available to invertebrates and ultimately to fish, through the food chain.

Several authors have cited the feasibility of using benthic algae as qualitative indicators for heavy metal pollution (Bryan and Hummerstone, 1973; Preston *et al.*, 1972; Foster, 1976; Haug *et al.*, 1974). Melhuus *et al.* (1978) and Foster (1976) attempted to use algae as a quantitative indicator of the state of heavy metal pollution in an area. Foster stated that in highly polluted aquatic environments, concentrations of contaminants in algae do not reflect the water concentrations to which they have been exposed. However, Melhuus *et al.* (1978) found that although zinc and cadmium seemed to bioconcentrate as a function of ambient concentration or distance from the point source, copper and lead behaved differently. They stated that there are two prerequisites for the use of contaminant residues in algae as a quantitative indicator of environmental pollution: (1) there must be an empirically determined bioconcentration ratio for a species or community of algae for a contaminant over a range of concentrations in the water, and (2) the concentration of the contaminant in the water must be relatively stable or else variations in concentration through time will make the bioconcentration ratio less interpretable.

Miyamoto *et al.* (1979) showed that in a model laboratory ecosystem the bioaccumulation ratio of algae was much greater than for other representative aquatic organisms for DDT and total DDT residue, which included the degradation products DDD and DDE (Table 6). When comparing bioaccumulation ratios for the organophosphorus insecticide Fenitrothion through time, however, they found that initially it was consistently higher in fish and invertebrates than in algae. Invertebrates and fish were better able to degrade or eliminate the compound so that after several weeks the algae had a greater bioaccumulation ratio than initially, whereas the concentration in other organisms had generally decreased. There is, therefore, an association between bioaccumulation ratio and a

Table 6   Bioaccumulation of DDT and metabolites in an aquatic ecosystem.

Organism	Bioaccumulation ratio		Percentage DDT in total
	DDT	DDT-R[a]	
Algae (7 species)	4 720	4 900	96
Daphnia pulex	2 560	1 270	
Snail	3 660	5 820	63
Carp	2 390	3 600	66

Data from Miyomoto *et al.* (1979)
[a] DDT-R is total of DDT, DDD and DDE

greater degradation/elimination rate in more active organisms. Thus the bioaccumulation ratio may increase with time and algae may represent a chronic reservoir of contamination by the prolonged addition of a chemical through a food chain to fish.

Warner and Morschel (1978) have shown that the uptake of organic compounds by algae varies greatly among species. They examined the uptake of dieldrin in 15 different species and found that absorption was completed in one hour for some species and apparently incomplete and still continuing after 30 hours for other species. They reported that the bioaccumulation ratio of dieldrin in algae varied from 1600 to 10 000 based on a volume-to-volume ratio.

Current or movement of the water past algal cells increases contaminant uptake. Rose and McIntire (1970), who exposed colonies of *Spirogyra* to various concentrations of dieldrin in artificial laboratory streams, found that increasing the current from still water resulted in about an order of magnitude increase in the bioaccumulation ratio (Table 7). It appears that in still water an organism absorbs the contaminants quickly from water immediately adjacent and that the current keeps the concentration of the contaminant renewed near the surface of the cells.

D. Biodegradation

The degradation of organic contaminants is influenced by several environmental variables including temperature, pH, light, and aerobic and anaerobic microorganisms. Recent studies have implicated aquatic algae as a factor mediating the metabolism of several types of contaminants and pesticides. Vallentyne and Bingham (1974) studied the effect of several species of algae and different chemical and physical factors on the degradation of 2,4-D in water. They showed that *Scenedesmus quadricauda* absorbed more 2,4-D from the water in the dark than under lighted conditions and absorbed more at pH 6 than above. Low temperature slowed uptake. Several 2,4-D metabolites were found that were produced by the alga itself. Several other species of algae of the genera *Chlorella* and *Chlamydomonas* did not take up 2,4-D, nor did they affect its metabolism. Butler *et al.* (1975a) isolated 21 different morphological and physiological strains of algae from eight different genera and found that the pesticides 2,4-D, diazinon, and methoxychlor were differentially degraded by various of the algal cultures, but that atrazine was not. Kar and Singh (1977, 1979) showed that blue-green algae repeatedly inoculated into cultures containing the insecticides benzene hexachloride, carbofuron, and hexachlorocyclohexane took up and detoxified the medium and allowed algal growth.

Matsumura and Esaac (1979) pointed out that many organic macromolecules such as chlorophyll and aromatic amines formed in algal cells are known to sensitize and enhance photochemical reactions of a number of synthetic organics. In a series of experiments they found that flavoprotein extracted from blue-green algae in the presence of light greatly increased the photodegradation of DDT, lindane, dieldrin,

Table 7   Comparison of the bioaccumulation ratio of *Spirogyra* exposed to dieldrin with
and without current.

Current velocity (cm s^{-1})	Dieldrin concentration (μg l^{-1})	Bioconcentration ratio
28	0·44	2 500
0	0·38	420
28	6·69	11 200
0	4·28	116

Data from Rose and McIntire (1970)

toxaphene, parathion and mexacarbate. Flavoproteins also appeared to enhance
reductive degradation under anaerobic conditions for mexacarbate, DDT, and
toxaphene.

E. Relative Vulnerability of Algae to Toxic Contaminants

Algae composing the primary producer level are of initial importance in providing
the energy that sustains invertebrates and fish in most aquatic ecosystems.
However, there have been few attempts to evaluate the relative vulnerability of
algae to toxic stress with that of other aquatic organisms. Patrick *et al.* (1968)
compared the acute toxic response of a diatom, a snail, and bluegill sunfish, and
found that there was a differential sensitivity among populations of these three
organisms to 20 industrial wastes. Morgan (1972) ranked the toxicity of a PCB and
DDT to a species of alga, two species of invertebrates and one species of fish.
Cultures of the alga, *Chlamydomonas*, showed the most initial acute sensitivity to the
PCB but completely recovered after two weeks. *Daphnia* exhibited the most
sensitive response to long-term exposures.

Bringham and Kuhn (1980) compared toxic response thresholds of contaminants
which stops multiplication in bacteria, algae, and protozoans. Of 156 inorganic and
organic chemicals, 23 had a toxic effect on the bacterium, *Pseudomonas putida*, 47 on
the alga *Scenedesmus quadricauda* and 43 on the flagellate protozoan, *Entosiphon
sulcatum*. The authors stress the importance of using a battery of test organisms to
assess a wide range of biological pollutant effects. Walsh *et al.* (1980) compared the
response of freshwater and estuarine algae, crustaceans, and fish to 23 different
textile mill effluents. They found that the algae, *Selenastrum capricornutum* and
Skeletonema costatum, were the most sensitive organisms tested and responded to all
effluents by either growth inhibition or stimulation.

Kenaga and Moolenaar (1979), however, presented a somewhat conflicting view
of the relative sensitivity of algae and aquatic plants. They stated that of 27 781

chemicals tested on aquatic plants, and 49 000 tested on the alga, *Chlorella*, only 33 caused complete mortality at concentrations below 1 mg l^{-1}, and only three of these compounds were also not toxic to *Daphnia* and fish at the same concentrations that affected the alga. This analysis supports their hypothesis that algae are the least sensitive of aquatic organisms to toxic contaminants.

III. Future Research

Algal species vary widely in their response to toxic chemicals. Hollister and Walsh (1973) demonstrated that there was a differential photoinhibitory response to various herbicides among single species of algae and that the sensitivity of various species of algae in four different phyla exposed to a single herbicide varied by nearly two orders of magnitude. Variation among species within phyla was lower, although there was still a tenfold difference from species to species.

Tests on single species of algae are therefore of limited applicability in assessing the effects of environmental contaminants on algal communities that are composed of an array of species with different sensitivities. Even at concentrations that are sublethal, toxicants can change the structure of algal communities (Mosser *et al.*, 1972a,b). However, only recently have there been research efforts addressing the question of how contaminant-induced changes in the structure and function of algal communities affect the aquatic ecosystem (DeNoyelles and Kettle, 1980).

Several interdisciplinary teams of scientists have outlined the complexity of environmental contaminant problems and assessment needs (National Science Foundation, 1976; Environmental Studies Board, 1981).

Future research should be directed toward four problem areas: (1) A toxicity test or series of tests needs to be devised that realistically assesses the impact of environmental contaminants on both structural and functional aspects of algal communities. (2) Additional research should be conducted to better establish the relationship between changes in community structure and changes in community function due to contaminant impact. (3) The magnitude to which algae are able to bioconcentrate and biodegrade contaminants is important in understanding how contaminants move through a food chain and also how algae provide some sort of protection to other organisms in the aquatic ecosystem. Future research in this area should focus on the identification of the kinds of algae that take up different contaminants and whether contaminants sorbed onto algae are degraded or are passed on to consumer organisms. (4) Ecosystem level studies should be conducted to determine how contaminant-induced changes in structure and function of algal communities affect other components of the aquatic ecosystem. This research should include measurements of community stability and how rapidly algal communities are able to recover from contaminant stress. Table 8 contains a conceptual hierarchical set of tests that consider the effects of contaminant stress on different levels of biological organization.

Table 8   Different levels of biological organization and qualities included in contaminant assessment.

---

1 SINGLE SPECIES TEST
   Range finding test, e.g. LC_{50} or EC_{50}

2 POPULATION TEST
   Mortality
   Reproductivity
   Growth rate
   Phenotype or genotype change
   Interaction of physical, chemical and biological factors with stress

3 MULTIPLE SPECIES TEST
   Competition
   Predation

4 ARTIFICIAL ECOSYSTEM TEST
   Change in community structure and function

5 ENCLOSURE OF NATURAL ECOSYSTEM
   Change in community structure and function

6 FIELD ASSESSMENT
   Critical ecosystem structure and function

IV. Conclusions

Environmental contaminants may act on a number of cellular and biochemical processes critical to growth and production of aquatic algae. These processes include photosynthesis, nucleic acid production, protein and lipid biosynthesis, uptake of inorganic nutrients, and nitrogen fixation. The direct effects of a particular toxic contaminant is very dependent on the species of algae and water quality conditions.

Indirect or secondary effects due to contaminant impact on organisms other than algae are chiefly from alteration of aquatic macrophyte communities and reduction of zooplankton communities. Death of large masses of aquatic macrophytes may release large pulses of nutrients but only when accompanied by severe de-oxygenation of the water. Complete suppression of macrophyte growth results in a reduction of nutrient concentrations in the water which results in less phytoplankton. Reduction in nutrients is due either to the loss of nutrient release of the macrophytes or to the increase of the oxidation–reduction potential of the mud–water interface which renders forms of nitrogen and phosphorus less soluble and reduces nutrient resuspension from the sediments.

The secondary effect on algae due to contaminant stress on the zooplankton community appears to be an increase in the algal standing biomass, photosynthetic rate, and nitrogen fixation.

Algae take up and bioconcentrate a variety of environmental contaminants present in the water. The rate of chemical bioconcentration and the final amount are functions of the particular species of algae, a number of physical and chemical environmental factors, and the nature of the contaminant.

Aquatic algae have recently been implicated in the degradation of organic contaminants. Algae appear to act directly in the degradation of organic chemicals or mediate photolysis. Thus, algae may be an important factor in the mitigation of contaminant effects on other organisms by sorption or rendering the toxic chemical unavailable by degradation to a harmless form.

References

Batterton, J. C., G. M. Boush and F. Matsumura (1971). *Bull. Environ. Contam. Toxicol.* **6**, 589–594.

Biggs, D. C., R. G. Rowland and C. F. Wurster (1979). *Bull. Environ. Contam. Toxicol.* **21**, 196–201.

Blinn, D. W., T. Tompkins and L. Zaleski (1977). *J. Physiol.* **13**, 58–61.

Bowes, G. W. (1972). *Plant Physiol. Leurcostes* **49**, 172–176.

Boyle, T. P. (1979). Responses of experimental lentic aquatic ecosystems to suppression of rooted macrophytes. *In* "Aquatic Plants, Lake Management and Ecosystem Consequences of Lake Harvesting" (J. E. Breck, R. T. Prentki and O. L. Loucks, eds), pp. 269–283. University of Wisconsin, Madison.

Boyle, T. P. (1980). *Environ. Poll.* **21**, 35–49.

Bringmann, F. and R. Kuhn (1980). *Water Res.* **14**, 231–241.

Bristow, J. M. and M. Whitcombe (1971). *Amer. J. Bot.* **58**, 8–13.

Brooker, M. P. (1974). *J. Institutional Water Engineers* **28**, 206–210.

Brooker, M. P. and R. W. Edwards (1975). *Water Res.* **9**, 1–15.

Brooks, A. S. and N. E. Liptak (1979). *Water Res.* **13**, 49–52.

Brown, J. R., L. V. Chow and C. B. Deng (1976). *Bull. Environ. Contam. Toxicol.* **15**, 437–441.

Brungs, W. A. (1976). Effects of Wastewater and Cooling Water Chlorination on Aquatic Life. U.S. EPA. Ecological Research Series, EPA-600/3-76-098, 46 pp.

Bryan, G. W. and L. G. Hummerstone (1973). *J. Mar. Biol. Ass. U.K.* **53**, 705–720.

Butcher, J. E., M. G. Boyer and C. D. Fowle (1977). *Bull. Environ. Contam. Toxicol.* **17**, 752–758.

Butler, G. L., T. R. Deason and J. C. O'Kelley (1975a). *Bull. Environ. Contam. Toxicol.* **13**, 149–152.

Butler, G. L., T. R. Deason, and J. C. O'Kelley (1975b). *Brit. Physiol. J.* **10**, 371–376.

Chiou, C. T., V. H. Frud, D. W. Schmedding and R. L. Kohnert (1977). *Environ. Sci. Technol.* **11**, 475–480.

Claeys, R. R., R. W. Claldwell, N. H. Cutshall and R. Holton (1975). *Pesticides Monitoring J.* **9**, 2–10.

DaSilva, E. J., L. E. Henricksson and E. Henricksson (1975). *Arch. Environ. Contam. Toxicol.* **3**, 193–204.

DeMarte, J. A. and R. T. Hartman (1974). *Ecology* **55**, 188–194.

Dennington, V. N., J. J. George and C. H. E. Wyborn (1975). *Environ. Poll.* 8, 233–237.

DeNoyelles, F., Jr and W. D. Kettle (1980). Herbicides in Kansas waters—Evaluations of the Effects of Agricultural Runoff and Aquatic Weed Control on Aquatic Food Chains. Kansas Water Resources Project Completion Report, Project No. A-092-Kan, Agreement No. 14-34-0001-8087.

Derby, S. B. and E. Rucker (1970). *Bull. Environ. Contam. Toxicol.* 5, 553.

Divavin, I. A., O. G. Mironov and I. M. Tsymbal (1978). *Soviet J. Mar. Biol.* 542–544.

Environmental Studies Board. Committee to Review Methods for Ecotoxicology (1981). "Testing for the Effects of Chemicals on Ecosystems." National Academy Press, Washington, 102 pp.

Fish, G. R. (1966). *Weed Res.* 6, 350–358.

Fisher, U. S. (1975). *Science* 180, 463–464.

Fitzgerald, G. P. and S. L. Faust (1963). *Appl. Microbiol.* 11, 345–351.

Fogg, G. E. (1971). *Plant Soil* 971, 343–401.

Foster, P. (1976). *Environ. Poll.* 10, 45–53.

Gentile, J. H., J. Cardin, M. Johnson and S. Sosnowski (1976). Power plants, chlorine and estuaries. U.S. EPA, Ecological Research Series No. EPA-600/3-76-055, Washington, D.C.

Glooschenko, V., M. Holdrinst, J. N. A. Lott and R. Frank (1979). *Bull. Environ. Contam. Toxicol.* 21, 515–520.

Goulder, R. (1969). *Oikos* 20, 300–309.

Hansen, D. J. (1976). PCB's: Effects On and Accumulation by Estuarian Organisms. National Conference on PCB's, EPA-560/6-75-004, pp. 282.

Harding, L. W. (1976). *Bull. Environ. Contam. Toxicol.* 16, 559–566.

Haug, A., S. Melsom and S. Omang (1974). *Environ. Poll.* 7, 179–192.

Hawxby, K., B. Tubea, J. Ownby and E. Bosler (1977). *Pesticide Biochem. Physiol.* 7, 203–209.

Hollibaugh, J. T., D. L. R. Seibert and W. H. Thomas (1980). *Estuar. Coastal Mar. Sci.* 10, 93–105.

Hollister, T. A. and G. E. Walsh (1973). *Bull. Environ. Contam. Toxicol.* 9, 291–295.

Horne, O. J. and C. R. Goldman (1974). *Science* 183, 409–411.

Hurlbert, S. H. (1975). *Res. Rev.* 57, 81–148.

Hurlbert, H. L., M. S. Mulla and H. A. Willson (1972). *Ecol. Monogr.* 42, 269–278.

Jewell, W. J. (1971). *J. Wat. Poll. Control Fed.* 43, 1457–1467.

Kar, S. and P. K. Singh (1977). *Bull. Environ. Contam. Toxicol.* 20, 707–714.

Kar, S. and P. K. Singh (1979). *Microbios Lett.* 10, 111–114.

Keith, L. H. and W. A. Telliard (1979). *Environ. Sci. Technol.* 13, 416–423.

Kenaga, E. E. (1972). *Res. Rev.* 44, 73–113.

Kenaga, E. E. and R. J. Moolenaar (1979). *Environ. Sci. Technol.* 13, 1479–1480.

Kirkwood, R. C. and W. W. Fletcher (1970). *Weed Res.* 10, 3–10.

Krzeczkowska-Wolosyzn, L. (1979). *Acta Hydrobiol.* 21, 139–147.

Lakshminarayana, J. S. S. and H. Bourque (1980). *Bull. Environ. Contam. Toxicol.* 24, 389–396.

Laube, V., S. Ramamoorthy and D. J. Kushner (1979). *Bull. Environ. Contam. Toxicol.* 21, 763–770.

Leland, H. V., S. N. Luoma and J. M. Fielden (1979). *J. Wat. Poll. Control Fed.* 51, 1592–1615.

Liden, L. H., D. T. Burton, L. H. Bongers and A. F. Holland (1980). *J. Wat. Poll. Control Fed.* **52**, 173–182.

MacFarlane, R. B., W. A. Glooschenko and R. C. Harriss (1972). *Hydrobiologica* **39**, 373–382.

Matthews, L. J. (1967). Further results of spraying lake weeds. *In* "Rotura and Waikato Water Weeds: Problems and the Search for a Solution" (V. J. Chapman and C. A. Bell, eds), pp. 76. University of Auckland, New Zealand.

Matsumura, F. and E. G. Esaac (1979). *ACS Symposium Series* **99**, 317–379.

McRoy, C. P., R. J. Barsdate and M. Nebert (1972). *Limnol. Oceanogr.* **17**, 52–67.

Melhuus, A., K. L. Seip, H. M. S. Seip and S. Myklestad (1978). *Environ. Poll.* **15**, 101.

Miyamoto, J., Y. Takimoto and K. Mihara (1979). *ACS Symposium Series* **99**, 3–20.

Morgan, G. B. and J. B. Lackey (1958). *Sewage Industr. Wastes* **30**, 283–286.

Morgan, J. A. (1972). *Bull. Environ. Contam. Toxicol.* **7**, 129–137.

Mortimer, C. H. (1971). *Limnol. Oceanogr.* **16**, 387–404.

Mosser, J. L., N. S. Fisher, T. Teng and C. F. Wurster (1972). *Science* **175**, 191–192.

Mosser, J. L., N. S. Fisher and C. F. Wurster (1972). *Science* **176**, 533–535.

Mulligan, H. F. and A. Baranowski (1969). *Verh. Internat. Verein. Limnol.* **17**, 802–810.

Mulligan, H. F., A. Baranowski and R. Johnson (1976). *Hydrobiologia* **48**, 109–116.

National Science Foundation (1976). Ecosystem Processes and Organic Contaminants: Research Needs and Interdisciplinary Perspective. US Government Printing Office, Washington, (NSF-RA-760008) 44 pp.

Nichols, D. S. and D. R. Keeny (1973). *Hydrobiologia* **42**, 509–525.

Ohkawa, H., R. Kikuchi and J. Miyamoto (1980). *J. Pesticide Sci.* **5**, 11–22.

Papst, M. H. and M. G. Boyer (1980). *Hydrobiologia* **69**, 245–250.

Patrick, R., J. Carins and A. Scherer (1968). *Progressive Fish-Culturist* **30**, 137–140.

Pokorny, J., L. Mentberger, B. Losos, P. Hartman and J. Hetesa (1971). Changes in biochemical and hydrobiological relations occurring when *Elodea* was controlled with paraquat. Proceedings of European Weed Research Council, 3rd International Symposium of Aquatic Weeds, pp. 217–229.

Powers, C. D., R. G. Rowland and C. F. Wurster (1977). *Environ. Poll.* **12**, 17–25.

Preston, A., D. F. Jefferies, J. W. A. Dutton, B. R. Harvey and A. K. Steele (1972). *Environ. Poll.* **3**, 69–82.

Rice, C. P. and H. C. Sikka (1973). *Bull. Environ. Contam. Toxicol.* **9**, 116–123.

Rose, F. L. and C. P. McIntire (1970). *Hydrobiologia* **35**, 481–493.

Ryan, J. B., D. N. Riemer and S. J. Toth (1972). *Weed Sci.* **20**, 482–486.

Schauberger, C. W. and R. B. Wildman (1977). *Environ. Contam. Toxicol.* **17**, 534–541.

Simpson, R. L. and P. Pimental (1972). *Search Agric.* (Cornell University) **2**, 1–39.

Sodergren, A. (1968). *Oikos* **19**, 126–138.

Sodergren, A. (1971). *Oikos* **22**, 215–220.

Stoermer, E. F., L. Sicko-Goad and D. Lazinsky (1980). Synergistic effects of phosphorus and heavy metals loading on Great Lakes phytoplankton. Proceedings of the Third USA-USSR symposium on the effects of pollutants upon aquatic ecosystems (W. R. Swain and V. R. Shannon, eds), pp. 171–186. EPA-600/9-80-034.

Strange, R. J. (1976). *J. Appl. Ecol.* **13**, 889–897.

Sumida, S. and N. Ueda (1975). *Plant Cell Physiol.* **14**, 781.

Toetz, D. W. (1971). *Limnol. Oceanogr.* **16**, 819–822.

Toetz, D. W. (1973). *Hydrobiologia* **41**, 275–290.

Toetz, D. W., L. Varga and M. Pierce (1977). *Water Res.* 11, 253–258.

Vallentyne, J. P. and S. W. Bingham (1974). *Weed Sci.* 22, 358–363.

Walker, C. R. (1963). Toxicological Effects of Herbicides on the Fish Environment. Proceedings of the Annual Air and Water Pollution Conference 8, 17–34.

Walsh, G. E. (1972). *Hyacinth Control J.* 10, 45–118.

Walsh, G. E., C. W. Miller and P. T. Heitmuller (1971). *Bull. Environ. Contam. Toxicol.* 6, 279–288.

Walsh, G. E., L. H. Bahner and W. B. Horning (1980). *Environ. Poll. (Ser. A)* 21, 169–179.

Way, J. M., J. F. Newman, N. W. Moore and F. W. Knaggs (1971). *J. Appl. Ecol.* 8, 509–532.

Werner, D. and E. Morschel (1978). *Bull. of Environ. Contam. Toxicol.* 20, 313–319.

Wetzel, R. G. (1975). "Limnology." Saunders, Philadelphia. 743 pp.

Wetzel, R. G. and B. A. Manny (1972). *Verh. Internat. Verein. Limnol.* 18, 162–170.

Wong, P. T. S., G. Barnison and Y. K. Chau (1979). *Bull. of Environ. Contam. Toxicol.* 23, 487–490.

Wurstbaugh, W. A. and C. S. Apperson (1978). *Bull. Environ. Contam. Toxicol.* 19, 641–647.

Wurster, C. F. (1968). *Science* 159, 1474–1475.

9 Algae as Monitors of Heavy Metals in Freshwaters

B. A. Whitton

Department of Botany
University of Durham
Durham, England

I. Introduction

During the past decade an increasing number of people concerned with environmental monitoring have gained access to sophisticated and rapid techniques for the analysis of heavy metals. At the same time there has been a relatively large number of accounts of biological methods for assessing the environmental impact of these metals, including some with freshwater algae. Although several of the methods appear very promising, none have been subjected to critical evaluation at a wide range of different research institutes. This chapter will review the various methods, assess which ones can be put to practical use now and suggest which might be considered suitable for ultimate development into a "standard" method.

To avoid confusion, the terms "monitor" and "heavy metal" need to be explained. "Indicator" has in the past often been used to describe an organism which can suggest the presence or absence of a particular environmental variable, while "monitor" has been used for quantitative assessment. As most types of environmental studies nowadays are quantitative, the term "monitor" is used for most of the present chapter.

The term "heavy metal", although often not rigidly defined, is generally held to refer to those metals having a density greater than 5, about 40 elements in all (Passow *et al.*, 1961). As pointed out by Nieboer and Richardson (1980), the term is often used where there are connotations of toxicity, and so data on lighter elements such as Be and Sr are sometimes included in general accounts of heavy metals. Because many authors use the term without defining it, Nieboer and Richardson (1980) proposed that it be abandoned entirely in favour of a classification separating metal ions into those which are oxygen-seeking, those which are nitrogen/sulphur-seeking, and those which are intermediate. Although Nieboer and Richardson (1980) demonstrated the biological relevance of their classification, the present

ALGAE AS ECOLOGICAL INDICATORS
ISBN 0 12 640620 0

author advocates that the term "heavy metal" should continue to be used in pollution studies. Scientific accuracy may be gained by avoiding the term, but much more will be lost in ability to communicate with administrators and the wider public.

Several reviews provide a guide to the extensive literature. Rai *et al.* (1981) outline the contents of some two hundred papers on algae and heavy metal pollution, the majority of which deal with freshwaters. Gadd and Griffiths (1978) helped to put algae into perspective with other microorganisms, while Whitton and Say (1975) did the same for other types of organisms in flowing waters. The influence of copper was discussed in detail by Stokes (1979) and that of zinc by Whitton (1980). Although not directly relevant to the present chapter, the account of Davies (1978) is valuable because it dealt critically with the literature on uptake and toxicity of heavy metals. Nieboer *et al.* (1978) reviewed mineral uptake and release by lichens, while Phillips's book (1981) deals at length with seasonal changes in heavy metal levels in animals.

II. Metal Composition of Field Populations

The capacity of algae and other aquatic plants to take up heavy metals from the water, producing an internal concentration greater than in their surroundings has been shown for many species. As a consequence it has frequently been suggested that chemical analyses of these submerged plants may give valuable information about contamination in the surrounding water. Because direct chemical analyses of water are already carried out so widely, it may be helpful to summarize the potential advantages of using aquatic plants before considering the results already obtained for particular algae.

(a) Plant monitors give an integrated picture of pollution within a particular system (Adams *et al.*, 1973; Empain, 1976; Trollope and Evans, 1976). This may be especially important where pollution is intermittent (Say *et al.*, 1981).

(b) Because of the high levels of accumulation found in most instances, measurement of the levels in plants increases the sensitivity of detection (Dietz, 1973).

(c) It is reasonable to assume that metal accumulation by a plant gives a better indication of the fraction of the metal in the environment which is likely to affect the aquatic ecosystem than most types of direct chemical analysis (Empain *et al.*, 1980).

(d) Dried plant materials are much easier to keep than water samples. Because they are compact and stable (with respect to most elements), they are suitable for long-term storage. Provided minimal precautions are taken, water management bodies can retain samples indefinitely, permitting further elements to be analysed years later. The samples are easy to send via the post for inter-laboratory comparisons of methods or for analysis by techniques, such as X-ray fluorescence spectrometry (Satake *et al.*, 1981), usually not available in laboratories responsible for routine pollution monitoring.

(*e*) The information obtained may also prove useful in the development of practical systems for the removal of heavy metals from mine and industrial effluents by encouraging the growth of plants (see Wixson, 1977; Jennett *et al.*, 1977).

Several authors have pointed out the value of algae (and bryophytes) over rooted plants in that they reflect only the properties of the ambient water rather than a combination of both water and sediments. Other potential advantages and disadvantages of algae will become apparent in the following discussion (see also Whitton *et al.*, 1981).

Probably because it is so widespread, *Cladophora* has been assessed more than any other alga (except perhaps *Chlorella* in the laboratory) for its ability to accumulate metals. The data, however, are fragmentary, and in no case has a standard method been developed which can be applied to routine monitoring without further research. The following account summarizes the methods which have been used.

Several reports deal with the use of *Cladophora* for concentrating radioisotopes (e.g. Kulikova, 1960; Angelovic, 1965; Williams, 1970). These include little information on heavy metals, but it is clear that elements may be concentrated at very high levels in comparison with the external environment. Burkett (1975) found in the laboratory that live *C. glomerata* accumulated more [^{203}Hg]-methylmercury than dead alga, at all test concentrations. The rate of uptake was apparently independent of methylmercury concentrations in the medium. Loss from the alga in uncontaminated water was slow, even when it had been killed with formalin. However, as pointed out by Burkett, the ability of *Cladophora* to retain mercurials may be lower under natural conditions where metabolic activity is greater. Gileva (1964) found little loss of inorganic mercury (Hg II) when contaminated *C. fracta* was placed in a desorbing solution.

Keeney *et al.* (1976) measured the levels of Cu, Zn, Cd and Pb in *C. glomerata* from two sites in Lake Ontario; their report includes a relatively detailed account of practical methods. Samples of fresh algae were washed thoroughly in distilled water and then air dried at 90°C. These were then digested with HNO_3, $HClO_4$ and H_2O_2, followed by HF to remove siliceous material. Metal analysis was carried out with differential pulse anodic stripping voltammetry, even though, as pointed out by the authors, the levels of metals accumulated were sufficient to permit analysis by (routine) atomic absorption spectroscopy. The authors summarized their results as the ratio between the concentration of metal in alga (µg metal per g dry weight) to that in "filtered" water (mg l^{-1}). (Following the conventions of Wilson, 1976, the term filtrable will be used through the rest of the chapter for a solution which has passed through a filter.) Keeney *et al.* compared their results with those for reports for Western Lake Erie and the upper Spokane River. The results for Cu are particularly consistent: Ontario; $2\cdot2 \times 10^3$, $1\cdot9 \times 10^3$; Erie, $1\cdot0 \times 10^3$; Upper Spokane, $2\cdot5 \times 10^3$. The ratios for the other metals fall within the range 10^3–10^4. Whitton (1979, based on Lloyd, 1977) reported a linear relationship between concentrations of Zn and Pb in *C. glomerata* and in the surrounding water of one river system at the time of collection, with an enrichment ratio for Zn of approximately $1\cdot3 \times 10^3$ over a range of aqueous Zn concentrations from $0\cdot01$ to $0\cdot35$ mg l^{-1}. However, two populations analysed by Trollope and Evans (1976) showed considerably higher enrichment ratios for Zn ($4\cdot0 \times 10^3$, 12×10^3). Presumably the concentration of Zn in the alga is dependent not only on the external concentration, but also other environmental variables.

The account of Empain *et al.* (1980) provides a comparison between the levels of metals accumulated by *C. glomerata* and by twelve other organisms (the alga *Lemanea fluviatilis*, ten bryophytes and the flowering plant *Lemna minor*). In comparison with the mean result obtained for all species, the authors concluded that the two algae were good for estimating Cd, but considerably underestimated Mn, Co, Zn and Pb. *Cladophora* also underestimated Cu, while *Lemanea* underestimated Fe and Cr. For this study whole plants of *Cladophora* were used, washed as free as possible of epiphytes (A. Empain, pers. comm.). Another useful comparison between *Cladophora* and other plants is given by Gale and Wixson (1979). This deals with accumulation of Mn, Zn and Pb in a meander system used to trap metal waste and the influence of various washing techniques on removal of metals from these plants.

There are a few other reports on heavy metal concentrations or enrichment ratios in freshwater *Cladophora*, but none gives much information on methods of sampling, preparation of material or analysis. Because of its widespread occurrence and the relative ease with which it can be recognized macroscopically, it is certainly a useful organism for monitoring both heavy metals and other toxic substances such as DDT (Meeks and Peterle, 1967; Ware *et al.*, 1968). Under favourable conditions *C. glomerata* can grow very rapidly, so the young tips are particularly suitable for differentiating between recent pollution and longer-term pollution; whole plants of *Cladophora* and bryophytes are more suitable for the latter. The use of *C. glomerata* is restricted by the seasonality of its growth, material suitable for harvesting being present for only just over half the year in most temperate regions. Its use as a monitor is also restricted at many sites where monitoring is important, because *Cladophora*, although tolerant of organic and nutrient pollution, is among the most sensitive of all algae to heavy metals (see Whitton, 1970). The presence of obvious growths of the alga at a site is in fact good evidence that heavy metal pollution is not having a marked influence on the ecosystem (Whitton, 1979).

Lemanea fluviatilis, included in the interspecies comparison of Empain *et al.* (1980), has been the subject of more detailed study by Harding and Whitton (1981).

This alga, which, like *Cladophora*, is easy to recognize macroscopically, is widespread and often abundant in temperate regions, but nevertheless overall is much less common than *Cladophora*. Like *Cladophora* it is suitable for harvesting for about half the year, but this includes late winter and early spring, when *Cladophora* is usually not obvious. It has the particular advantage that it appears to be tolerant of at least some types of heavy metal pollution and is often conspicuous in rivers subject to the influence of old lead mines. Harding and Whitton (1976) used only the terminal 2-cm tips of the filaments, as they wanted to establish the relationship between levels of metals in the plant and the river water. Since metal levels in many rivers are known to fluctuate markedly, it was assumed that the younger the part of the filament used for analysis, the more closely the algal composition was likely to reflect that the water collected at the same time. Significant positive correlations were found between the logarithm of the concentrations of Zn, Cd and Pb in alga taken from a wide range of sites and the logarithm of the concentrations of these elements in the water. The enrichment ratios for each metal (Fig. 1) decreased with increasing aqueous concentrations, an observation apparently contrasting with the result for *Cladophora* and Zn mentioned above, but agreeing with that for the same three metals in the liverwort *Scapania undulata* (Whitton *et al.*, 1982).

Analysis of the data on *Lemanea fluviatilis* from all sites indicates that other chemical features of the water probably influence the enrichment ratio of Zn for any particular level in the environment. In particular, high levels of aqueous Ca and/or

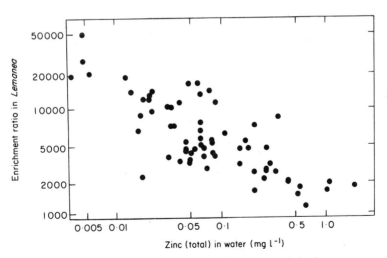

Fig. 1 Relationship between enrichment ratios for zinc in *Lemanea* and zinc in water at a range of stream and river sites ($r = -0.840$; $P = <0.0001$).

Mg appear to lead to reduced Zn in the alga, though experimental studies would be needed to confirm this.

Lemanea is usually relatively easy to transplant from one site to another, and experiments carried out by Harding and Whitton (1977) showed that transplanted populations eventually stabilize at a Zn concentration similar to that in filaments native to a site. Initial rapid changes were followed by periods of slower change, reaching an equilibrium by about 1000 h (in mid-winter). Deliberate transplants may therefore be made to a site where the alga can grow for part of the year, even though it can not maintain permanent populations.

Deb *et al.* (1974) analysed the mineral composition of *L. australis* collected from Manipur State, India, where the alga is eaten locally. Silver was present at detectable levels, in contrast to other analyses for Ag made on Indian plants. The authors pointed out that the presence of the metal indicated its existence in local rocks and suggested that a thorough survey in the region was desirable. Unfortunately there have been no subsequent reports to indicate whether this algal study has led to the location of a new silver mine!

Other algae which promise to be useful for monitoring heavy metals are *Nitella* and other charophytes which are not heavily calcified. *Nitella* is most widespread in shallow lakes, although it also occurs in flowing waters. There is extensive literature on ion uptake in this alga which, though not concerned with heavy metals, provides a useful background for anyone wanting to quantify its response to heavy metals. Harding and Whitton (1978) reported that the Zn and Pb contents of *Nitella flexilis* in a reservoir polluted by mining activities were correlated with the contents of these elements in the water but not in the underlying sediments. The rate and extent of Cd uptake in *Nitella* sp. were investigated by Kinade and Erdman (1975); both depended on the levels of Ca and Mg in solution, with about twice as much Cd accumulated in soft as opposed to hard water.

Few other freshwater algae are sufficiently widespread, easy to recognize macroscopically and easy to harvest without contaminating materials, that they can be recommended for general use for monitoring heavy metals, although no doubt many species can be utilized for local purposes. At least in north-west Europe freshwater forms of *Enteromorpha* are sometimes abundant, often in waters where there are few other plants suitable for analysis. Although there are no reports of the heavy metal composition of *Enteromorpha* in freshwater, there are many for estuarine populations (e.g. Klumpp and Peterson, 1979; Harding, 1980) and marine populations (e.g. Stenner and Nickless, 1974, 1975; Seeliger and Edwards, 1977; Sivalingam, 1978).

The study of Trollope and Evans (1976) of algae near zinc smelter wastes in the Lower Swansea Valley, Wales, included analyses for a range of genera besides *Cladophora (Coccomyxa, Microspora, Mougeotia, Oedogonium, Oscillatoria, Spirogyra, Ulothrix, Zygnema)*. The authors give no details of the methods by which the materials were harvested and cleaned of larger particles, but later stages of preparation were similar to those used by many other researchers: the algae were washed three times with distilled water and dried at 100°C. The mean value for Zn enrichment ratios adjacent to the smelter waste (mean aqueous $Zn = 16.7$ mg l^{-1}) was 1.4×10^3, while those for water distant from the waste (mean aqueous $Zn = 0.12$ mg l^{-1}) was 6.7×10^3. The mineral composition (including Mn, Fe, Zn) of a range of algal genera from non-polluted sites was reported by Boyd and Lawrence (1967), but without details of the composition of the water.

Several authors have used whole communities rather than trying to obtain unialgal materials. Denny and Welsh (1979) compared the Pb content of plankton samples from three lakes in the English Lake District. Ullswater, with an old lead mine in its catchment, had much more Pb in the plankton than the other lakes, reaching up to 722 µg Pb g^{-1} organic dry weight. This represents an enrichment ratio of more than 10^5. A *Melosira* dominated population had more Pb than an *Asterionella* one. The authors suggested that this might be because the former depended on cells which had been resting on Pb-rich bottom sediments, whereas the latter depended on growth of populations which had remained in the main water column.

Periphyton has also been used for monitoring heavy metals. Laboratory studies on periphyton from the Columbia River (Cushing and Rose, 1970; Rose and Cushing, 1970) showed that ^{65}Zn uptake by the whole community was directly proportional to the external metal concentration, but that most of the initial sorption took place on the upper surface of the community. The results of short-term exposure should therefore be expressed on an areal rather than a gravimetric basis. A marked decrease in the radionuclide content of the periphyton of this river has recently been reported (Cushing *et al.*, 1981) since the closure of the Hanford reactors.

Uptake of metals by diatom communities on artificial substrates (diatometers) has been followed by Patrick *et al.* (1975) and Friant and Koerner (1981). The latter found significantly higher levels of Cr on the diatometer below an effluent than above it, even though this was not detected by direct analysis of river sediments. The authors suggested that there was excellent potential for using the diatometer for routine analysis of river sediments. Racks of periphyton developing on glass slides were also employed by Johnson *et al.* (1978) for monitoring levels of metals at various sites around two highly contaminated lakes in Indiana. They concluded that periphyton may be a useful monitor where great differences exist in metal levels in the water, but seemed less convinced than Friant and Koerner (1981) of the widespread suitability of this approach.

A rather different use of algae has been initiated by E. Fjerdingstad, father and son, and other researchers from Copenhagen. They have analysed heavy metals by proton-induced X-ray spectrometry in samples of the red snow alga *Chlamydomonas* from Greenland (Fjerdingstad, 1973; Fjerdingstad *et al.*, 1974), Switzerland (Fjerdingstad and Kemp, 1975) and Spitsbergen (Fjerdingstad *et al.*, 1978). It was suggested by Fjerdingstad *et al.* (1978) that the higher Pb content of algae in Switzerland and Spitsbergen compared to Greenland (Table 1) was due to

Table 1 Ca and Pb contents (μg g^{-1}) of *Chlamydomonas nivalis* at three different locations: Ca is relatively constant, while Pb appears to be influenced by man's activities.[a]

	Contents	
	Ca$_{\bar{x}}$	Pb$_{\bar{x}}$
East Greenhead	2850	13·2
Spitsbergen	2640	42·1
Switzerland, Steingletscher Glacier	2915	133

[a] Data of Fjerdingstad and Kemp (1975) and Fjerdingstad *et al.* (1978)

Switzerland being an industrialized country, with a main road only a few kilometres from the sample site, and to Spitsbergen being involved in coal mining.

It will be apparent from the above accounts that the various researchers have had to be opportunistic in their selection of field materials for chemical analysis. Algae have often proved excellent monitors of heavy metals, but suitable species or communities are not always present. If plants are to be used routinely, it will frequently be necessary to include both algae and bryophytes in the sampling programme. These will usually provide one or more suitable species at most times of year in most rivers and shallower parts of lakes. If the information is to be of more than local interest, it is essential that more effort is put into the development of standard methods for sampling, harvesting, washing and digestion.

III. Metal Composition of Laboratory Populations

Numerous studies on heavy metal accumulation have been made in the laboratory, some of which are summarized by Rai *et al.* (1981). Many give information helpful in interpeting field results. Although none has been developed especially as a bioassay for assessing what fraction of a metal in a sample of natural water is available for biological uptake, several report on the selectivity of the algae for the less coarse fractions. *Synedra ulna*, for example, accumulates 0·45 μm filtrable Hg II rather than particulate Hg II (Fujita and Hashizume, 1975). The presence of clay particles (illite and montmorillonite) has been found (Keulder, 1975) to increase the uptake of ^{65}Zn by *Scenedesmus obliquus*, apparently by the removal of ions normally competing with Zn uptake. Algae themselves may also mobilize heavy metals bound to sediments, as shown by Laube *et al.* (1979) for Cu and Cd by *Anabaena* sp. and *Ankistrodesmus braunii*.

Most algae show an increase in metal accumulation over the range of increasing external concentrations which do not have a marked effect on growth. For instance, Cain *et al.* (1980) found that Cd accumulation increased over the range 0·01–1·0 mg l^{-1} aqueous Cd, but declined at higher concentrations, presumably owing to the severe toxic effects on the alga. However, Pb uptake by *Pediastrum tetras* was found to be independent of the external Pb concentration by Wettern *et al.* (1976), although there was a linear relationship between Mn in the medium and the alga. In their review, Gadd and Griffiths (1978) concluded that the amount of metal taken up by passive mechanisms and bound on the surface of algal cells is quite low in comparison with that taken up by metabolic or energy-dependent processes. In one example, Fujita and Hashizume (1975) found that 20% of the ^{203}Hg II supplied to dividing cells of *Synedra ulna* was taken up by passive adsorption, whereas 50% could not be eliminated even with cysteine solution and was accumulated in the inner part of the cell. However studies of ^{203}Hg II uptake over a 20-h period by *Pediastum boryanum* (Richardson *et al.*, 1975) showed that senescent coenobia accumulated more than young ones and the authors attributed uptake to adsorption by various cell components. Empty skeletons of the coenobia accumulated about one-third as much as complete but senescent ones, indicating that passive adsorption by cell walls is particularly important. The authors suggested that the silica component of the wall might be adsorbing much of this fraction. Filip and Lynn (1972) found over a 2-h period with *Selenastrum capricornutum* that as much Hg II was taken up by dead cells as by live, whether the latter were in the light or dark. Cd uptake differed in the two planktonic diatoms, *Asterionella formosa* and *Fragilaria crotonensis* (Conway and Williams, 1979). A range of experiments, including comparisons of light and dark and live versus dead cells, indicated that sorption of Cd by *Asterionella* was partially an active process, whereas that by *Fragilaria* appeared to be passive. Differences probably exist also between elements. In a study of ^{65}Zn, ^{109}Cd, ^{203}Hg and ^{210}Pb uptake by 14 algal strains (Hassett *et al.*, 1981) none of them removed Zn, whereas significant uptake took place in some instances with all the other metals. The authors suggested that this might be because cells were kept in the dark during the labelling period and light might be essential for Zn uptake.

Further discussion of the relative importance of passive and active uptake mechanisms lies outside the scope of this chapter, but it will be important to establish the key facts for useful indicator organisms like *Cladophora*, *Lemanea* and *Enteromorpha*. At present we do not even know whether these organisms respond differently to a pulse of heavy metal during the night as opposed to daylight.

IV. Laboratory Assays of Toxicity

There are even more studies on the toxicity of heavy metals to algae than on their accumulation, many of them with information useful in the design of laboratory assays, but only a small fraction of them deal with the development of assays for practical purposes.

The earlier use of algae in laboratory assays was concerned mostly with algicides for swimming pools or blue-green algal blooms in reservoirs. Assays on a range of possible algicides, including Cu and Ag, were reported during the mid-1960s by G. P. Fitzgerald. His studies used straightforward batch culture techniques, with liquid media (25 ml in 50-ml flasks), but without shaking or aeration (Fitzgerald and Faust, 1963). The organisms tested were mostly ones isolated specifically for the project from relevant sites, including a blue-green alga from a swimming pool which appeared to have considerable resistance to copper sulphate (Fitzgerald and Faust, 1963). *Microcystis aeruginosa* and *Chlorella pyrenoidosa* were used (Fitzgerald and Faust, 1963) to assay both the algicidal and algistatic properties of Cu supplied in various ways. When Cu was supplied from five different sources, including three commercial products, its toxicity in any one medium was quite similar. In contrast, differences in the composition of the medium led to marked differences in Cu toxicity. Algal biomass, growth form and extracellular materials can all at times influence assay results (Fitzgerald, 1964). Young and Lisk (1972) modified Fitzgerald's methods slightly, in particular by measuring growth over a period rather than just comparing algal biomass at one time.

The most detailed study of Cu as an algicide is that of Elder and Horne (1978), which was designed to compare the likely effects of Cu at two sites in California, a new reservoir and Clear Lake. Their account could well serve as a model for future projects where accurate information is important for specific bodies of water, since it combines both laboratory and field observations. The bioassays were carried out in large (4 litre) conical flasks with water (3 litre) from the relevant lake. The cultures were incubated under conditions of light and temperature that simulated as closely as possible those of the lake, and the flasks were shaken gently each day. Algal responses were measured as chlorophyll a content and ^{14}C uptake; both processes were depressed significantly at only 5–10 μg l^{-1} Cu. The authors pointed out that blue-green algae are especially susceptible to Cu toxicity, primarily because of the inhibition of nitrogen fixation.

Cu has also been reported to be highly toxic to the (presumably non-nitrogen-fixing) blue-green alga *Spirulina platensis* in Lake Nakuru, Kenya, as a result of laboratory assays using natural lake water and its associated algal population (Källqvist and Meadows, 1978). In spite of this being a soda lake, 50 μg l^{-1} Cu reduced the growth rate of *Spirulina* to 40% of the control; net oxygen production of the community was much less sensitive to Cu. The authors suggested that the concentrations of Cu that could develop in the lake as a result of factory pollution might change the algal species composition before it influenced the rate of primary production or the total algal biomass.

A number of authors have used a modification of the US Environmental Protection Agency's Algal Assay Procedure with *Selenastrum capricornutum* (Miller et al., 1978) for assaying the effects of heavy metals, though only one study (Greene et al., 1978) deals with a specific lake (Long Lake, Washington). Parrish and Burks (1977) made slight modifications to the method to measure the influence of Cd on the specific growth rate of *S. capricornutum*. Micronutrients other than Fe were omitted during the actual experiments; Fe was added in equimolar amounts with ethylenediaminetetra-acetic acid (EDTA); inocula were prepared by the method of Cain and Trainor (1973) to ensure that cells were in exponential growth. The authors found the alga to be slightly less sensitive to Cd than reported previously by Bartlett *et al.* (1974). Other studies involving the use of *S. capricornutum* as an assay organism include those of Hendricks (1978) on interactions between Zn and sulphide and of Kuwabara (1981) on the effects of adding Cu over different time periods.

Parrish and Burks (1977) suggested that, although the Algal Assay Procedure was good for comparative experiments within one laboratory, interlaboratory variation made predictive experiments questionable. This seems an unnecessarily gloomy view for the long run, but the method certainly needs much research before it can be recommended for general use with heavy metals. Because of contamination from even high-grade laboratory chemicals, it is difficult to prepare media with identical trace element composition in different laboratories. For instance, Zn is usually present at levels of 20–60 μg l^{-1} in media where it has been excluded from all trace metal stocks (observations in author's laboratory). As Zn and Cd can interact markedly in their effects on algae, though not necessarily always in the same way (Whitton and Shehata, 1982) differing levels of Zn contamination in the medium could easily affect the results of assays on Cd toxicity. Another problem which is in urgent need of study is the genetic stability of the widely used strain of *S. capricornutum*. The fact that this has not been looked at critically may prove to be harmless as far as studies on eutrophication are concerned, but there is too much evidence for genetic differences within species (see Section VI) with respect to metal tolerance for the problem to be ignored in this case.

A number of other strains of Chlorococcales have been used in laboratory assays on metal toxicity. Many of these use *Chorella vulgaris*, though, as there has been little attempt to concentrate work on a particular strain, it is hard to compare one set of results with another. Examples of studies which are useful in designing assays are listed in Table 2.

Rai *et al.* (1981) include an extensive table summarizing the contents of papers on the influence of environmental factors on heavy metal toxicity. Only one other account will be discussed here: the use of *Scenedesmus quadricauda* by Bringmann and Kühn (1980), which is particularly valuable because it compares the effects of 156 different substances (including Ni, Cu, Ag, Cd, Hg, Pb) on organisms from three different phyla (*Pseudomonas putida*, *Scenedesmus quadricauda*, *Entosiphon sulcatum*). Sufficient practical detail is included that their account can be treated as a standard method, although no information is given about the original sources of three organisms. The alga was more tolerant of all the heavy metals (except Cd) than either of the other two organisms. The difference in sensitivity between the alga and the bacterium were particularly marked for Ni and Cu, and between the alga and the protozoan for Pb.

Table 2 Literature on use of freshwater algae for laboratory bioassays and important physiological studies not discussed elsewhere in this chapter

Metal	Organism	Observation	Reference
Cr, Zn	*Nitzschia linearis*	Toxicity v. snail and bluegill	Patrick and Cairns (1968)
Cu	*Chlorella pyrenoidosa*	Photosynthesis	Steemann Nielsen et al. (1969)
Cu, Zn, Pb	Filamentous greens	*Cladophora* and *Oedogonium* spp. sensitive	Whitton (1970)
^{226}Ra	2 blue-greens, 5 greens	Accumulation	Havlik and Robertson (1973)
Mn, Ag, Pb	*Microthamnion kuetzingianum*	Toxicity and uptake	Lorch (1974)
Cr	*Chlorella pyrenoidosa* *Nitzschia palea*	Photosynthesis and growth	Wium-Anderson (1974)
Cu, Zn	*Hormidium rivulare*	Influence of pH	Hargreaves and Whitton (1976)
Pb	*Selenastrum capricornutum* + *Chlorella* strains	0·5 mg l^{-1} causes 50% inhibition of growth	Monahan (1976)
Cu, Zn, Cd	*Selenastrum capricornutum*	Development of method for toxicity assays	Miller et al. (1976)
Cd	*Asterionella formosa*	10 µg l^{-1} stops growth	Conway (1978)
Cu	*Lyngbya nigra*	0·45 µM inhibits growth	Gupta and Arora (1978)
Hg	*Scenedesmus acutus*	Development of Hg "buffer"	Huisman and Ten Hoopen (1978)
Cu	8 algae from 3 phyla	*Gloeocystis gigas* produces extracellular compounds reducing Cu^{++} activity	Swallow et al. (1978)
Pb	5 greens	Effect of complexing on toxicity	Weber et al. (1978)
Cr, Ni, Cu, Zn, Hg, Pb	*Scenedesmus quadricauda*	Metals tested individually and as mixture	Wong et al. (1978)
Cr	*Chlorella pyrenoidosa* *Chlorella* sp.	Comparison of Cr sources on toxicity	Meisch and Schmitt-Beckmann (1979)

Metal	Organism		Reference
Cu	21 algae from 5 phyla	Extracellular complexing agents	McKnight and Morel (1979)
Hg	*Chlorella vulgaris*	Reversal of toxicity by absorbic acid and glutathione	Rai (1979)
Cu, Pb	*Diatoma tenue* var. *elongatum*	Influence on polyphosphate bodies	Sicko-Goad and Stoermer (1979)
Hg	*Anabaena inaequalis*	$8 \ \mu g \ l^{-1}$ stops growth	Stratton *et al.* (1979)
Cu	filamentous greens	*Hormidium* most tolerant	Francke and Hillebrand (1980)
Zn, Cd, Hg	*Euglena gracilis*	Division and movement particularly affected	de Filippis *et al.* (1981)
Zn, Cd	*Coelastrum probiscoideum*	Influence of light regime on interaction between metals	Müller and Payer (1980)
Zn	blue–green algae	Resistant strains, especially *Calothrix*	Shehata and Whitton (1981)
Cu, Zn	*Eunotia exigua* v. *Pinnularia acoricola*	Both very resistant, but *Eunotia exigua* more so	Whitton and Diaz (1981)

V. Field Assays of Toxicity

A development of the approach of trying to simulate natural conditions in the laboratory by using large containers, natural lake water and a suitable temperature and light regime is to carry out the assays actually at the field site. Blinn et al. (1977) monitored the effects of Hg in clear plastic chambers in Lake Powell, Arizona. These held 6000 l of lake water and were open at the top, although the side extended sufficiently far above the water to prevent exchange between the experimental system and the lake. The influence of Hg II was tested on photosynthetic activity (^{14}C uptake) over 24-h periods. A reduction of at least 40% occurred with concentrations as low as 60 μg l^{-1} Hg, but the detailed responses of the spring and summer phytoplankton populations were rather different. Chambers (of slightly smaller volume) have also been employed by Kerrison et al. (1980) to test the influence of Cd in Lake Comabbio, a highly eutrophic lake in Northern Italy. With this study, however, only four of the chambers were closed at the bottom; another four were left open to permit exchange of Cd between the water and the sediments. At concentrations between 10 and 100 μg l^{-1}, Cd had marked effects on the enclosed ecosystems. These included decreases in phytoplankton and zooplankton biomass, decrease in oxygen concentration and increase in ammonia.

The most detailed studies on the influence of heavy metals in confined bodies of lake water are those of the MELIMEX experiment carried out by researchers from the EAWAG institute in Lake Baldegg, Switzerland (Schweiz. Z. Hydrol. 41, 165–314). Chambers 12 m in diameter and 10 m deep, termed limno–corrals (Gächter, 1979), permitted not only contact with the sediments, but also a limited flow through of water. Three limno–corrals were used for the experiment on phytoplankton (Gächter and Máreš, 1979), two being enriched with metals and one being a control. The metal-enriched water included a mixture of Cu, Zn, Cd, Hg and Pb, all at the legally tolerated limits for Switzerland. The account of the experiment is well worth reading in full, since it was carried out over a period of 17 months, but the most obvious effects were depressed photosynthetic activity, reduced phytoplankton species number, and major shifts in community structure. The authors concluded that the metal concentration limits set for flowing waters entering Swiss lakes were not low enough to prevent adverse effects in the receiving waters. In particular the limit set for dissolved Zn ($3 \cdot 1 \times 10^{-6}$ mol l^{-1} = $0 \cdot 2$ mg l^{-1}) should be lowered.

Although there are numerous comments in the literature (see Fitzgerald and Faust (1963) for earlier references and Whitaker et al. (1978) about the use of copper sulphate and other copper-containing algicides to control algal blooms, surprisingly few describe in any detail the fate of the copper and its influence on the alga. There are, for instance, apparently no reports on whether weather conditions or time of day influence the susceptibility of the algae. The study by Elder and Horne (1978) on Lake Perris, California, however, does provide much useful information. These authors pointed out that alkaline pH and high organic carbon levels in the water may detoxify Cu, so that it is not only an ineffective algicide but it may actually stimulate growth. Such conditions are common in lakes that have excessive algal growth, but the sensitivity of nitrogen fixation to Cu suggests that Cu treatment is suitable for lakes such as Clear Lake, which are nitrogen-limited and subject to large blooms of nitrogen-fixing blue-green algae.

VI. Adaptation

Streams and ponds with high levels of heavy metals often show differences in the amounts and types of algae present in comparison with environments lacking such enrichment (see Section VII). It seems reasonable to suspect that differences often also occur within species, either as environmental or genetic responses to metal enrichment. Evidence for environmental adaptation comes mostly from the laboratory. As mentioned above, Kuwabara (1981) tested the possibility of adding Cu to *Selenastrum capricornutum* at a rate slow enough to allow adaptation within one cell generation. In laboratory cultures of *Stigeoclonium tenue*, basal growth increasingly predominates over "upright" growth as Zn is increased (Harding and Whitton, 1977). At the field site with the highest Zn level (20 mg l^{-1}), it existed in a predominantly basal form; in culture at lower Zn levels it developed typical upright growth (Harding and Whitton, 1976). The proportion of mucilaginous species of algae appears to be greater in high Zn streams than similar streams lacking enrichment. An increased frequency of "knee-joints" has been noted in *Hormidium rivulare* populations tested in the laboratory, both as an environmental and a genetic response to high Zn (Say *et al.*, 1977). In spite of such observations, however, morphology in general provides little clue at present to the heavy metal status of a body of water.

A number of examples have been reported of marked differences in metal resistance within a single species which almost certainly reflect genetic differences. All the reports so far from freshwater deal with green algae. These include Ni-tolerant *Scenedesmus* (Stokes *et al.*, 1973), Ni-tolerant *Chlorella* (Stokes, 1975), Cu-tolerant *Scenedesmus* (Mierle and Stokes, 1976) and Cu-tolerant *Chlorella vulgaris* (Foster, 1977; Buttler *et al.*, 1980). Laboratory assays of Zn tolerance in populations of three *Hormidium* spp. (Say *et al.*, 1977) and *Stigeoclonium tenue* (Harding and Whitton, 1976) from many different streams showed a marked correlation between the experimentally determined tolerance and the level of aqueous Zn found in a stream. Other filamentous green algae (*Microthamnion strictissimum*, *Mougeotia* sp.: Whitton, 1980) are known to show similar differences in Zn tolerance depending on the Zn status of the sites from which they were removed, but no evidence for this could be found in the red alga *Lemanea fluviatilis* (Harding and Whitton, 1981). Good evidence for intraspecific differences in Zn tolerance have however been found in marine populations of a diatom (Jensen *et al.*, 1974) and a brown alga (Russell and Morris, 1970; Hall *et al.*, 1979).

Laboratory assays of the genetic tolerance of field populations provide one of the best ways of quantifying the toxicity of Zn at a site and allow estimates to be made of the influence of other factors such as Ca and P on this toxicity, but the method does require improvement in two main ways. It is at present difficult to design laboratory media where the level of metal bringing about a particular toxic response is the same as that found in the field, though helpful information on media suited for toxicity tests is given by Hutchinson and Stokes (1975). Resistance is often, but not always, found to be higher in the laboratory. This presumably reflects in large part the extent to which features of environmental chemistry other than the heavy metal under test resemble closely those in the natural environment. This is difficult to simulate where it is desired to use standard test media, especially where the field populations come from sites with very low levels of phosphate.

The other difficulty in assaying the tolerance of field populations in the laboratory is the possibility of adaptation taking place during the period of assay. Visual observation should be adequate to rule this out where filamentous materials are tested within a few days of being harvested, but more care is needed where dead cells can be seen only with the microscope. Obvious increases in tolerance within a relatively few subcultures have been reported for Al in *Chlorella pyrenoidosa* (Foy and Gerloff, 1972), Ni and Cu in *Chlorella* and *Scenedesmus* (Stokes, 1975) and Co, Ni, Cu, Zn and Cd in "*Anacystis nidulans*" (Whitton and Shehata, 1982).

Experiments on tolerant strains provide a further use besides their ability to assay the present toxicity of a metal at a site. The influence of environmental factors needs to be quantified on tolerant rather than sensitive strains, if the long-term effects of future heavy metal effluents are to be predicted accurately. At least in the case of Zn, there is evidence from a range of organisms that the relative importance of various factors on toxicity differs between tolerant and sensitive strains (see Whitton and Shehata, 1982). Mg, in particular, appears to play a much greater role with tolerant strains.

VII. Species and Community Composition

It would clearly be useful to be able to comment on the metal composition of a water body simply by observing its flora. The presence of any species whose environmental limits are understood could in theory be used as an indicator, but in practice the approach is of limited use, either because the ecological ranges of species are too broad or too little is known about their ecology. There are apparently no algal equivalents of the zinc violet, *Viola calaminaria*, a species more or less restricted to zinc-rich soils. The absence of sensitive species may prove to be a more useful indication of heavy metal enrichment than the mere presence of resistant ones, but, with the exception of a few normally conspicuous forms, this requires too thorough a floristic survey to be of widespread use. The absence of *Cladophora glomerata* in eutrophic rivers may however suggest sites influenced by heavy metal pollution, especially where abundant growths of *Stigeoclonium* tenue are present

(Whitton, 1979). *Cladophora* is apparently always especially sensitive to heavy metals (Whitton, 1970), but the absence of other algae may be because they are at a marked selective disadvantage rather than being killed outright by the elevated level of metal. This is likely to lead to conflicting observations. For instance, Williams and Mount (1965) concluded from an experimental study that *Synedra ulna* is highly tolerant of Zn, whereas Say and Whitton (1980) concluded that this species was probably absent in part of a stream system owing to the presence of elevated Zn.

Copper enrichment in reservoirs and ponds leads to the selective killing of planktonic blue-green algae as a group with other algae, especially greens, replacing them (e.g. Januszko, 1976, 1978). There is no evidence for a similar selective effect of any other metal on a particular phylum. In the case of Zn, a floristic comparison of a wide range of streams (Whitton and Diaz, 1980) has shown that not only does the number of species decrease with increasing environmental levels of metal, but that the percentage decrease is about the same for each of the major phyla. Although increased Zn leads to a reduction in number of species, sites with levels of Zn, which by most criteria would be regarded as very high, can sometimes be relatively species rich. A stream with a gradient of aqueous Zn from 25·6 to 1·2 mg l^{-1} (Say and Whitton, 1980) showed an increase in number of species on passing down the gradient, but only from 41 to 25; an unpolluted tributary had 61 species, 11 of which were absent in the main stream. In contrast, Say (1978) found only two algae (*Gomphonema parvulum, Achnanthes minutissima*) at a site polluted not only with Zn (22·8 mg l^{-1}) and Cd (0·44 mg l^{-1}), but also with very high levels of suspended matter.

It will be apparent that it is still difficult to generalize on the influence of heavy metal pollution of algal species and community composition. The individual situations in nature are too diverse, ranging, for instance, from lakes to fast-flowing rivers, from one metal to mixtures of many, often with suspended matter present as well and from extremely nutrient poor to nutrient rich waters. A list of useful floristic papers is given in Table 3. These provide some guidance for anyone needing to predict the short- and long-term effects of heavy metal pollution at particular sites.

VIII. Conclusions

Algae have been employed as monitors (or indicators) of heavy metals in freshwaters using many different approaches, yet none of these approaches has been developed so well that it can be applied routinely for water management purposes. Many studies of heavy metal pollution appear to have made use of algae more because of the particular interest of the researchers than because algae were obviously the most suitable organisms for the project. There is an urgent need to develop two or three standard methods which could lead to the use of algae becoming routine tools for monitoring metals in freshwaters. Two are particularly

Table 3 Floristic studies of the influence of heavy metals

Metal	Location	Observations	Reference
Cr, Cu	USA	List based on various observations; various details given	Palmer (1959)
Cu	England, R. Dove	*Chlorococcum* and *Achnanthes affinis* replaced other algae	Hynes (1960)
Cu, Zn, Cd, Pb	India, Udaipur, Zawar mines	No algae in effluent, but various algae when waste dried	Rana *et al.* (1971)
Cu, Zn	Canada, NW. Miramichi River System	Waters also low pH: *Achnanthes microcephala, Eunotia exigua, Pinnularia interrupta f. biceps, Fragilaria virescens* all very resistant	Besch *et al.* (1972)
Zn, Cd, Pb	USA, New Lead Belt	Floristic notes on polluted sites	Gale *et al.* (1973); Whitton (1980)
Many	England, various sites	Mine drainages; pH ⩽3·0 and with elevated metals	Hargreaves *et al.* (1975)
Cu, Zn, Pb	Wales, R. Ystwyth	*Hormidium* spp. resistant	McLean and Jones (1975)
Cu	Poland, experimental fish ponds	Increased overall biomass; stimulated greens and diatoms	Januszko (1976)
Cu, Zn, Cd, Hg	Canada, Flin Flon wastes	Includes floristic lists	Jackson (1978)
Cu, Zn, Cd	France, rivers near Decazeville	*Cladophora* replaced by *Stigeoclonium* etc.	Say (1978)
Cu	Canada, prairie lakes	*Aphanizomenon* eliminated, replaced mostly by greens, diatoms and chrysophytes	Whitaker *et al.* (1978)
Cu, Zn, Cd, Hg, Pb	Switzerland, L. Baldegg	Various blue-green algae, diatoms and 5 species of green listed as more abundant in metal-polluted water	Gächter and Máreš (1979)

Cu, Pb	Canada, subarctic lakes	Mines wastes on species composition and diversity	Moore et al. (1979)
Cu, Ni etc.	Canada, Sudbury lakes	*Peridinium inconspicuum* dominant, whereas chrysophytes in uncontaminated. Low pH probably controlling variable	Yan (1979)
Cu, Zn, Hg, Pb	Canada, eutrophic subarctic lake	Epipelic communities (resistant include *Anomoeoneis vitrea* and *Pinnularia brebissoni*)	Moore (1980)
Zn	England, N. Pennine stream	Gradient of concentrations on flora	Say and Whitton (1980)
Cu	Massachusetts reservoir	*Ceratium* replaced by *Nannochloris* and *Ourococcus*	McKnight (1981)

promising. A set of assays parallel to those developed for predicting and interpreting nutrient eutrophication would be useful in many situations; it should, however, be clear from comments made in Section IV that the methods already available can not be applied uncritically for heavy metals. The use of metal content in field populations to monitor events in water bodies is also an approach which could be used routinely, especially for flowing waters. It has the advantage that the methods involved can be regarded as merely an extension of standard chemical analyses of the water. Interpretation of the data may eventually prove to be straightforward, but much more information is needed from both the field and the laboratory about metal accumulation by the potentially most useful organisms, *Cladophora*, *Lemanea*, *Enteromorpha* and *Nitella*.

References

Adams, S., H. Cole, Jr and L. B. Massie (1973). *Environ. Pollut.* 5, 117–147.

Angelovic, J. W. (1965). "Some effects of Accumulated Radium on the Productivity of Algae." Michigan, Ann Arbor Univ. Microfilm 65–12, 593.

Bartlett, L., F. W. Rabe, and W. H. Funk (1974). *Wat. Res.* 8, 179–185.

Besch, W. K., M. Ricard and R. Cantin (1972). *Int. Rev. Ges. Hydrobiol.* 57, 39–74.

Blinn, D. W., T. Tompkins and L. Zaleski (1977). *J. Phycol.* 13, 58–61.

Boyd, C. E. and J. M. Lawrence (1967). *Proceedings of Annual Conference of S.-E. Game & Fish Commissioners* 20, 413–424.

Bringmann, G. and R. Kühn (1980). *Wat. Res.* 14, 231–241.

Burkett, R. D. (1975). *J. Phycol.* 11, 55–59.

Butler, M., A. E. J. Haskew and M. M. Young (1980). *Plant Cell Environ.* 3, 119–126.

Cain, J. R., D. C. Paschal and C. M. Hayden (1980). *Archs Environ. Contam. Toxicol.* 9, 9–16.

Cain, J. R. and F. R. Trainor (1973). *Phycologia* 12, 227–232.

Conway, H. L. (1978). *J. Fish. Res. Bd Canada* 35, 286–294.

Conway, H. L. and S. C. Williams (1979). *J. Fish. Res. Bd Canada* 36, 579–586.

Cushing, C. E. and F. L. Rose (1970). *Limnol. Oceanogr.* 15, 762–767.

Cushing, C. E., D. G. Watson, A. J. Scott and J. M. Gurtisen (1981). *Health Physics* 41, 59–67.

Davies, A. G. (1978). *Adv. Mar. Biol.* 15, 381–508.

Deb, D. B., B. Krishna, K. Mukherjee, S. Bhattacharya, A. N. Choudhary, H. B. Das and Sh. T. Singh (1974). *Curr. Sci.* 43, 269.

de Filippis, L., R. Hampp and H. Ziegler (1981). *Z. Pflanzenphysiol.* 101, 37–47.

Denny, P. and R. P. Welsh (1979). *Environ. Pollut.* 18, 1–9.

Dietz, F. (1973). The enrichment of heavy metals in submerged plants. *In* "Advances in Water Pollution Research, Proceedings 6th International Conference" (S. H. Jenkins, ed.), pp. 53–62. Pergamon, Oxford.

Elder, J. F. and A. J. Horne (1978). Copper cycles and $CuSO_4$ algicidal capacity in two California lakes. *In* "Environmental Management, Vol. 2, No. 1", pp. 17–30. Springer, New York.

Empain, A. (1976). *Mém. Soc. R. bot. Belg.* 7, 141–156.

Empain, A., J. Lambinon, C. Mouret and R. Kirchmann (1980). Utilisation des bryophytes

aquatiques et subaquatiques comme indicateurs biologiques de la qualité des eaux courantes. *In* "La Pollution des Eaux Continentales, 2nd Edn" (P. Pesson, ed.), pp. 195–223. Gauthier-Villars, Paris.
Filip, S. D. and R. I. Lynn (1972). *Chemosphere* 6, 251–254.
Fitzgerald, G. P. (1964). *Appl. Microbiol.* 12, 247–253.
Fitzgerald, G. P. and S. L. Faust (1963). *Appl. Microbiol.* 11, 345–351.
Fjerdingstad, E. (1973). *Schweiz. Z. Hydrol.* 35, 247–251.
Fjerdingstad, E., K. Kemp, E. Fjerdingstad and L. Vanggaard (1974). *Arch. Hydrobiol.* 73, 70–83.
Fjerdingstad, E., L. Vanggaard, K. Kemp and E. Fjerdingstad (1978). *Arch. Hydrobiol.* 84, 120–134.
Foster, P. L. (1977). *Nature, Lond.* 269, 322–323.
Foy, C. D. and G. C. Gerloff (1972). *J. Phycol.* 8, 268–271.
Francke, J. A. and H. Hillebrand (1980). *Aquatic Bot.* 8, 285–289.
Friant, S. L. and K. Koemer (1981). *Wat. Res.* 15, 161–167.
Fujita, M. and K. Hashizume (1975). *Wat. Res.* 9, 889–894.
Gächter, R. (1979). *Schweiz. Z. Hydrol.* 41, 169–176.
Gächter, R. and A. Máreš, (1979). *Schweiz. Z. Hydrol.* 41, 228–246.
Gadd, G. M. and A. J. Griffiths (1978). *Microbial Ecology* 4, 303–317.
Gale, N. L., M. G. Hardie, J. C. Jennett and A. Aleti (1973). "Trace Substances in Environmental Health VI", pp. 95–106. Univ. Missouri, Columbia.
Gale, N. L. and B. G. Wixson (1979). Removal of heavy metals from industrial effluents by algae. *In* "Development in Industrial Microbiology Volume 20", pp. 259–273. Society for Industrial Microbiology.
Gilera, É. A. (1964). *Fiziologiya Rasteni* 11, 581–586.
Greene, J. C., W. E. Miller, T. Shiroyama, R. A. Soltero and K. Putman (1978). *Mitt. int. Verh. theor. angew. Limnol.* 21, 372–384.
Gupta, A. B. and A. Arora (1978). *Physiol. Plant* 44, 215–220.
Hall, A., A. H. Fielding and M. Butler (1979). *Mar. Biol.* 54, 195–199.
Harding, J. P. C. (1980). "Concentrations of Metals in Plant Samples from the Mersey Estuary November 1979—February 1980" North West Water Authority TS-BS-80-5. Warrington, England.
Harding, J. P. C. and B. A. Whitton (1976). *Brit. phycol. J.* 11, 417–426.
Harding, J. P. C. and B. A. Whitton (1977). *Brit. phycol. J.* 12, 17–21.
Harding, J. P. C. and B. A. Whitton (1978). *Wat. Res.* 12, 307–316.
Harding, J. P. C. and B. A. Whitton (1981). *Wat. Res.* 15, 301–319.
Hargreaves, J. W., E. J. H. Lloyd and B. A. Whitton (1975). *Freshwat. Biol.* 5, 563–576.
Hargreaves, J. W. and B. A. Whitton (1976). *Oecologia, Berl.* 26, 235–243.
Hassett, J. M., J. C. Jennett and J. E. Smith (1981). *Appl. Environ. Microbiol.* 41, 1097–1106.
Harlik, B. and E. Robertson (1973). *J. Hyg. Epidem. Microbiol. Immnol.* 17, 393–405.
Hendricks, A. C. (1978). *J. Wat. Pollut. Control Fed.* 50, 163–168.
Huisman, J. and H. J. G. Ten Hoopen (1978). *Wat. Air Soil Pollut.* 10, 325–333.
Hutchinson, T. C. and P. M. Stokes (1975). "Heavy Metal Toxicity and Algal Bioassays." American Society for Testing and Materials Tech. Publ. 573, 320–343.
Hynes, H. B. N. (1960). "The Biology of Polluted Waters." Liverpool U. P., England.
Jackson, T. A. (1978). *Environ. Geol.* 2, 173–189.
Januszko, M. (1976). *Pol. Arch. Hydrobiol.* 23, 95–103.

278   B. A. Whitton

Jennett, J. C., J. M. Hassett and J. E. Smith (1977). Proc. Annual Conference on Trace Substances in Environmental Health II, pp. 448–458. Univ. Missouri, Columbia.

Jensen, A., B. Rystad and S. Melsom (1974). *J. Exp. Mar. Biol. Ecol.* 15, 145–157.

Johnson, G. D., A. W. McIntosh and G. J. Atchison (1978). *Bull. Environ. Contam. Toxicol.* 19, 733–740.

Källqvist, T. and B. S. Meadows (1978). *Wat. Res.* 12, 771–775.

Keeney, W. L., W. G. Breck, G. W. Van Loon and J. A. Page (1976). *Wat. Res.* 10, 981–984.

Kerrison, P. H., A. R. Sprocati, O. Ravera and L. Amantini (1980). *Environ. Technol. Lett.* 1, 169–176.

Keulder, P. C. (1975). *J. Limnol. Soc. S. Africa* 1, 33–35.

Kinkade, M. L. and H. E. Erdman (1975). *Environ. Res.* 10, 308–313.

Klumpp, D. W. and P. J. Peterson (1979). *Environ. Pollut.* 19, 11–20.

Kulikova, G. M. (1960). *C. R. Acad. Sci. U.S.S.R.* 135, 978–980.

Kuwabara, J. S. (1981). *J. Phycol., Suppl.* 17, 6.

Laube, V., S. Ramamoorthy and D. J. Kushner (1979). *Bull. Environ. Contam. Toxicol.* 21, 763–770.

Lloyd, E. J. H. (1977). "Accumulation of Metals by Aquatic Plants in the River System." Ph.D. Thesis, University of Durham, England.

Lorch, D. (1974). "Toxizität, Aufnahme und Speicherung der Schwermetalle Blei, Mangan und Quecksilber durch die Süsswassergrunalge *Microthamnion kuetzingianum* Naeg. (Chaetophoraceae)" Doctoral Dissertation, Univ. Hamburg, Germany.

McKnight, D. M. (1981). *Limnol. Oceanogr.* 26, 518–531.

McKnight, D. M. and F. M. M. Morel (1979). *Limnol. Oceanogr.* 24, 823–837.

Meeks, R. L. and T. J. Peterle (1977). R. F. Project 1794, Report No. COO-1358-3. Ohio State University Research Foundation, Columbus.

Meisch, H.-U. and I. Schmitt-Beckmann (1979). *Z. Pflanzenphysiol.* 94, 231–239.

Mierle, G. and P. M. Stokes (1976). *Trace Substances in Environmental Health X*, pp. 113–122. Univ. Missouri, Columbia.

Miller, W. E., J. C. Greene and T. Shiroyama (1978). "The *Selenastrum capricornutum* Printz Algal Assay Bottle Test." Corvallis Environmental Research Laboratory, Office of Research and Development, US EPA, Corvallis.

Monahan, T. H. (1976). *J. Phycol.* 12, 358–362.

Moore, J. W. (1981). *Wat. Res.* 15, 97–105.

Moore, J. W., D. J. Sutherland and V. A. Beaubien (1979). *Wat. Res.* 13, 1193–1202.

Muller, K. W. and H.-D. Payer (1980). *Physiol. Plant.* 50, 265–268.

Nieboer, E. and D. H. S. Richardson (1980). *Environ. Pollut. (Ser. B)*, 3–26.

Nieboer, E., D. H. S. Richardson and F. D. Tomassini (1978). *Bryologist* 81, 226–246.

Palmer, C. M. (1959). "Algae in Water Supplies". U.S. Dept. Health, Education & Welfare, Cincinnati, Ohio.

Parrish, D. and S. L. Burks (1977). "A Comparison of the Effects of Chelated and Non-chelated Cadmium on the Specific Growth Rate of *Selenastrum capricornutum*." US Dept. of Commerce National Technical Information Service. PB-275 775.

Passow, H., A. Rothstein and T. W. Clarkson (1961). *Pharmac. Rev.* 13, 185–224.

Patrick, R., T. Bott and R. Larson (1975). "The Role of Trace Elements in Management of Nuisance Growths." Environmental Protection Technology Series, EPA 660/2-75-008. 250 pp. Corvalis, Oregon.

Patrick, R., J. Cairns Jr and A. Scheier (1968). *Progressive Fish-Culturist* 137–140.

Phillips, D. J. H. (1981). "Quantitative Aquatic Biological Indicators." 48 pp. Applied Science Publishers, London.

Rai, L. C. (1979). *Phykos* 18, 105–109.

Rai, L. C., J. P. Gaur and H. D. Kumar (1981). *Biol. Rev.* 56, 99–152.

Rana, B. C., T. Gopal and H. D. Kumar (1971). *Environ. Health* 13, 138–143.

Richardson, T. R., W. F. Millington and H. M. Miles (1975). *J. Phycol.* 11, 320–323.

Rose, F. L. and C. E. Cushing (1970). *Science, N.Y.* 168, 576–577.

Russell, G. and O. P. Morris (1970). *Nature, Lond.* 228, 288–289.

Satake, K., P. J. Say and B. A. Whitton (1981). Use of X-ray fluorescence spectrometry to study heavy metal composition of aquatic bryophytes. In "Heavy Metals in Northern England. Environmental and Biological Aspects" (P. J. Say and B. A. Whitton, eds), pp. 135–146. Department of Botany, University of Durham.

Say, P. J. (1978). *Ann. Limnol.* 14, 113–131.

Say, P. J., B. M. Biaz and B. A. Whitton (1977). *Freshwat. Biol.* 7, 377–384.

Say, P. J., J. P. C. Harding and B. A. Whitton (1981). *Environ. Pollut. (Ser. B)* 2, 295–307.

Say, P. J. and B. A. Whitton (1980). *Hydrobiologia* 76, 255–262.

Seelinger, U. and P. W. Edwards (1977). *Mar. Pollut. Bull.* 8, 16–19.

Shehata, F. H. A. and B. A. Whitton (1981). *Verh. Int. Verein. Theor. Angew. Limnol.* 21, 1466–1471.

Sicko-Goad, L. and E. F. Stoermer (1979). *J. Phycol.* 15, 316–321.

Sivalingam, P. M. (1978). *Botanica Marina* 21, 327–330.

Steemann Nielsen, L. Kamp-Nielsen and S. Wium-Anderson (1969). *Physiol. Plant.* 22, 1121–1133.

Stenner, R. D. and G. Nickless (1974). *Wat. Air Soil Pollut.* 3, 279–291.

Stokes, P. M. (1975). Adaptation of green algae to high levels of copper and nickel in aquatic environments. In "International Conference on Heavy Metals in the Environment" (T. C. Hutchinson, ed.), pp. 137–154. Toronto, Canada.

Stokes, P. M. (1979). Copper accumulations in freshwater biota. In "Copper in the Environment. I" (J. P. Nriagu, ed.), pp. 357–381. J. Wiley, New York.

Stokes, P. M., T. C. Hutchinson and K. Krauter (1973). *Can. J. Bot.* 51, 2155–2168.

Stratton, G. W., A. L. Huber and C. T. Corke (1979). *Appl. Environ. Microbiol.* 38, 537–543.

Swallow, K. C., J. C. Westall, D. M. McKnight, N. M. Morel and F. M. M. Morel (1978). *Limnol. Oceanogr.* 23, 538–542.

Trollope, D. R. and B. Evans (1976). *Environ. Pollut.* 11, 109–116.

Ware, G. W., M. K. Dee and W. P. Cahill (1968). *Bull. Environ. Contam. Toxicol.* 3, 333–338.

Weber, A., M. Melkonian, D. W. Lorch and M. Wettern (1978). *Mitt. Int. Verein. Limnol. Theor. Angew.* 21, 254–260.

Wetton, M., D. W. Lorch and A. Weber (1976). *Arch. Hydrobiol.* 77, 267–276.

Whitaker, J., J. Barica, H. Kling and M. Buckley (1978). *Environ. Pollut.* 15, 185–194.

Whitton, B. A. (1970). *Wat. Res.* 4, 457–476.

Whitton, B. A. (1979). Plants as indicators of river water quality. In "Biological Indicators of Water Quality" (A. James and L. Evison, eds), pp. 5.1–5.35. J. Wiley, Chichester.

Whitton, B. A. (1980). Zinc plants in rivers and streams. In "Zinc in the Environment. Part II. Health Effects" (J. P. Nriagu, ed.), pp. 363–400. J. Wiley, New York.

Whitton, B. A. and B. M. Diaz (1980). *Trace Substances in Environmental Health* XIV, 457–463, Univ. Missouri, Columbia.

280   *B. A. Whitton*

Whitton, B. A. and B. M. Diaz (1981). *Verh. Int. Verein. Theor. Angew. Limnol.* **21**, 1459–1465.

Whitton, B. A. and P. J. Say (1975). Heavy metals. *In* "River Ecology" (B. A. Whitton, ed.), pp. 286–311. Blackwell, Oxford.

Whitton, B. A., P. J. Say and B. P. Jupp (1982). *Environ. Pollut. (Ser. B).* **3**, 299–316.

Whitton, B. A., P. J. Say and J. D. Wehr (1981). Use of plants to monitor heavy metals in rivers. *In* "Heavy Metals in Northern England. Environmental and Biological Aspects" (P. J. Say and B. A. Whitton, eds), pp. 147–152. Department of Botany, University of Durham, England.

Whitton, B. A. and F. H. A. Shehata (1982). *Environ. Pollut. (Ser. A).* 275–281.

Williams, L. G. (1970). *J. Phycol.* **6**, 314–316.

Williams, L. G. and D. I. Mount (1965). *Amer. J. Bot..* **52**, 26–34.

Wilson, A. L. (1976). "Concentration of Trace Metals in River Waters." Water Research Centre Technical Report TR 16. Water Research Centre, Stevenage, Herts.

Wium-Anderson, S. (1974). *Physiol. Plant.* **32**, 308–310.

Wixson, B. G. (ed.) (1977). "The Missouri Lead Study." 1108 pp. National Science Foundation, New York.

Wong, P. T. S., Y. K. Chau and P. L. Luxon (1978). *J. Fish. Res. Bd Canada* **35**, 479–481.

Yan, N. D. (1979). *Water Air Soil Pollut.* **11**, 43–55.

Young, R. G. and D. J. Lisk (1972). *J. Wat. Pollut. Control Fed.* **44**, 1643–1647.

10 Algal Assays for Vitamins

D. G. Swift

Graduate School of Oceanography
University of Rhode Island
Kingston, Rhode Island, USA

I. Introduction

A. *Auxotrophy*

It is common for one or more vitamins to be required absolutely for continued growth by an alga in axenic culture. Vitamin B_{12}, thiamine, and biotin, individually or in combination, are vitamins that are important for phytoplankton. These substances are required at low external concentration (10^{-13} to 10^{-10} M) and function as cofactors in biochemical pathways. Requirement for an organic substance at a micronutrient level by an organism that otherwise grows autotrophically is termed auxotrophy. Provasoli (1963) includes, as an alternative to a particular organic substance, "its physiological equivalent", to allow for the differences in chemical structure that may occur between the substance as added to culture medium, the form in which it is taken up by the cells, and the form which is active in biochemical pathways or the actual end product of the pathway. Because many species of algae require vitamins for growth, it is likely that vitamins have some role in determining which species develop and dominate in a phytoplankton assemblage.

B. *Vitamins and Natural Populations*

Auxotrophic algae are included among the various species which are dominant in particular regions or become dominant at certain times during the annual sequence of phytoplankton in the ocean or in lakes. This suggests that vitamin concentrations are often sufficient for good growth; otherwise auxotrophs could not compete so successfully with autotrophs. Conditions with regard to other nutrients and physical factors must also be satisfactory in order for significant populations to develop. It is important to recognize that other nutrients might be limiting when conditions are such that a vitamin would otherwise limit growth. In addition,

ALGAE AS ECOLOGICAL INDICATORS
ISBN 0 12 640620 0

competition between auxotrophs for coexistence or sequential dominance occurs, such as in the spring diatom blooms or summer diatom populations common in some lakes and estuaries or coastal regions.

Vitamins can be measured with good sensitivity by bioassay, but there is significant cost involved in effort, time, and facilities. Using general knowledge about vitamins and phytoplankton growth, together with data concerning the phytoplankton species present or the nutrient cycles occurring, one might then hypothesize an ecological role for a vitamin in a particular region. Following this, vitamin bioassay can be undertaken to test or modify the hypothesis, and it may be desirable or necessary to examine also the vitamin requirements of pure cultures of the important algal species from the region.

C. Occurrence within Major Algal Groups

Among 400 clones from all major algal groups that were tested, 44% required B_{12}, 21% required thiamine, and 4% required biotin (Provasoli and Carlucci, 1974). Within a genus there may be species with and without a vitamin requirement, and within a species there may be individual clones with different vitamin requirements (Lewin and Lewin, 1960; Guillard, 1968; Provasoli, 1958, 1963; Hargraves and Guillard, 1974). While a B_{12} or thiamine requirement often occurs alone, organisms requiring biotin usually also require one of the other two vitamins. The occurrence of auxotrophy is not related to energy metabolism. Facultative and obligate heterotrophs also comprise species with and species without a need for vitamins. In addition, existence of a vitamin requirement is not particularly related to whether the environment from which the species was isolated is normally rich in vitamins or low in vitamins. The review of Provasoli and Carlucci (1974) includes a comprehensive listing of vitamin requirements of algal species and patterns of requirements in algal groups. Other useful reviews are those of Provasoli (1963), Thomas (1968), Carlucci (1974), Berland et al. (1978), Swift (1980), and Bonin et al. (1981).

Distribution of vitamin requirements among algal groups shows certain trends. Table 1 shows that vitamin B_{12}, the most common vitamin requirement, is more likely to be required by species of some groups than by others. Among centric diatoms, the majority require B_{12} only or else no vitamin at all. Skeletonema costatum (Droop 1955a), and Thalassiosira pseudonana ($= Cyclotella nana$) and T. weissflogii ($= T. fluviatilis$) are species with a B_{12} requirement (Guillard and Ryther, 1962; Guillard, 1968). The first is very important in phytoplankton populations of many coastal and estuarine regions, and the latter two organisms have been the subject of a number of laboratory studies on the physiology of algae. T. nordenskioeldii (Swift and Guillard 1978) and Detonula confervacea (Guillard and Ryther, 1962) are examples of species that do not require B_{12}. These species are also ecologically important in temperate and boreal regions. About half of pennate diatoms require B_{12}. Several clones (10%) require thiamine only, and several more

Table 1 B$_{12}$ requirement in algae[a]

	%
Dinoflagellates	80
Chrysophytes	80
Diatoms	
Centric	70
Pennate	50
Haptophytes	50
Green algae	30
Cyanobacteria	20

[a]Approximate frequency of occurrence in clones reported from each group. Calculated from data reported in Provasoli and Carlucci (1974) and Swift and Guillard (1978).

require thiamine plus B$_{12}$, which is true also for one centric diatom. The seven clones designated *Bellerochea* (Hargraves and Guillard, 1974), a centric diatom genus, actually show morphological characteristics intermediate between centric and pennate diatoms (Stosch, 1977) and include clones with requirements of B$_{12}$ only, thiamine only, and B$_{12}$ plus thiamine, nutritional patterns more like pennate than centric diatoms. No diatom species has yet been shown to have a biotin requirement (Provasoli and Carlucci, 1974). Vitamin requirements are not linked to food value; many of the diatoms that are important as food for raising shellfish are B$_{12}$ requirers, such as *T. pseudonana*, *S. costatum*, and species of *Chaetoceros*.

About 80% of dinoflagellates studied require B$_{12}$, alone or in combination with thiamine or biotin or both (Provasoli and Carlucci, 1974; Loeblich, 1967). Examples are the requirement of *Gonyaulax polyedra* for B$_{12}$ only, *Prorocentrum micans* for B$_{12}$ plus biotin, and of *Amphidinium carteri* and *Gymnodinium breve* for all three vitamins. Marine ciliates studied by Stoecker *et al.* (1981) selectively feed on dinoflagellates.

Virtually all species of the Prymnesiophyceae (=Haptophyceae) have some vitamin requirement, and a requirement for thiamin occurs in the majority (Provasoli and Carlucci, 1974). Most require thiamine only or thiamine plus B$_{12}$, but several need B$_{12}$ only. None requires biotin. The coccolithophore *Emiliana huxleyi* (=*Coccolithus huxleyi*) which is widespread as an oceanic phytoplankter, usually in oligotrophic subtropical or tropical regions, needs thiamine only (Guillard, 1963). *Isochrysis galbana* and *Pavlova lutheri* (=*Monochrysis lutheri*), which have been well studied in the laboratory (Turner, 1979; Swift and Taylor, 1974; Droop, 1955b, 1957a, 1961, 1968, 1973) and are very effective as food in mariculture (Ryther and Goldman, 1975), require B$_{12}$ + thiamine.

Among the Chrysophyceae, excluding the *Pavlova* species now considered as prymnesiophytes, responses include one autotrophic species and another ten or so species requiring two or three vitamins in various combinations, most needing B_{12} + thiamine, with or without biotin (Provasoli and Carlucci, 1974; Turner, 1979).

Among the Chlorophyceae and Cyanobacteria, most are autotrophic, with a few species in each group being auxotrophic for vitamin B_{12} and a few in the former group requiring thiamine only or B_{12} plus thiamine. Neither group includes species reported to have a biotin requirement (Provasoli and Carlucci, 1974). *Dunaliella tertiolecta*, with no vitamin requirement (Carlucci and Bowes, 1970a), and *Brachiomonas submarina*, with B_{12} required and thiamin stimulatory (Droop and Scott, 1978; Turner, 1979), are some green flagellates used as food for herbivores in culture.

Among the three vitamins cited, the greatest amount of scientific investigation has been undertaken concerning vitamin B_{12}. Thus many of the details and examples presented about physiological responses, ecological roles, or bioassay details will refer to vitamin B_{12}.

II. Biological Responses

A. Cell Yield

At sufficiently low vitamin concentration, the cell crop is a function of initial B_{12} concentration (Fig. 1). This is the type of response on which most bioassay procedures depend, and it can also be used to calculate the minimum cellular B_{12} requirement, which is 2–18 molecules per $(\mu m)^3$ of cell volume for almost all auxotrophic algae or bacteria (Guillard and Cassie, 1963; Droop, 1957; Bradbeer, 1971), indicating that physiological activity occurs at a very low concentration (1000 molecules per cell of *T. pseudonana* 3H). A response reflecting cell density, such as chlorophyll *a* content or *in vivo* fluorescence or light scattering, can also be used in quantitative work or assays (Wood, 1963; Riley, 1966; Pintner and Altmyer, 1979; Brand *et al.*, 1981). A B_{12} concentration of 0·1 ng l^{-1}, which is a low concentration for lakes or for oceanic coastal regions, is sufficient to support a cell density in cultures of over 10^6 cells l^{-1} of most species, which greatly exceeds most natural population levels. The low density of natural phytoplankton populations relative to that of laboratory cultures and the small cellular requirement for B_{12} means that B_{12} is not likely to have a limiting role on the basis of supplying minimum cell vitamin content (Droop, 1957b).

B. Growth Rate

At sufficiently low concentrations, initial B_{12} concentration can control growth rate of algae in batch cultures (Fig. 2). A Michaelis–Menten type of kinetics is observed.

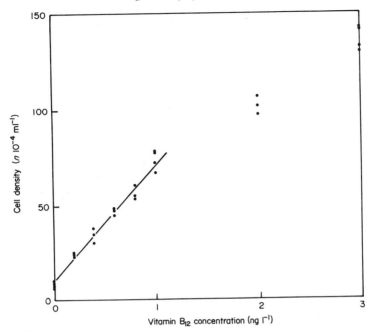

Fig. 1a Final cell crop of *Thalassiosira pseudonana* (clone 3H) (= *Cyclotella nana*) as a function of initial vitamin concentration. Growth in artificial sea water medium. From Swift (1967).

The primary ecological influence of vitamin B_{12} (or other vitamins) in the environment is probably through its effect on growth rate: vitamin-requiring species can develop in the population if there is sufficient B_{12} for rapid growth rate (Daisley, 1957). Rate responses of some algae are shown in Table 2. The half saturation constants for processes dependent on B_{12} vary for different species, from about 0.1 ng l^{-1} to 4 ng l^{-1} for photoautotrophic species, so that growth rates for various species are affected over a wide range of B_{12} concentration. The effect on growth rate allows use of $^{14}CO_2$ uptake rate in bioassays as a measure of response, an alternative to using parameters that are proportional to cell density (Gold, 1964; Carlucci and Silbernagel, 1966a,b, 1967).

The values of kinetics constants are the same in batch and chemostat growth for some species such as *T. pseudonana* 3H (Swift and Taylor, 1974), and different for others such as *P. lutheri* (Droop, 1968, 1973; Swift and Taylor, 1974) or *S. costatum* (Wood, 1963; Riley, 1966; Droop, 1970). Cells can excrete a vitamin B_{12} binding factor which complicates computation of kinetics constants (Droop, 1966, 1968).

C. Stimulation

Careful experiments to deplete the medium and cells of excess B_{12} carryover usually

D. G. Swift

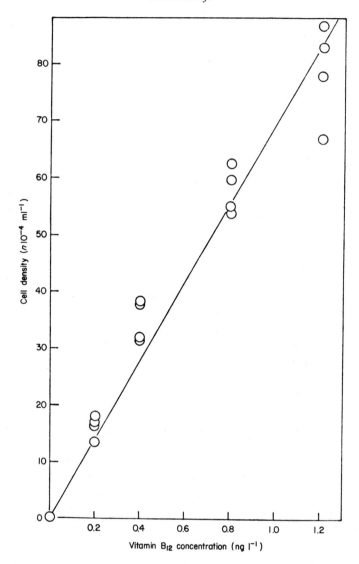

Fig 1b Standard curve for a vitamin B_{12} assay using *T. pseudonana* 3H. External standard in charcoal treated Gulf of Maine sea water. Slope, or cell yield, is 7×10^5 cells per pg B_{12}. From Swift and Guillard (1978).

result in observation of an absolute requirement in cultures that were previously only stimulated by B_{12} (Guillard, 1963). However, stimulation of growth by B_{12} occurs in a number of autotrophic clones of centric diatoms, such as *T. nordenskioeldii* isolated from the Gulf of Maine (Swift and Guillard, 1978). In

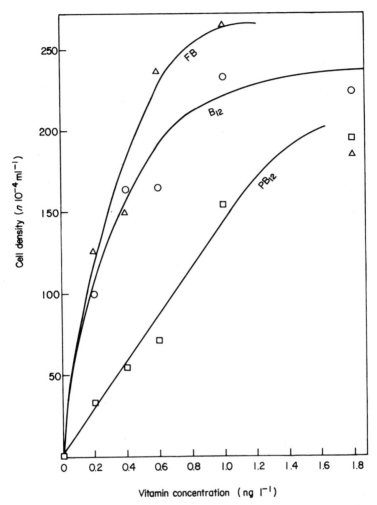

Fig. 1c Cell crop produced by *Bellerochea polymorpha* (clone 675d) in response to vitamin B_{12} and two analogs. B_{12} curve is for vitamin B_{12}, FB curve is for Factor B, and PB_{12} is for pseudovitamin B_{12}. From Swift and Guillard (1978).

culture cells grow faster and with shorter lag period if B_{12} is present (and they take it up). If no B_{12} is available initially, cells eventually synthesize it and grow (B_{12} can be detected in their medium).

Normally, culture media used for algae are made up with vitamin concentrations greatly in excess of the minimum requirements of auxotrophs (Guillard and Ryther, 1962). This allows for longer viability of cells in the light for both auxotrophs and autotrophs. Two or more successive transfers into vitamin-free medium are usually required to see effects of vitamin depletion on auxotrophs.

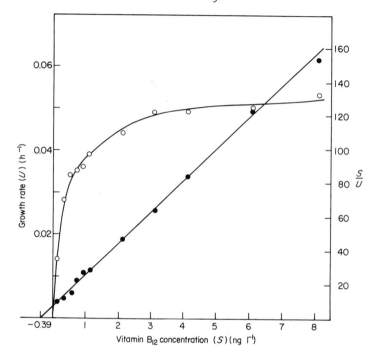

Fig. 2 The specific growth rate (*U*) of *Thalassiosira pseudonana* (clone 3H) as a function of initial vitamin B_{12} concentration (open circles, curved line). A linear transformation (solid circles) was used to calculate kinetics constants, such as in Table 2. From Swift and Taylor (1974).

III. Structure, Properties, and Roles of Vitamins

A. Forms of Vitamin B_{12}

1. Vitamin B_{12}   Figure 3 shows vitamin B_{12} in one of its coenzyme forms. It is a porphyrin-like ring structure containing a cobalt atom at the centre. The important parts of this structure in considering biological activity are the groups bonded to cobalt above and below the plane of the large ring (Table 3). The group below the plane is involved in the specificity of organisms for different analogues; the cobalt and group above it are involved in the enzymatic reactions of the vitamin. There is a nucleotide-like sequence below the planar structure, bound between the cobalt and the side chain of one pyrrol ring. This is 5,6-dimethylbenzimidazole in vitamin B_{12}; other structures may be present in vitamin B_{12} analogues which are synthesized and used by some microorganisms. In the coenzyme form shown here, 5′-deoxyadenosine is bound to cobalt above the plane of the large ring. Another coenzyme form contains a methyl group in this position. In reagent form, vitamin

Table 2 Rate responses to vitamins in batch cultures (unless otherwise stated)

Organism	Response	Vitamin	K_s (ng l^{-1})	Reference
FLAGELLATES				
Pavlova lutheri	Growth rate	B$_{12}$	2·8	Swift and Taylor (1974)
P. lutheri	No growth rate change	B$_{12}$	<0·1	Droop (1961)
P. lutheri (chemostat)	B$_{12}$ uptake	B$_{12}$	2·6	Droop (1968)
P. lutheri (chemostat)	Complex growth response	B$_{12}$	0·14	Droop (1968)
P. lutheri	C uptake	Thiamine	125	Carlucci and Silbernagel (1969)
Amphidinium carteri	C uptake	Biotin	4·0	Carlucci and Silbernagel (1969)
Ochromonas malhamensis	Growth rate	B$_{12}$	13	Ford (1958)
DIATOMS				
Thalassiosira pseudonana 3H	Growth rate	B$_{12}$	0·39	Swift and Taylor (1974)
T. pseudonana 3H (chemostat)	Growth rate	B$_{12}$	0·26	Swift and Taylor (1974)
T. oceanica 13-1[a]	Growth rate	B$_{12}$	0·2–0·4	Carlucci and Silbernagel (1969)
T. oceanica 13-1[a]	C uptake	B$_{12}$	2·9	Carlucci and Silbernagel (1969)
Skeletonema costatum (chemostat)	Growth rate	B$_{12}$	0·3	Droop (1970)
S. costatum	Growth rate	B$_{12}$	~2	Wood (1963), Riley (1966)
OTHER				
Natural populations, Lake Washington	B$_{12}$ uptake	B$_{12}$	6–110	Parker (1977)

[a] Formerly *Thalassiosira pseudonana*, clone 13-1 (Hasle, 1983)

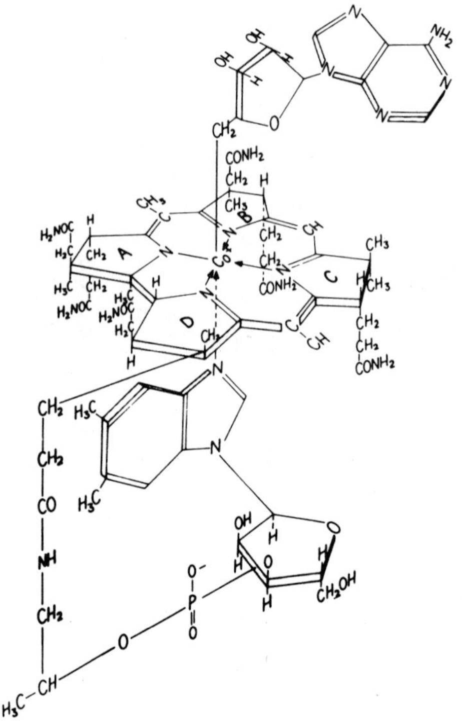

Fig. 3 The spatial formula of coenzyme B_{12}. From Smith (1965) and Swift (1980).

Table 3 Cobamide structure

SIDE CHAIN BETWEEN COBALT AND RING
B_{12}: 5,6 dimethylbenzimidazole (Fig. 3)
Pseudo B_{12}: adenine
Factor B: none

BOUND TO COBALT
-CN: in commercial reagent
-OH: in slightly alkaline solution
5′ deoxyadenosine: in Coenzyme B_{12} (Fig. 3)
methyl: in Methyl-B_{12} (a coenzyme)

B_{12} is sold as cyanocobalamin which contains a cyanogen group bonded to cobalt above the large ring. The cyanogen form is a result of the chemical purification process: treatment of samples converts cobalamins with other moieties in this position to cyanocobalamin, because it has good stability (Smith, 1965). In solutions at slightly alkaline pH, such as sea water or culture media, the natural form of the vitamin is hydroxocobalamin, with a hydroxyl group replacing the cyanogen or 5′-deoxyadenosyl moiety. Smith (1965) includes detailed information on structure, properties and activity of vitamin B_{12} and related substances.

2. *B_{12} analogues* Cobalamin, containing 5,6-dimethylbenzimidazole, can meet the B_{12} requirement of higher animals and any auxotrophic microorganisms. Mammals have a strict requirement for this form. Vitamin B_{12} analogues (Table 3) containing different moieties bound to cobalt below the planar structure are active for some microorganisms (Table 4). Forms with slightly substituted dimethylben-zimidazole have high activity for many organisms. Forms with the large ring structure identical (cobamides) but containing moieties such as adenine (pseudo vitamin B_{12}) or guanine in the side-chain are active for the bacterium *Lactobacillus leichmannii*. The bacterium *Escherichia coli* can utilize B_{12} or any of the analogue forms, including those lacking the side-chain altogether.

Table 4 Response of organisms to B_{12} and analogues

	B_{12}	Pseudo B_{12}	Factor B
MAMMALIAN SPECIFICITY PATTERN			
Mammals	100	0	0
Algae:			
Thalassiosira pseudonana 3H	100	0	0
Amphidinium carteri	100	0	0
LACTOBACILLUS SPECIFICITY PATTERN			
Algae:			
Euglena gracilis	100	100	0
Pavlova lutheri	100	40	0
Bacterium:			
Lactobacillus leichmannii	100	50–100	0
COLIFORM SPECIFICITY PATTERN			
Algae:			
Thalassiosira oceanica 13-1[a]	100	10–60	30–65
Bellerochea polymorpha 675d	100	40	120
Bacterium:			
Escherichia coli	100	100	100

[a] Formerly *Thalassiosira pseudonana* 13-1 (Hasle, 1983)

The specificity pattern for an alga is determined by data such as that shown in Fig. 1c. For a quantitative response, data should be taken from the linear portion of a dose–response curve (Fig. 1c, 1a). Otherwise the concentration at which the test is done can influence the result; a very low efficiency of utilization will appear much higher if cell crop at a single high concentration of analogue is compared to that for B_{12}. All of the specificity patterns outlined previously have been found in algae

(Guillard, 1968; Provasoli and Carlucci, 1974). Different clones of the same species can have different B_{12} specificity (Guillard, 1968). Table 4 includes some of the algal species useful for bioassay. In undertaking bioassays or comparing results from the literature, it is necessary to know whether the assay organism responds to B_{12} alone or to additional substances. Two assay organisms might give different results for the same water sample because of different analogue specificity. To interpret ecological implications of vitamin assays, it is desirable to know the vitamin requirements of the significant species in the region, including B_{12} specificity if there is a B_{12} requirement. Such information is not easily obtained, for it is necessary to culture the algae to study vitamin requirements; they must be free of bacteria as well, because bacteria can use or produce vitamins.

3. Bound B_{12}   Used algal culture medium contains a substance which binds B_{12}, making it unavailable for uptake by cells (Kristensen, 1955, 1956; Ford, 1958; Droop, 1968). This binding factor is produced by many of the species that have been tested, both those requiring B_{12} as well as those with no requirement (Droop, 1968; Pintner and Altmyer, 1973). Greatest release occurs in stationary phase cultures at very high population density. The binder is characterized as a protein (Daisley, 1970; Pintner and Altmyer, 1979) and is stable under refrigeration but loses its ability to bind B_{12} in several weeks at $15°C$ or $20°C$. Proteolytic agents or autoclaving cause total destruction.

Intracellular proteins also bind B_{12} (Sarhan *et al.*, 1980; Daisley, 1970), so death and breakage of cells are a source of bound B_{12} or binding factor and may contribute also to that observed during culture growth. Bound B_{12} can be observed in natural water samples with seasonal variation in its importance (Swift, 1981). Such binders for B_{12} are probably of bacterial, as well as as algal, origin and include proteins to which B_{12} is normally bound in cells as well as proteins or fragments which bind B_{12} non-specifically.

B. B_{12} Biochemistry

As is apparent from its complex structure, B_{12} is a larger molecule (molecular weight 1365) than most vitamins and cofactors. B_{12} is stable in solution at pH 4–7. Crystalline B_{12} contains water of hydration, which should be allowed for in calculating concentration (Smith, 1965). Concentrated solutions can be made and sterilized by filtration. They can be kept frozen when not in use. Dilutions should be made when needed because concentrations for experiments or assays at natural levels are very low (less than 10^{-12} M), and vitamin solutions are more stable when concentrated (Smith, 1965).

B_{12} is important for animals and microorganisms only and is synthesized by the latter. Higher plants lack B_{12} and use different cofactors for the corresponding biochemical reactions (Sebrell and Harris, 1968; Jaenicke, 1964). Animals must obtain B_{12} from their food sources or microbial flora. Thus bacteria, fungi, and algae are the primary sources of B_{12}, not only for themselves but also for higher organisms, as ingested food.

Vitamin B_{12} is a cofactor in over a dozen reactions, but there is little evidence to single out any one as being primarily significant for algal physiology (Swift, 1980). Many of the reactions are in heterotrophic organisms and probably do not occur in algae (Jaenicke, 1964; Stadtman, 1971). B_{12} is a cofactor in reactions involving intramolecular shifts of C or H and in group transfer reactions. Three reactions are linked to some algae. The nucleotide reductase reaction, which reduces the ribose in ribonucleotides to deoxyribose to furnish precursors for DNA synthesis, utilizes a B_{12} coenzyme in *Euglena gracilis* (Gleason and Hogenkamp, 1970). B_{12} deficiency leads to abnormally large cells in *Euglena* because nuclear and cell division cease while RNA and protein synthesis continue (Epstein *et al.*, 1962; Lefort-Tran *et al.*, 1980).

Formation of methionine in the bacterium *E. coli* utilizes a methylated B_{12} and folic acid derivatives, although some strains contain an alternative pathway or both pathways (Stadtman, 1971; Wagner, 1966). Cobalamin is linked to methionine synthesis in *Ochromonas malhamensis* (Barker, 1972) and to stimulation of methionine synthesis in *Prymnesium parvum* (Jaenicke, 1964). It may be involved in methionine synthesis in *Nechloris pseudoalveolaris*, a green flagellate not auxotrophic for B_{12}, although DNA involvement is possible also (Easley, 1969).

B_{12} coenzyme is a cofactor in the isomerization of the branched 4-carbon fatty acid derivative methyl malonyl Co A, to the straight chain succinyl Co A. This could be involved in breakdown of fatty acids or amino acids or (in reverse direction) for synthesis of isoleucine or leucine. Part of the B_{12} requirement of *O. malhamensis* is used in this reaction (Wagner, 1966).

Thus there is no certain evidence of the particular biochemical role for B_{12} in planktonic algae. Those species for which particular information is available, such as *O. malhamensis* and *E. gracilis*, are atypical in nutrition and physiology because of heterotrophic capability; hence it is not wise to extrapolate to other algal species groups. Regarding gross effects, those species with rigid cell walls or plate structures do not show the same morphological changes with B_{12} starvation as *E. gracilis*.

Algae not requiring an external supply of vitamin B_{12} use it in their biochemical pathways after synthesizing it in the cell (Carlucci and Bowes, 1970a; Swift and Guillard, 1978; Droop, 1968). These organisms can also take up B_{12} from the medium, as do the auxotrophs.

Minimum cell requirement for auxotrophs, as discussed in the section on cell yield, is only about 2–18 molecules per $(\mu m)^3$, computed for batch cultures of algae or bacteria (Guillard and Cassie, 1963; Droop, 1957; Bradbeer, 1971). Assay of cell extracts gives results of similar magnitude (Carlucci and Bowes, 1972a). Cells can take up B_{12} very rapidly and accumulate 10^3 to 10^4 times the minimum content (Di

Girolamo and Bradbeer, 1971; Droop, 1968, 1973b) so that cells in early stages of batch culture, or in chemostat culture, may have cell vitamin content greatly in excess of the minimum.

C. Thiamine

Thiamine is shown in Fig. 4. It has a pyrimidine and a thiazole portion with a CH_2 bridge. Table 5 shows response of some assay organisms and algal species studied in the laboratory. Among algal groups, a thiamine requirement is common in haptyphytes (prymnesiophytes) and chrysophytes. Species in these groups which have been tested for specificity need the pyrimidine moiety (Turner, 1979; Provasoli and Carlucci, 1974; Carlucci and Silbernagel, 1966b). Some green flagellates show a stimulatory response to thiamine, which can be replaced by the thiazole moiety (Turner, 1979).

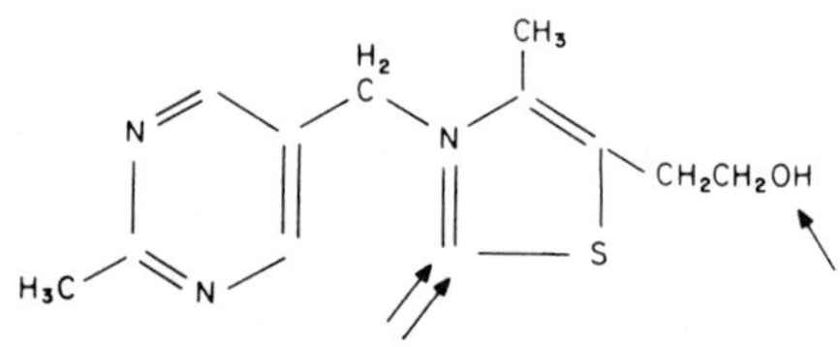

Fig. 4 Structure of thiamine. Single arrow indicates where pyrophosphate attaches. Double arrow indicates site of substrate attachment.

The reagent form, thiamine hydrochloride, has a characteristic odour and is readily soluble. Solutions are stable at pH ≤ 5, and a solution of at least 0.1% forms an acid solution. Thiamine is found in all organisms, free thiamine in plants and thiamine pyrophosphate in animals (Sebrell and Harris, 1972). Thiamine is easily destroyed by high temperature and alkaline pH (Gold et al., 1966; Gold, 1968). However, amino acids and chelators inhibit destruction whereas cupric ion promotes it (Gold, 1968; Wada and Suzuki, 1965). Thiamine itself is a weak metal complexer. Retention of thiamine activity in sea water, normally slightly alkaline, is not predicted according to properties of its aqueous solutions. However, the same holds true for mammalian body fluids which are also alkaline, and the amino acids present are thought to promote stability (Wada and Suzuki, 1965). Thiamine in a sea water culture medium retains 85% activity for *P. lutheri* assay even after autoclaving (Swift and Guillard, 1977). This medium contained a trace metal mix of metal salts and chelator.

The biochemical roles of thiamine and biotin in algae can probably be accurately deduced from studies with other microorganisms or with higher plant cells, which do synthesize thiamine and biotin. Animals need an external supply from food or bacterial flora, so phytoplankton supply the initial vitamin in the food chain in aquatic ecosystems (Provasoli, 1971).

Thiamine pyrophosphate is a cofactor in the decarboxylation of pyruvic and other α-keto acids. The pyruvate dehydrogenase complex of several enzymes and cofactors catalyses decarboxylation and reduction of pyruvate to acetyl Co A in mitochondria. Several other similar dehydrogenase complexes occur in the citric acid cycle and in breakdown or synthesis of some amino acids. Thiamine pyrophosphate is also a cofactor for two transketolase step in the Calvin cycle in chloroplasts of higher plants (Hatch, 1976).

Table 5 Specificity of thiamine response

Organism	Requirement	Reference
ALGAE		
Pavlova lutheri	Pyrimidine	Carlucci and Silbernagel (1966b)
Isochrysis galbana	Pyrimidine	Turner (1979)
Ochromonas malhamensis	Intact thiamine	Provasoli and Carlucci (1974)
FUNGI		
Phycomycete S-3	Intact thiamine	Vishniac and Riley (1961)
Cryptococcus albidus (yeast)	Thiazole	Ohwada and Taga (1972)

D. Biotin

Structure of the biotin molecule is shown in Fig. 5. It is soluble in hot water, slightly soluble in cold water (usually enough for a stock solution). Green plants synthesize biotin, but roots can also take it up. It is assumed that most marine animals derive biotin from microorganisms. There are some structural analogues for biotin lacking the S atom or substituting oxygen, but little is known of their biological activity.

In a study of effect of temperature and solar radiation on vitamins in sterile sea water, Carlucci *et al.* (1969) found biotin activity to increase at most temperatures, indicating breakdown is slight or that more active products are produced. Sunlight (on to large quartz tubes of solution) destroyed $75^{\circ}{}_{0}$ of B_{12} and biotin in two weeks, with further decrease occurring over two months. The assays were performed with the dinoflagellate *Amphidinium carteri*. Table 6 lists some bioassay organisms.

Biotin is a cofactor in some carboxylations and transcarboxylations (Hatch, 1976; Sebrell and Harris, 1968). An example is the acetyl Co A carboxylase system, which

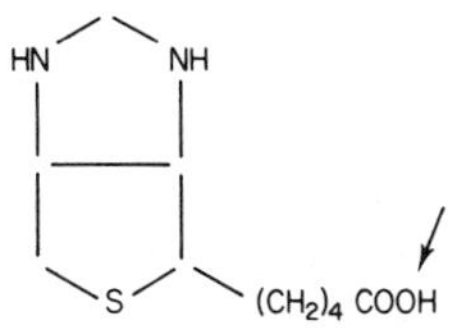

Fig. 5   Structure of biotin. Arrow indicates where protein binds.

D. G. Swift

Table 6   Bioassay organisms for biotin. The first two listed are more sensitive than the others.

Organism	Reference
ALGA	
Amphidinium carteri (dinoflagellate)	Carlucci and Silbernagel (1967)
BACTERIA	
Achromobacter sp.	Ohwada (1972)
Serratia marinorubra (bacterium)	Litchfield and Hood (1965)
Unidentified marine bacterium	Antia (1963)

is a complex of several proteins, one of which binds biotin. The conversion of acetyl Co A and CO_2 (with ATP needed also) to malonyl Co A is an initial step in fatty acid synthesis. Similar enzymes are propionyl Co A carboxylase (to form methyl malonyl Co A) and pyruvate carboxylase (to form oxaloacetate in mitochondria). Biotin is not a cofactor for the initial steps in autotrophic CO_2 fixation, which are carried out by the enzymes ribulose diphosphate carboxylase and phosphoenolpyruvate carboxylase.

IV. Bioassay for Vitamins

A. Basic Features of Algal Bioassay for Vitamins

Bioassay is the only method at present that is sufficiently sensitive and specific for quantitative determination of vitamin concentration in natural waters. (Various chemical, radioactive, and physical methods and their current shortcomings will be discussed below, as will the enrichment response type of bioassay experiment which is directed at identifying limiting factors or suitability of water for growth of a particular species.) In the bioassays described here, an organism with an absolute requirement for a particular vitamin is used to measure the concentration of the vitamin by allowing the organism to grow in the solution being assayed, with the final crop or some other measure giving a calibrated response to the vitamin. Although most of the assay organisms are photoautotrophic algae, techniques are basically the same for assays utilizing an alga growing heterotrophically in bright light, dim light or dark, or a bacterium or fungus growing heterotrophically. Handling of assays using heterotrophs differs only in the omission of light and in the provision of a suitable organic substrate (not contaminated by vitamins). Although a longer incubation period is required, the algal assays with phototrophic growth have several advantages relative to those utilizing heterotrophic microorganisms. Algal assays have greater sensitivity, generally, so are suitable for samples from either oligotrophic waters or eutrophic waters (with suitable dilution). Photoautotrophic growth avoids the possibility of vitamin contamination of the organic substrate or other essential organic nutrients, and also avoids the problem of response by the assay organism to an end product of the pathway using the vitamin. Such organic substances could be present in the water sample or the organic additions, but they are not generally taken up by the photoautotrophic algae.

Bioassays require certain manipulations in sample collection, storage, and preparation, but large volumes of water or extraction procedures are not necessary. The essential features of bioassays are as follows:

(1) Water samples must be taken in apparatus which has been cleaned to avoid vitamin contamination (from organisms growing on its surface or material shed from the ship surface) or contamination of the sample with substances, such as rust particles or oil, which might inhibit growth of the assay organism (usually an alga).

(2) Samples must be kept cold after collection or handled appropriately to prevent bacterial activity which can change vitamin concentration.

(3) The natural population must be removed. This is usually done by filtration. After this step samples can be stored frozen. Vitamin activity will be stable for long periods (Carlucci *et al.*, 1969).

(4) Nutrients necessary for growth of the assay organism, except for the factor being assayed, are added in excess to the sample.

(5) The prepared sample is sterilized by autoclaving or filtration.

(6) Samples are inoculated with cells of the assay organism which have already been depleted of the vitamin being assayed.

(7) Samples are incubated under appropriate conditions. Final cell density or some other measure of response is determined and then compared quantitatively to that in controls made with known vitamin additions.

Various aspects of the above steps can be considered in greater detail. The B_{12} assay developed by Ryther and Guillard (1962) uses *T. pseudonana* 3H. Determination of response after growth levels off in standards (about five days of incubation) can be made using cell counts, optical density, or other measures of growth (see Cell Yield section). Gold (1964) developed the assay further by using a short exposure period with ^{14}C-labelled CO_2 after inoculation and initial incubation. Carlucci and Silbernagel (1966a) modified the technique further by selecting the optimum incubation time before ^{14}C addition and adding controls to check for inhibition. An assay organism having different specificity from the species (or clone) of the earlier methods was used—*T. pseudonana* clone 13-1, now *T. oceanica* (see Table 4).

B. Standards

1. Types of standard curves Standards are normally made in sea water freed of vitamins by charcoal treatment (Fig. 1). Careful washing of the charcoal is necessary to remove inhibitory substances prior to use (Ryther and Guillard, 1962). Less charcoal can be used to further avoid the possibility of adding inhibitory substances (Haines and Guillard, 1974; Swift and Guillard, 1977). Usually standards are made by addition of vitamins to vitamin-free sea water from the same region, termed external standards. If the sea water vitamin concentration is at an appropriately low level, then internal standards can be made by adding vitamins to sea water used without charcoal treatment (Fig. 6, upper line). Another type of internal standard usually used is the addition of small amounts of vitamin to some of the replicates of a sample in order to detect inhibition in the sample (Carlucci and

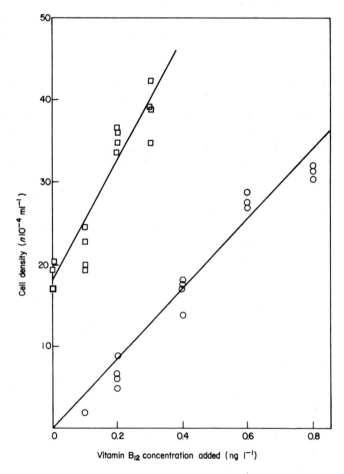

Fig. 6   External and internal standards. Circles: external standards in charcoal-treated sea water from the Gulf of Maine. Squares: internal standards using untreated water sample. Normally less difference is observed between slopes of the two types of standard curves.

Silbernagel, 1966a). In the presence of inhibition, the increment of growth from the added vitamin is less than that for the same amount of vitamin in the external standards. Such inhibition can often be circumvented by dilution of the sample. A desirable feature of using internal standards is the avoidance of a procedure, such as charcoal treatment to provide vitamin-free sea water, to which standards but not samples are exposed. This may have unknown effects. Such treatment also removes other organic substances and trace metals in addition to vitamins, but modern culture media formulations add back metals plus chelators in excess. Overall, vitamin-free sea water is preferable to artificial sea water as a matrix for standards or as a diluent in vitamin assay, probably because of good matching of major and minor ionic constituents and their ratios.

Many sea water samples must be diluted to remain within the linear response portion of an assay. Droop (1955b) describes an assay using *Pavlova lutheri* in which there is not a separate standard response curve for calibration. Instead, the water sample and sample diluted with vitamin-free sea water receive graded additions of B_{12} (Fig. 7). The distance between the parallel line responses is used in calculating the vitamin concentration in the unknown. (If the sample B_{12} concentration exceeds the linear response range, the assay can be made using two different dilutions of the water sample.) While highly suitable for statistical analysis as pointed out by Droop (1955b) and M. Fiala (Banyuls, France, unpub. manuscript), at least six cultures are required for an individual assay. Another technique (Fig. 8) is the dilution of the sample with graded amounts of vitamin-free sea water (Guillard and Ryther, 1962).

The simplest type of assay set-up utilizes a standard curve for the linear portion of the response, an internal standard if *in situ* vitamin concentration is sufficiently low but otherwise an external standard. Samples are prepared in replicate (usually triplicate) with dilutions made also if concentrations are likely to exceed the linear range (Carlucci and Silbernagel, 1966a; Swift and Guillard, 1977). Additional samples should be added to check for inhibition.

2. Problems with standards   Certain problems can occur with any of the types of standards described if calibration requirements are not met or at least known in sufficient quantitative detail so that valid computation of sample concentration can be made. (1) If the vitamin-free sea water used in various dilutions is not totally free of vitamins, then the background level present can be allowed for in calculations. (2) A more serious level of contamination can come from glassware or laboratory water. Initial detergent or acid cleaning of all utensils and use of pure water (such as deionized and distilled water) for cleaning laboratory equipment and making medium reagents is necessary to avoid random contamination of assays. Contaminants containing biological materials (such as dust, culture residues, and moulds in stored water or solutions) will add vitamins. In addition, materials which inhibit growth of algae, such as chromic acid cleaning solutions should be avoided except when it can be verified that they can be washed away. (3) Sometimes a standard curve in charcoal-treated sea water will have a different slope from that

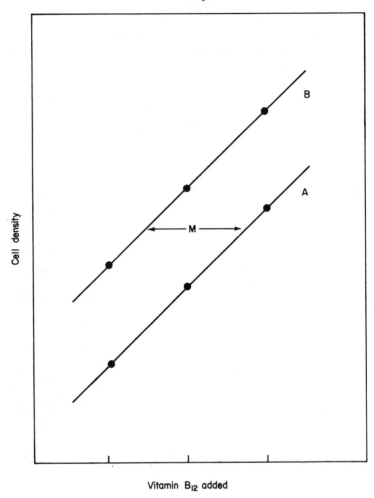

Cell density

Vitamin B₁₂ added

Fig. 7 Diagram of linear parallel line assay. Line B is response with graded additions of vitamin to sample, containing original concentration Z (unknown). Line A is response with graded additions to sample diluted quantitatively to $1/r$ of original concentration. Distance M between the regressions is equal to $Z(r - 1)/r$. Thus Z can be calculated from M and r; $Z = Mr/(r - 1)$. From Droop (1955b).

done in untreated sea water (Fig. 6). It should be verified that there is not a great discrepancy. Otherwise assays done by the parallel line method will show divergence of the lines; those with graded dilution of the sample will yield a curved line (Fig. 9); and systematic errors will occur in computations for undiluted samples when an external standard is the reference (Droop, 1955b; Swift, 1973, 1981; Swift and Guillard, 1977). (4) It is desirable to have as little vitamin carryover as possible

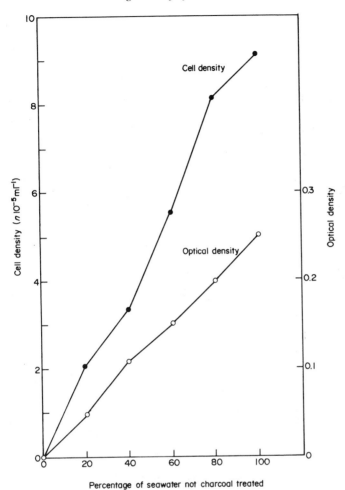

Fig. 8 Assay of sea water for B_{12} by graded dilution of sample with charcoal-treated sea water. Organism: *T. pseudonana* 3H. An external standard curve was used with this assay. From Ryther and Guillard (1962).

in the inoculum. A zero or very low response in the vitamin-free controls verifies that the inoculum was depleted of vitamin, provides a general check on avoidance of contamination in the procedure overall, and leads to greater precision and sensitivity. This is fairly simple with *T. pseudonana* but with some other species a much larger inoculum of less depleted cells is required in order to avoid a long lag period (Carlucci and Silbernagel, 1966b, 1967).

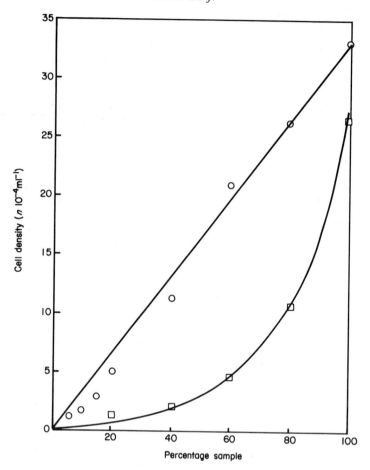

Fig. 9 Assay of two different sea water samples for vitamin B_{12} using graded dilution with charcoal-treated sea water. Gulf of Maine samples. Circles: 10 m sample; response is suitable for assay. Squares: 30 m sample; data for the dilutions (less than 100% sample) are not useful because of non-linearity. From Swift (1973).

C. *Organisms to be Used For Bioassay*

The algae listed in Table 4 are useful for bioassay of B_{12}. Assays utilizing the small *Thalassiosira* species are rapid and sensitive. Two species, *T. pseudonana* clone 3H and *T. oceanica* clone 13-1 (formerly considered *T. pseudonana* also) are in use (Carlucci and Silbernagel, 1966a; Ryther and Guillard, 1962; Hasle, 1983). The wide tolerance of salinity and temperature of clone 3H makes it suitable also for assay of freshwater samples after dilution with vitamin-free sea water or artificial medium. Most freshwater samples have sufficiently high vitamin B_{12} that dilution is necessary anyway. Other freshwater organisms such as *E. gracilis* or *O. malhamensis* could also be used (Parker, 1977; Ohwada and Taga, 1972b). *Pavlova lutheri* is also suitable for assay of sea water (Droop, 1955b), although a longer incubation period is necessary.

Bioassay organisms for thiamine are listed in Table 5. *Pavlova lutheri* (Carlucci and Silbernagel, 1966b) is probably the organism of choice, based on the observation that few samples assayed showed no thiamine activity (Swift, 1981), unlike ones assayed with the yeast *Cryptococcus albidus* (Ohwada and Taga, 1972a; Natarajan and Dugdale, 1966). It is possible that inhibition of the assay organism may be responsible in some cases for the negative assay result.

Bioassay organisms for biotin are listed in Table 6. Inhibition to the assay organism can be a problem with *Amphidinium carteri* (Swift, 1981) and the bacterium *Serratia marinorubra* (Natarajan, 1968; Litchfield and Hood, 1965), although in some cases this may be caused during sample preparation (Swift, 1981).

D. Preparation of Inoculum

Cells must be sufficiently depleted of excess stored vitamin in the inoculum to give a quantitative response in assay. However, they must have sufficient vigour to commence growth without a long lag. Cells are much more easily depleted of nitrogen or phosphorus than of vitamins (or other micronutrients or trace elements). Culture media such as f/2 (Guillard and Ryther, 1962) contain vitamins in excess. The actual amount depends to some extent on the care with which the vitamin stock solution is maintained in order to avoid loss of activity. In general it takes at least one transfer into vitamin-free medium to deplete thiamine and biotin. The second sequential transfer should show some response, but total depletion of stored cellular vitamin may not have yet occurred. For B_{12}, because of lower minimum requirement and greater stability, an additional transfer is required.

For convenience while doing assays, it is desirable to keep stocks of each assay culture growing at a low level of the vitamin for which it is used. For example, *T. pseudonana* 3H with 2–5 ng l^{-1} B_{12} can be transferred every week or two. It will grow at almost maximum growth rate and will produce a cell crop of about 10^6 ml^{-1}. (Swift and Taylor, 1974; Swift and Guillard, 1977). This is just beyond the linear portion of a dose–response curve (Fig. 1). Cells in log or stationary phase can be diluted in vitamin-free sea water to provide initial density of 200–500 ml^{-1} when assay cultures are started. Such cells are in a physiological state to commence growth again without a long lag, but the B_{12} carryover will still be small. Most of these cells from stationary phase will not divide again without more B_{12}. Carryover can be estimated on the basis of the dilution as about 0·001 ng l^{-1}.

E. Specificity

The specificity of the assay is relevant for B_{12} and thiamine (Tables 4 and 5) in order

D. G. Swift

to interpret results. Little data exist regarding presence of individual thiamine moieties in natural waters. All of the thiamine assay organisms respond to intact thiamine. Results of assays having different specificity will be directly comparable only if the concentrations of the individual moieties are equal (or some are absent). With regard to vitamin B_{12}, it is important to know the specificity of assays in order to compare results, because the assays which include analogues in their response will yield a higher value for the same sample (if analogues are present). Use of two different assay organisms has allowed estimation of analogues in several studies. Swift (1981) found mean ratios of cobamides (B_{12} plus biologically utilized analogues, particularly factor B and pseudo B_{12}) to B_{12} to be 2·3 in winter and summer samples from the offshore Gulf of Maine, and 1·4 in spring. The two assays utilized *T. pseudonana* 3H and *Bellerochea polymorpha* 675d (Table 4). Allowing for the latter clone not having the same efficiency in utilizing analogues as it does B_{12}, one can calculate that mean analogue concentration in winter and summer is 160% of B_{12} concentration, and 50% of B_{12} in spring. Most samples from the mixed layer in spring had less than this because the mean was influenced by a higher value at 100 m depth. Bruno *et al.* (1981) assayed B_{12} in water samples from the Peconic Bay estuary, Long Island, using *T. pseudonana* 3H and *T. oceanica* 13-1. Means of samples at different stations over seasonal cycles showed analogue concentration 150% of B_{12} concentration. However, individual peaks tended to occur at different times for analogues and B_{12}, and the ratio for individual samples showed extremely wide variation.

Sediments are extremely rich in vitamin B_{12} and related substances. This is probably because of high density of bacteria and the importance of B_{12} in bacterial biochemical pathways. In sediment samples and bacterial cultures, there is greater cobamide activity in compounds other than B_{12} (Starr *et al.*, 1957; Burkholder and Burkholder, 1956, 1958). Using this information and data for analysis of water samples, Swift (1981) and Bruno *et al.* (1981) suggest that a high concentration of analogues is indicative of cobamide production from bacterial sources, particularly in sediments. Cobamide concentrations will be higher when there is an entirely mixed water column or when mixing from deeper water occurs or when bacterial activity *in situ* is significant.

V. Inhibition and Non-specific Responses

Growth responses which are less in magnitude than expected from the amount of vitamin present can occur as a result of inhibition in the medium. B_{12} binding factor will prevent use of vitamin which is bound to it, but usually an assay is designed to destroy or to retain binding factor so that results can be interpreted properly. The presence of inhibitory substances not directly related to vitamin concentration can be detected and assessed by use of internal standards (as discussed previously) if they are not of great magnitude. Such inhibition can occur in sea water samples

(Ohwada, 1972; Litchfield and Hood, 1965; Swift, 1981). Pintner and Altmyer (1979) studied inhibition due to B_{12} binder. Culture media in which one species had grown was assayed with other organisms. Addition of B_{12} reversed the inhibition caused by binding factor. However in a number of instances inhibition occurred which was not reversed by B_{12}. Similarly Carlucci and Bowes (1970a,b) sometimes observed growth inhibition of an assay organism by medium in which another clone had produced vitamin. The factors involved are heat stabile and very inhibitory (Pintner and Altmyer, 1979). Thus non-specific inhibition (or possible allelopathic effects) can occur to prevent a suitable growth response.

Inhibition can also occur from conditions which are not favourable for growth, such as deficiency or a required element or presence of a toxic substance. This can come about by contact of the sample with objects made of materials such as rubbers or plastics which are unsuitable for culturing purposes because toxic substances leach from them (Blankley, 1973).

VI. Chemical Methods

Although methods other than bioassay may become practical in the future, only bioassay has sufficient sensitivity (for detecting such low concentration) and specificity (for responding to the biologically active molecules and not to structurally similar breakdown products. High performance liquid chromatography (Beck and Brink, 1976, 1978) is highly discriminative but is insufficiently sensitive to apply without an extraction procedure to concentrate B_{12}. Radioisotope methods utilize binding of B_{12} to a B_{12} binder (intrinsic factor from mammalian cells). Natural B_{12} competes with added radioactive B_{12} (^{57}Co label). Techniques have been described with extraction (Beck, 1978) and without it (Sharma *et al.*, 1979). There is a problem with not knowing the accuracy and specificity of these methods relative to bioassay. Sharma *et al.* (1979) found results from the isotope assay that were four to ten times greater than bioassay results, indicating that water samples contain substances that bind to intrinsic factor (or affect the binding) but which lack cobamide activity for organisms.

Radioactive B_{12} has been of use as a tracer in such areas as chemostat growth in the laboratory (Droop, 1969, 1968) and uptake rates of natural populations (Parker, 1977). Usually a known B_{12} standard (laboratory studies) or a bioassay (field studies) is employed to yield more than relative measurements.

VII. Ecological Aspects

A. Vitamin Production and Utilization

Bacteria are known to be a source of vitamins in water and sediments (Starr *et al.*,

1957; Berland *et al.*, 1978a; Lebedeva *et al.*, 1971; Vacelet, 1975; Martin and Vacelet, 1975; Haines and Guillard, 1974). Vitamin content of sediments and suspended material is significant (Burkholder and Burkholder, 1956), so land drainage or mixing with deeper water can provide a source of vitamins. Water samples from just above the bottom in Lake Sagami showed higher concentrations than those less deep in the same layer (Ohwada and Taga, 1972b).

Algae that do not require an external source of a particular vitamin synthesize it and may release a detectable amount during growth or senescence (Carlucci and Bowes, 1970a).

Mixed laboratory cultures of a B_{12}-requiring alga and a B_{12}-producing alga or bacterium have been grown successfully (Carlucci and Bowes, 1970b; Haines and Guillard, 1974).

Algae not requiring vitamins are capable of vitamin uptake (Droop, 1968). Obviously such ability is necessary so that a mutation leading to auxotrophy is viable. Diatom clones which are stimulated by vitamin B_{12} (but lack an absolute requirement take up the vitamin if it is present but produce detectable B_{12} when grown without it (Swift and Guillard, 1978)).

Zooplankton feeding probably has a role as well. In studies with radioactive B_{12}, Droop and Scott (1978) observed the conversion of some of the algal B_{12} to soluble B_{12} after grazing. The bacterial flora of zooplankton and their faecal pellets probably acts to produce more B_{12}, because their environment is rich in organic material.

Uptake of B_{12} by natural populations varies seasonally. In Lake Washington, concentration in winter was about 10 ng l^{-1} and uptake rate was about 0·2 ng l^{-1} d^{-1}. In summer, the concentration was around 4 ng l^{-1} with uptake rate of 1 ng l^{-1} d^{-1} (Parker, 1977). Input processes (liberation by phytoplankton and bacteria, mixing from deep water, sediments, run-off or advection) acts constantly to balance uptake and losses, so that net concentration changes are small relative to the potential for uptake. Algae and bacteria are both significant in the uptake of B_{12} by natural populations, according to studies with size fractionation of populations (personal communication, L. Herold, University of Southern California).

VIII. Vitamin Concentrations in Particular Regions

A. Vitamin B_{12} in Sea Water

Vitamin concentrations have been studied in a number of regions. In marine waters, concentrations decrease as one goes offshore and show a general correlation with nutrient concentration, productivity level etc. A sampling from the literature on B_{12} serves to show the ecological variation. Values are lowest in the Sargasso Sea, generally 0·01–0·1 ng l^{-1} (Menzel and Spaeth, 1962) with a general pattern suggesting that B_{12} as well as inorganic nutrients increase during winter, when stratification breaks down and mixing occurs. After a small spring diatom bloom, B_{12} drops rapidly to the lowest values. The dominant phytoplankter in summer is *Emiliana huxleyi* which requires only thiamine. Samples from the North Pacific average $0\cdot1$ ng l^{-1} B_{12} in the upper 60 m with greater values in deep water (Carlucci and Silbernagel, 1966c). In the Antarctic most samples contained 1 ng l^{-1} or less, but 20% of samples had no detectable B_{12} and the average was only $0\cdot2$ ng l^{-1} in the upper 75 m. B_{12} values are low in relation to the high inorganic nutrient concentrations (Carlucci and Cuhel, 1975). In an inlet of the Barents Sea, the range of values is 0 to 2 ng l^{-1} (Propp, 1970). Minimum values occur during the spring bloom.

In a number of regions studied that are closer to continents the maximum values are greater. The offshore Gulf of Maine with $0\cdot1$–$1\cdot6$ ng l^{-1} B_{12} has rather low concentrations for a coastal region (Swift, 1981). The range of values in Long Island Sound is 4–16 ng l^{-1} (Vishniac and Riley, 1961). B_{12} increases during the winter, along with nitrate and phosphate, and undergoes a strong drop after the spring bloom develops. B_{12} concentrations in coastal regions at eastern Long Island are often less than expected relative to Long Island Sound; Napeague Bay (Block Island Sound) samples contain $<0\cdot05$ to 11 ng l^{-1} and in the nearby Peconic Bay estuary the range is 0 to 30 ng l^{-1} (Bruno and Staker, 1978; Bruno *et al.*, 1981). In the Strait of Georgia, British Columbia, the range of values is around 1–12 ng l^{-1}. The pattern of high values before the spring bloom and a decrease during the bloom is supplemented by summer peaks associated with particles in river runoff (Cattell, 1969, 1973). In Sagami Bay, Japan, the average values in different seasons varied between $0\cdot6$ ng l^{-1} and $4\cdot8$ ng l^{-1} (Ohwada and Taga, 1972a).

B. *Vitamin B_{12} in Lakes*

Maximum values in lakes resemble those in some of the marine coastal regions. Lake Washington (Parker, 1977) has B_{12} from 4 to 10 ng l^{-1}. Cyanobacteria are dominant in the plankton there, but greatest B_{12} uptake rates occurred when diatoms were present. Richer lakes in the English Lake District had similar B_{12} levels, such as 4–16 ng l^{-1} annual range, e.g. from n.d. to 5 ng l^{-1}. In lakes with higher B_{12}, the particulate B_{12} was rarely more than 10–20% of total B_{12}, whereas in the clearer lakes there was often particulate B_{12} present when dissolved B_{12} was not detected (Daisley, 1969).

Lake Sagami, Japan (Ohwada and Taga, 1972b) contained 0–12 ng l^{-1} in the upper layer, with lowest values occurring during diatom dominance and highest values when a cyanobacterium dominated. In two French subalpine lakes, B_{12} showed a pattern of high concentration during winter, decrease during a spring algal bloom and gradual increase in summer (Pommel, 1975). In Lake Leman B_{12} values are approximately 1–6 ng l^{-1} and spring bloom species are diatoms and flagellates. In Lake Nantua the range of values is 4–30 ng l^{-1}, and cyanobacteria dominate.

C. Thiamine and Biotin

There are less distinct seasonal patterns for thiamine and biotin. There is perhaps a trend towards following the pattern of phytoplankton populations and of gradual increase in concentration during summer so that winter values, while high, are not the yearly maxima.

Concentrations of thiamine and biotin in sea water are sometimes so low as to be undetectable. Levels tend to correlate with the chlorophyll a present. Generalization from several studies (see Swift, 1980, for citations) in Pacific and South Polar waters indicates values of around 5 ng l^{-1} for thiamine in offshore waters, up to 40 ng l^{-1} in coastal water and double that or greater in inlets and shallow regions. Levels for biotin (when detectable) are around 1–4 ng l^{-1} in offshore waters and 6–20 ng l^{-1} in coastal regions and inlets.

Levels of thiamine and biotin in lakes can encompass the full range of values cited for marine waters and can rise to even higher levels in eutrophied systems.

IX. Vitamins in Enrichment Response Experiments

Although it seems possible that vitamin limitation might be detectable by enrichment response experiments, there are several factors that make this less promising than hoped for. First, the lack of vitamins will have had an effect on species selection in the existing populations so that the cells which might have responded to vitamin enrichment do not exist in the population. Secondly, in regions where vitamins might be limiting, other nutrients are likely to be limiting also, so that effects occur as part of multiple nutrient effects. Some of the other nutrients have a more rapid and dramatic effect. For example, vitamin B_{12} enrichment of Sargasso Sea water did not increase ^{14}C uptake (Menzel and Ryther, 1961); iron and silicate sometimes did (Ryther and Guillard, 1959).

Use of a test organism in such experiments eliminates some of the difficulty with a natural population. An experimental difficulty is to effectively deplete the inoculum of vitamins as well as other nutrients. Vitamin depletion takes place more slowly than depletion of other nutrients. Minimum cell requirements are small, and a great excess is present in stock culture media. Because of this, it is surprising that even some secondary effect of vitamins has been observed in a few instances. Smayda (1964) enriched Sargasso Sea water with various nutrients singly, or omitted them singly or in groups from a suite of enrichments. Growth of *T. oceanica* 13-1 as a test organism was measured. The primary result was that presence of chelator improved growth, indicating that removal of a metal toxicity or deficiency was a primary step in improving growth. (Revealing a vitamin response in the presence of such inhibition or deficiency, even using an organism that should show a vitamin response, in water samples that one would expect to be improved by vitamins, is not a simple task.) In other experiments with test organisms, results showed very interesting differences in water from various locations for different test organisms, but there was no evidence that enrichments lacking vitamins were much worse for improving growth than enrichments lacking other nutrients (Smayda, 1970; Ignatiades and Smayda, 1970). This suggests that the inocula were more depleted of other nutrients than of vitamins.

Berland *et al.* (1978b) studied the effect of six enrichments, individually and in combination, on ^{14}C uptake in water from an oligotrophic inlet near Marseille. Natural populations and ten different cultured species were tested; the effect of individual factors was ordered. Phosphate was always the most important factor. Nitrate and chelators were next, followed by vitamins. Vitamins were considered second as a limiting factor for the diatom *Cylindrotheca closterium* (a B_{12} auxotroph) and the prymnesiophyte *Pavlova pinguis*, as a third limiting factor for *Asterionella japonica*, and essentially without effect for *Skeletonema costatum* (a B_{12} auxotroph) and *Phaeodactylum tricornutum* (an autotroph).

Thus the experimental approach of nutrient enrichment experiments to detect ecological effects of a vitamin is one that is likely to yield useful information only if it is designed for that specific purpose. Clearly the same techniques that are important for good bioassay of vitamin concentration are necessary in design of such enrichment experiments. Chapters 4 and 5 discuss concepts and techniques for such an approach in much greater detail.

X. Detecting an Ecological Role

Auxotrophic algae form a significant portion of phytoplankton populations in natural waters. This fact, together with the observation that vitamin concentrations do not frequently reach undetectable values, suggests that vitamin concentrations are often sufficient to allow near maximum growth rates so that auxotrophs can compete successfully with autotrophs. Minimum cellular B_{12} requirements are very small, so the primary influence from vitamins is probably on growth rate. If a vitamin is thought to be a possible limiting factor, such an hypothesis should be formulated for a particular species (or group of species) at a particular time or season. The appropriate time at which to put substantial effort into vitamin assays is when such species grow fast (as during initiation of their seasonal bloom) or when they might have been expected to grow but did not, possibly owing to lack of vitamin. In further testing of the hypothesis it is also useful to study algal isolates in the laboratory and to try to undertake some simple enrichment experiments.

Although a bloom may initially deplete the water of B_{12}, the concentration usually rises again through the activities of bacteria that develop after algal excretion and cell breakage increase organic substrates, and through the activities of succeeding algal species, which may be autotrophic for the vitamin. The existence of binding factor and of B_{12} analogues having activity for some species or clones complicates the consideration of B_{12} assay and utilization, as does the capacity of autotrophs to take up vitamins for which there is no absolute requirement. In the environment there is competition between auxotrophs, and the analogue specificity and growth kinetics are important in determining which clones develop populations more rapidly.

References

Barker, H. A. (1972). *Ann. Rev. Biochem.* **41**, 55–90.

Berland, B. R., D. J. Bonin and S. Y. Maestrini (1978b). *Int. Revue Ges. Hydrobiol.* **63**, 501–31.

Berland, B. R., D. J. Bonin, M. Fiala and S. Y. Maestrini (1978a). Importance des vitamines en mer. Consommation et production par les algues et les bactéries. *In* "Actualités de Biochimie marine." Colloque G.A.B.I.M.-C.N.R.S. Marseille 1976, pp. 121–146. Ed. speciale du C.N.R.S., Paris.

Berland, B. R., D. J. Bonin, J.-P. Durbec and S. Y. Maestrini (1976). *Hydrobiologia* **50**, 167–172.

Beck, R. A. (1978). *Anal. Chem.* **50**, 200–202.

Beck, R. A. and J. J. Brink (1976). *Environ. Sci. Technol.* **10**, 173–175.

Beck, R. A. and J. J. Brink (1978). *Environ. Sci. Technol.* **12**, 435–438.

Blankley, W. F. (1973). Toxic and inhibitory materials associated with culturing. *In* "Handbook of Phycological Methods" (Stein, J. R., ed.), pp. 207–229. Cambridge University Press, Cambridge.

Bonin, D. F., S. Y. Maestrini and J. W. Leftley (1981). The role of hormones and vitamins in species succession of phytoplankton. *In* "Physiological Bases of Phytoplankton Ecology" (T. Platt, ed.), pp. 310–322. *Can. Bull. Fish. Aquat. Sci.* **210**.

Bradbeer, C. (1971). *Arch. Biochem. Biophys.* **144**, 184–192.

Brand, L. E., R. R. L. Guillard and L. S. Murphy (1981). *J. Plankton Res.* **3**, 193–201.

Bruno, S. F. and R. D. Staker (1978). *Limnol. Oceanogr.* **23**, 1045–1050.

Bruno, S. F., R. D. Staker and L. L. Curtis (1981). *J. Mar. Res.* **39**, 335–352.

Burkholder, P. R. and L. M. Burkholder (1956). *Limnol. Oceanogr.* **1**, 202–208.

Carlucci, A. F. (1974). Production and utilization of dissolved vitamins by marine phytoplankton. *In* "Effect of the Ocean Environment on Microbial Activities" (Colwell, R. R. and Morita, R. Y., eds), pp. 449–456. University Park Press, Baltimore.

Carlucci, A. F. and P. M. Bowes (1970a). *J. Phycol.* **6**, 351–357.

Carlucci, A. F. and P. M. Bowes (1970b). *J. Phycol.* **6**, 393–400.

Carlucci, A. F. and P. M. Bowes (1972). *J. Phycol.* **8**, 133–137.

Carlucci, A. F. and R. L. Cuhel (1975). Vitamins in the South Polar Seas I. Distribution and significance of dissolved and particulate vitamin B_{12}, thiamin, and biotin in the Southern Indian Ocean. *In* "Adaptations within Antarctic Ecosystems." Proceedings of the Third S.C.A.R. Symposium on Antarctic Biology. (Llano, G. A., ed.), pp. 115–128. Gulf Printing Co., Houston.

Carlucci, A. F. and S. B. Silbernagel (1966a). *Can. J. Microbiol.* **12**, 175–183.

Carlucci, A. F. and S. B. Silbernagel (1966b). *Can. J. Microbiol.* **12**, 1079–1089.

Carlucci, A. F. and S. B. Silbernagel (1966c). *Limnol. Oceanogr.* **11**, 642–646.

Carlucci, A. F. and S. B. Silbernagel (1967). *Can. J. Microbiol.* **13**, 979–986.

Carlucci, A. F. and S. B. Silbernagel (1969). *J. Phycol.* **5**, 64–67.

Carlucci, A. F., S. B. Silbernagel and P. M. McNally (1969). *J. Phycol.* 5, 302–305.

Cattell, S. A. (1969). Dinoflagellates and vitamin B_{12} in the Strait of Georgia, British Columbia. Ph.D. Thesis. Univ. of British Columbia, Vancouver, B.C.

Cattell, S. A. (1973). *J. Fish. Res. Bd. Can.* 30, 215–222.

Daisley, K. W. (1957). *Nature* 180, 1042–1043.

Daisley, K. W. (1969). *Limnol. Oceanogr.* 14, 224–228.

Daisley, K. W. (1970). *Int. J. Biochem.* 1, 561–574.

DiGirolamo, P. M., R. J. Kadner and C. Bradbeer (1971). *J. Bact.* 106, 751–757.

Droop, M. R. (1955a). *J. Mar. Biol. Ass. U.K.* 34, 229–231.

Droop, M. R. (1955b). *J. Mar. Biol. Ass. U.K.* 34, 435–440.

Droop, M. R. (1957a). *J. Gen. Microbiol.* 16, 286–293.

Droop, M. R. (1957b). *Nature* 180, 1041–1042.

Droop, M. R. (1961). *J. Mar. Biol. Ass. U.K.* 41, 69–76.

Droop, M. R. (1966). *J. Mar. Biol. Ass. U.K.* 46, 659–671.

Droop, M. R. (1968). *J. Mar. Biol. Ass. U.K.* 48, 689–733.

Droop, M. R. (1970). *Helgolander Wiss. Meeresunters.* 20, 629–636.

Droop, M. R. (1973). *J. Phycol.* 9, 264–272.

Droop, M. R. and J. M. Scott (1978). *J. Mar. Biol. Ass. U.K.* 58, 749–772.

Easley, L. W. (1969). *J. Protozool.* 16, 289–294.

Epstein, S. S., J. B. Weiss, D. Causeley and P. Bush (1962). *J. Protozool.* 9, 336–339.

Ford, J. E. (1958). *J. Gen. Microbiol.* 19, 161–172.

Gleason, F. K. and H. P. C. Hogenkamp (1970). *J. Biol. Chem.* 245, 4894–4899.

Gold, K. (1964). *Limnol. Oceanogr.* 9, 343–347.

Gold, K. (1968). *Limnol. Oceanogr.* 13, 185–188.

Gold, K., O. Roels and H. Bank (1966). *Limnol. Oceanogr.* 11, 410–413.

Guillard, R. R. L. (1963). Organic sources of nitrogen for marine centric diatoms. *In* "Marine Microbiology" (Oppenheimer, C. H., ed.), pp. 93–104. C. C. Thomas, Springfield, Ill.

Guillard, R. R. L. (1968). *J. Phycol.* 4, 59–64.

Guillard, R. R. L. and V. Cassie (1963). *Limnol. Oceanogr.* 8, 161–165.

Guillard, R. R. L. and J. H. Ryther (1962). *Can. J. Microbiol.* 8, 229–239.

Haines, K. C. and R. R. L. Guillard (1974). *J. Phycol.* 10, 245–252.

Hargraves, P. E. and R. R. L. Guillard (1974). *Phycologia* 13, 163–172.

Hasle, G. R. (1983). *J. Phycol.* 19, 220–229.

Hatch, M. D. (1976). Photosynthesis: the path of carbon. *In* "Plant Biochemistry" (3rd ed.). (Bonner, J. and Varner, J. E. eds), pp. 797–844. Academic Press, New York and London.

Ignatiades, L. and T. J. Smayda (1970). *J. Phycol.* 6, 357–364.

Jaenicke, L. (1964). *Ann. Rev. Biochem.* 33, 287–312.

Kristensen, H. P. O. (1955). *Acta Physiol. Scand.* 33, 232–237.

Kristensen, H. P. O. (1956). *Acta Physiol. Scand.* 37, 8–13.

Lebedeva, M. N., Y. M. Markianovich and L. G. Gutveyb (1971). *Hydrobiological J.* 2, 14–19.

Lefort-Tran, J., M. H. Bre, J. L. Ranck and M. Pouphile (1980). *J. Cell Sci.* 41, 245–261.

Lewin, J. C. and R. A. Lewin (1960). *Can. J. Microbiol.* 6, 127–134.

Litchfield, C. D. and D. W. Hood (1965). *Appl. Microbiol.* 13, 886–894.

Loeblich, A. R. III (1957). *Phykos.* 5, 216–255.

Martin, J.-L. Y. and E. Vacelet (1975). *Cahiers de Biologie Marine* 16, 511–519.

Menzel, D. W. and J. P. Spaeth (1962). *Limnol. Oceanogr.* 7, 151–154.

Natarajan, K. V. (1968). *Appl. Microbiol.* **16**, 366–369.

Ohwada, K. (1972). *Mar. Biol.* **14**, 10–17.

Ohwada, K. and N. Taga (1972a). *Mar. Chem.* **1**, 61–73.

Ohwada, K. and N. Taga (1972b). *Limnol. Oceanogr.* **17**, 315–320.

Parker, M. (1977). *Limnol. Oceanogr.* **22**, 527–538.

Pintner, I. J. and Altmyer, V. L. (1979). *J. Phycol.* **15**, 391–398.

Pommel, B. (1975). *Ann. Hydrobiol.* **6**, 103–121.

Propp, L. N. (1970). *Oceanology* **10**, 676–681.

Provasoli, L. (1958). Growth factors in unicellular marine algae. *In* "Perspectives in Marine Biology" (Buzzati-Traverso, A. A., ed.), pp. 385–403. Univ. Calif. Press, Berkeley.

Provasoli, L. (1963). Organic regulation of phytoplankton fertility. *In* "The Sea" Vol. 2 (Hill, M. N., ed.), pp. 165–219. Interscience, New York.

Provasoli, L. (1971). Nutritional relationships in marine organisms. *In* "Fertility of the Sea" Vol. 2 (Costlow, J. D., ed.), pp. 369–382. Gordon and Breach, New York.

Provasoli, L. and A. F. Carlucci (1974). Vitamins and growth regulators. *In* "Algal Physiology and Biochemistry" (W. D. P. Stewart, ed.), pp. 741–787. Blackwell, Oxford.

Riley, G. A. (1966). *In* "Marine Biology II". Proceedings of the Second International Interdisciplinary Conference 1962 (Oppenheimer, C. H., ed.), pp. 101–104, 334. New York Academy of Sciences.

Ryther, J. H. and J. C. Goldman (1975). *Ann. Rev. Microbiol.* **29**, 429–443.

Ryther, J. H. and R. R. L. Guillard (1959). *Deep Sea Res.* **6**, 65–69.

Ryther, J. H. and R. R. L. Guillard (1962). *Can. J. Microbiol.* **8**, 437–445.

Sarhan, F., M. Houde and J. P. Cheneval (1980). *J. Protozool.* **27**, 235–238.

Sebrell, W. H. Jr and R. S. Harris (1968). "The Vitamins" (Vol. 2, 2nd ed.). Academic Press, New York and London.

Sharma, G. M., H. R. DuBois, A. T. Pastore and S. F. Bruno (1979). *Anal. Chem.* **51**, 196–199.

Smayda, T. J. (1964). Enrichment experiments using the marine centric diatom *Cyclotella nana* (clone 13-l) as an assay organism. Proceedings of a Symposium on Experimental Marine Ecology. Occasional Pub. 2, pp. 25–32. Grad. Sch. of Oceanogr., Univ. of Rhode Island, Kingston.

Smayda, T. J. (1970). *Helgoländer Wiss. meeresunters.* **20**, 172–194.

Smith, E. L. (1965). "Vitamin B$_{12}$" (3rd ed.). 180 pp. Methuen, London.

Stadtman, T. C. (1971). *Science* **171**, 859–867.

Starr, T. J., M. E. Jones and D. Martinez (1957). *Limnol. Oceanogr.* **2**, 114–119.

Stoecker, D., R. R. L. Guillard and R. M. Kavee (1981). *Biol. Bull.* **160**, 136–145.

von Stosch, H. A. (1977). Observations on *Bellerochea* and *Streptotheca*, including descriptions of three new planktonic diatom species. Nova Hedwigia Beih. **54**, 113–166. (Fourth Symp. on Recent and Fossil Marine Diatoms (1976).)

Swift, D. G. (1967). Growth of vitamin B$_{12}$-limited cultures: *Cyclotella nana*, *Monochrysis lutheri*, and *Isochrysis galbana*. M.A. Thesis. 85pp. Johns Hopkins Univ., Baltimore.

Swift, D. G. (1973). Vitamin B$_{12}$ as an ecological factor for centric diatoms in the Gulf of Maine. Ph.D. Thesis. 180 pp. The Johns Hopkins Univ., Baltimore.

Swift, D. G. (1980). Vitamins and phytoplankton growth. *In* "The Physiological Ecology of Phytoplankton" (Morris, I. ed.), pp. 329–368. Blackwell, Oxford.

Swift, D. G. (1981). *J. Mar. Res.* **39**, 375–403.

Swift, D. G. and R. R. L. Guillard (1977). *J. Mar. Res.* **35**, 309–320.

Swift, D. G. and R. R. L. Guillard (1978). *J. Phycol.* **14**, 377–386.

Swift, D. G. and W. R. Taylor (1974). *J. Phycol.* **10**, 385–391.

Thomas, W. H. (1968). Nutrient requirements and utilization: Algae, Table 27. *In* "Metabolism" (Altman, P. L. and Ditmer, D. S., eds), pp. 210–228. Fed. Am. Soc. Exp. Biol.

Turner, M. F. (1979). *J. Mar. Biol. Ass. U.K.* **59**, 535–552.

Vacelet, C. (1975). *Cahiers de Biologie Marine* **16**, 383–394.

Vishniac, H. S. and G. A. Riley (1961). *Limnol. Oceanogr.* **6**, 36–41.

Wada, S. and H. Suzuki (1965). Protective factors of thiamin. III. The protective action of amino acids against thiamin destruction *in vivo*. (Transl. fr. Japanese by Medical Literature, Inc., Philadelphia) Kasergaku Zasshi **16**, 326–328.

Wagner, F. (1966). *Ann. Rev. Biochem.* **35**, 405–434.

Wood, E. A. (1963). An Evaluation of the Role of Vitamin B_{12} in the Marine Environment. Ph.D. Thesis. Yale Univ., New Haven.

Part Five

Industrial Applications

11 Utilization of Algal Cultures and Assays by Industry

M. T. Elnabarawy and A. N. Welter

3M Company, Environmental Laboratory
Environmental Engineering and Pollution Control
St Paul, Minnesota, USA

I. Introduction

Significant environmental legislation has been enacted in the US during the past decade. These laws address the control of industrial wastes and/or chemical substances which might be associated with an increased risk to human health and/or the environment. These public laws include The Clean Water Act (CWA); the Federal Insecticide, Fungicide, and Rodenticide Act (FIFRA); the Resource Conservation and Recovery Act (RCRA); the Toxic Substances Control Act (TSCA, TOSCA); and the National Environmental Policy Act (NEPA). These laws, either directly or indirectly, are concerned with arresting the pollution of lakes, rivers, streams or groundwater and with the restoration of water quality.

The organisms more commonly used as indicator species in the assessment of water quality are representative species of various trophic levels, e.g. algae–phytoplankton, daphnids–invertebrates, and fish–vertebrates. Thus, these organisms serve as biological monitors and are used to assess both physical and chemical changes occurring in waterways.

This chapter provides an outline of possible uses of algae by industry, e.g. in assessing the environmental impact of new chemical substances or monitoring wastes which may enter the aquatic environment.

II. Algae as Biological Indicators

A. Evolution of Test Systems

In 1967 representatives of government and industry assembled to form a task force having as its objective the development of an algal assay method. As a result of this initiative, the Algal Assay:Bottle Test (AA:BT) procedure was developed, evaluated, and standardized (USEPA-CERL, 1971). In 1975, the American Public Health Association published the AA:BT procedure in its 14th edition of *Standard Methods for the Examination of Water and Wastewater.*

Miller *et al.* (1978) reported their results after intensive research seeking to improve and expand the understanding of results obtained using the AA:BT (1971). The test as described is intended primarily for use in the following general situations:

(1) Assessment of the nutrient status of a receiving water;
(2) Evaluation of materials whether they elicit stimulatory or inhibitory effects on algal growth;
(3) Assessment of complex wastes, defining their impact upon receiving waters (Leischman, 1979).

The test protocol "Marine Algal Assay Procedure:Bottle Test" (USEPA-1974), is a companion procedure to the Freshwater Algal Assay:Bottle Test (AA:BT) and is utilized to evaluate algal growth and productivity in estuarine and marine coastal situations.

Payne and Hall (1979) described a modification of the Bottle Test which consisted of a 5-day exposure of a unialgal culture to varying concentrations of a test material. This abbreviated exposure period is followed by a 9-day recovery period in an environment free of test material. This modified test method allows for the quantification of responses ranging from reduced growth, through algistatic and finally including algicidal responses.

Walsh *et al.* (1980a) recently described a marine algal bioassay method useful for the rapid screening of substances with single or multiple species tests. Briefly, algae are grown in optically matched spectrophotometric tubes, allowing for the direct measurement of population density while eliminating the necessity of removing sample aliquots for each density determination.

In 1981, the Organization for Economic Cooperation and Development (OECD) adopted the "Alga, Growth Inhibition Test" as a basic level test to indicate possible impact of a test substance on this functionally important organism (OECD, 1981). As is generally accepted, the test end-point will require the determination of an increase in cell numbers and/or biomass of a fast-growing species of green algae, including *Selenastrum capricornutum.*

In this brief review of the evolution of algal testing methods, the development of the algal assay-bottle test has been emphasized. This method or modifications of it allow the investigator to select that technique which will most effectively provide answers to the questions being posed.

B. Possibilities and Limitations of Algal Systems

The algal bottle test is utilized to identify materials affecting growth, biological availability of nutrients, and finally, to determine the dose–response curves for chemicals possessing growth-limiting characteristics. Unialgal cultures grown under specific laboratory conditions are utilized to define the growth response to individual or multiple nutrients and to correlate these effects with changes in conditions of algal growth. Miller (1978) has discussed criteria used in defining an appropriate algal test species: availability, uniformity in growth (responsiveness), ease of measurement, etc.

The types of information required dictate the experimental design of an algal test. Thus, these tests may be conducted while varying the exposure duration, number of species to be tested, etc.

Short-term tests, 14 days or less in duration, may be used to rank chemicals relative to their effect on algal growth. Because of the short life cycle of the algae these 14-day tests may be considered as chronic exposure tests which can assess effects over several generations. Therefore, for the purpose of ranking chemicals, these short-term exposures can provide useful subchronic toxicity information, a very important, but elusive, bioassay goal.

The document *Algal Assay Procedure: Bottle Test* (1971) has outlined steps which various laboratories, representing industry, academia, and government, undertook to develop a reliable test method. In these strictly controlled tests, excellent agreement was achieved by the participating laboratories. It was noted that differences in mean specific growth rate and the maximum standing crop during the first four days of a test were due to varying lag periods and exponential growth rates. Such differences could be minimized if the test period were long enough for several cell divisions to occur thus assuring the end of the log growth phase as opposed to arbitrarily setting an exposure period for the test.

In an effort to define the environmental impact of a test material on an algal culture more accurately, exposure periods of variable duration and subculturing of flasks having no visible algal population, are utilized. To define the nature of an observed algal response, an algal culture may be exposed to a toxicant for a period of five days, followed by a 9-day recovery period. The log growth response curve will then define the nature of the effects observed. Effects on algal growth whether stimulatory, inhibitory, biphasic, algistatic, algicidal, or no apparent effect can be recorded for the specific experimental conditions used.

Thus, the primary ecological significance of an algal test is demonstrable by the continuation of (1) growth inhibition or (2) an extensive algicidal response following removal of a toxicant. Subculturing in "toxicant"-free medium allows the investigator to differentiate between an algistatic or algicidal response, because in the former situation, a normal log growth phase will occur. Use of this test system can allow the investigator to predict whether phytotoxicity may become a concern under the exposure conditions being investigated. Another variable is the determination made during the experimental design phase of whether a single or multiple species is (are) to be exposed to the "toxicant". Decisions of this type are based on the physiological response of each test species, particularly since there are species variations relative to the period of maximum growth. Payne and Hall (1979) have described an algal assay in which test algal species were selected on the basis of their sensitivity and reactivity to: (1) nutrients, (2) trace metals, or (3) species-specific nutrient requirement.

C. Relation to Other Biological Methods

The general philosophy governing ecotoxicological testing is predicated on the use of multispecies tests representing various trophic levels which enable the investigator to determine the ecological significance of toxicant-induced effects. Thus, through proper selection of the test species, effects on parameters such as growth, reproduction, survival, photosynthesis, and physiologically induced changes may be studied.

Because of concern with the possible perturbation of the aquatic environment by pesticides, test development historically centred on methods designed to predict environmental impacts on this ecosystem. This methodology has become relatively sophisticated and utilizes aquatic plants, invertebrates, and vertebrates as the test species. Success of this approach has resulted in the publication of water quality criteria for priority pollutants (FR 45, 1980) and a handbook delineating available aquatic toxicology data for pesticides (USEPA, 1979).

One may readily predict the most sensitive test species if the mode of action of a chemical is known, thus algae would be highly sensitive to the presence of a herbicide, daphnids to insecticides, fish to piscicides, etc. The multitrophic level concept of testing can define a most sensitive species specifically in those cases where minimal toxicological information is available for a test substance. Furthermore, the ecological interactions to be impacted can be defined.

As a representative species of a lower trophic level, algae are ecologically important organisms because they serve as an important food source for other aquatic species. Their utility as a test organism is based upon their short life cycle, hence ease in performing multigeneration exposure studies.

As representative aquatic invertebrates, daphnids occupy an intermediate position on the trophic scale, utilizing algae as a food source, while themselves serving as food for higher organisms. The principal advantages in utilizing daphnids as a test species include: (1) ready availability, (2) relative ease of maintaining a colony, (3) comparatively short life cycle, (4) ease of determining toxicant effects on growth, reproduction and survival. Disadvantages to the use of daphnids as a test species include: (1) absence of standardized culture methods, (2) variation in sensitivity during their life cycle, (3) effect of previous exposure to toxicant on daphnids, brood release, reproduction, etc., (4) effect of nutrients on responsiveness. Walsh (1980b) has discussed the results of testing textile effluents utilizing the multitrophic level testing method. Variations in sensitivity to the effluent within a trophic level was an interesting finding of this study. Pomeroy

(1980) similarly concluded that daphnids were 96% effective as indicators in identifying "harmful" chemicals whereas algae and fish were but 56 and 62% effective, respectively.

A representative aquatic vertebrate is the fish. This organism is of economic importance, serving man both as a food source and as a means of recreational enjoyment. Perturbations in the aquatic ecosystem resulting in a decline in the availability of edible fish could have severe economic and social consequences. Thus, there is a definite need for performing multitrophic level investigations of possible effects of new chemical substances on this ecosystem.

III. Utilization of Algal Systems by Industry

A. Role of Algae in Assessment of Water Quality

Algae are directly affected by chemical wastes containing major nutrients such as phosphorus (P) and nitrogen (N). In the presence of excess nutrients, the algae are capable of rapid growth and multiplication. If unrestricted, these conditions may lead to a population shift, dominance, and/or algal bloom, conditions which are indicative of a deterioration in water quality. Therefore algal assays are highly useful in the determination of: (1) biological availability of nitrogen and phosphorus species, and sensitivity to changes in N and P loading; (2) assessment of the environmental impact of effluent discharges on the receiving stream; (3) evaluation of chemical substances and products as they might affect algal growth; and (4) monitoring of pesticides, metals, and their effect on algal species (Mitchell, 1974; USEPA, 1976).

Bioassays provide the only direct method for assessing the biological availability of metals in water. Algal assays for heavy metals can detect low levels in the environment, e.g. 0·01 p.p.m. silver (Hutchinson, 1973). The most common algal response to increasing levels of heavy metals such as copper, nickel, mercury, silver or cadmium is a decrease in growth rate.

Some of the algal test systems used to define the nature of various sample sources are listed in Table 1. The system cited most frequently is the Algal Assay Bottle Test or a variation, although this provides information relative to the toxicity of a toxicant in an artificial environment only. The laboratory stream/continuous-flow-through trough models provide more accurate "real world" assay conditions as opposed to static laboratory bioassay systems since they approximate algal growth conditions under natural settings. However, the latter system does have the advantage of being less expensive, providing meaningful and reproducible data within an acceptable time frame.

Algae have also proven useful in assessing water quality and are utilized as in-stream monitors. Palmer (1969) reported on algae which were tolerant to pollution, thriving in polluted areas. These algae are readily sampled and identified.

Table 1 Algal assay application

Bioassay sample source	Assay type[a]	Species	References
Evaluation of chelators	AA:BT	M. aeruginosa	Allen (1980)
Kraft mill effluent	Experimental C-F trough	Multispecies	Bothwell (1980)
Metal interaction	AA:BT	S. capricornutum C. stigmatophora	Christensen (1979)
Arsenic, herbicides	AA:BT	Chlamydomonas	Christensen (1980)
Municipal wastewater effluent	AA:BT	S. capricornutum	Greene (1975)
Industrial wastewater effluent			
Agricultural waste-water effluent			
Municipal wastes	AA:BT	S. capricornutum	Greene (1976)
Smelter wastes			
Model system: monitor	Modified	S. capricornutum	Guy (1980)
Chemical speciation	Am. Pub. Health Assoc.		
Oil dispersants	Plate test	Chlamydomonas Deinaliella	Heldal (1978)
Artificial refinery mixture	Laboratory stream	Multispecies	Honig (1980)
Wastewater effluents	AA:BT	S. capricornutum	Miller (1976)
Detergent materials	AA:BT	S. capricornutum	Payne (1978)
Chlorinated	Effluent assay	A. affinis C. pyrenoedosa	Simmons (1979)
Effluents	Modification		

[a] AA:BT—Algal Assay Bottle Test
C-F Trough—Continuous Flow Trough
Acute Toxic Effects—No mention of a specific test

Furthermore, Palmer has proposed an algal pollution index based on the genera of algae present in a sample. The less diverse the algal population, the greater the probability that the sample was obtained from a polluted area. Patrick (1973) also discussed the usefulness of algae in assessing water quality under both field and laboratory conditions. In the field it becomes imperative to observe, quantitate, and analyse the natural communities present in such a way that shifts in population can be detected and evaluated as a function of time. Schlichting (1980) described and discussed the use of an *in situ* biological monitoring system which, in part, utilizes algae as the indicator organism.

Honig (1980) utilized a laboratory stream system for observing the effects of an artificial refinery mixture on periphyton. This system can readily identify taxonomic shifts. The shift in periphyton species composition may be indicative of a shift in food quality which will be detectable at higher trophic levels.

Bothwell (1980) experimentally determined the effect of secondarily treated kraft mill effluent on the accumulation rate of algae using a continuous-flow trough. Generally this effluent increased the accumulation rate, although at higher concentrations, a biphasic response was observed, i.e. an initial transient inhibition followed by stimulation. Chlorophyll accumulation provided a consistent index of relative algal growth rate.

B. Stimulatory/Inhibitory Effects of Effluent Discharges

To assess the effects of selected concentrations of effluents on algal growth stimulation/inhibition, a standard or acceptable species is subjected to laboratory testing. These laboratory tests provide information on the relative toxicity of effluents. Only biologically active components of the effluent are responsible for the regulation of algal production. Thus the integration of many factors, chemical and physical, which contribute to growth effects in test algae are demonstrated by the algal growth response.

The effect of wastewater effluents upon algal growth has been determined using the algal assay bottle test wherein the maximum 14-day algal yield is correlated with the nutrient content of the test waters. Several reports (Miller, 1975, 1976; Greene, 1975) have addressed the utility of this test system in defining water pollution problems. Determination of changes occurring during the log phase of growth, e.g. day seven, could be attributed to availability of nutrients and/or metals, presence of toxic wastes, commercial poisons, etc. in the receiving water.

Simmons (1979) has discussed the implication of the chlorination procedure on sewage treatment plant effluent. Determining mean algal growth rates during the 9- to 15-day period of log phase growth indicated that inhibitory effects could be associated with the chlorination procedure. Two mechanisms were postulated: (1) inhibition of algal growth could be attributed to the presence of total residual chlorine in the effluent stream; (2) formation of chloro–organics could inhibit algal growth.

A modified laboratory test system can provide information on stimulation, inhibition, or algistatic and algicidal activities as induced by the effluent. Variability in toxicant exposure periods can be related to the degree of activity noted in the laboratory model. Similarly, subculturing will determine whether an apparent inhibitory response is merely algistatic or is, in fact, truly algicidal.

C. Algal Assay of Chemical Substances

Algal assays have been utilized in the assessment of aquatic toxic responses elicited by the introduction of chemicals to this ecosystem. The philosophy of multiple-species testing and algal toxicity responsiveness has been discussed by Payne (1978).

The system providing the most useful data has been that described by Payne (1978, 1979) and is a modification of the algal assay bottle test. In brief, different algal species are used to determine selective toxicity. Initially, these algae are exposed to varying concentrations of a toxicant. Minimally, biomass is determined at day five, and aliquots from flasks showing an inhibitory response are subcultured in a toxicant-free system for an additional nine days, biomass again being determined at selected intervals. A determination of the type of algal response can then be made, whether it is algistatic or algicidal. This test system provides the investigator with the advantage of flexibility wherein the protocol can be modified based on newly obtained results.

D. Biological Treatment of Wastewaters

A major concern- during the past decade has been the improvement and preservation of the aquatic environment, and this has led to an increase in the effectiveness of treatment processes to control levels of nitrogen and phosphorus, since altered nutrient inputs are known to affect water quality markedly. The recent literature has utilized model concepts in relating nutrient concentrations to phytoplankton concentrations (Schnoor, 1980; Vollenweider, 1980). Models of this type may serve as valid predictors, dictating the extent and duration of treatment and controlling the nutrient levels being released from waste treatment plants into receiving waters.

The role of algae in oxidation-stabilization ponds is primarily that of an oxygen source for aerobic and facultative bacteria. Stabilization of the organic material entering an aerobic pond is accomplished mainly through aerobic bacterial activity.

Three benefits of algae usage are readily apparent: (1) reoxygenation, (2) mineralization, and (3) contribution to the food chain. Three disadvantages may also occur: (1) algal toxicity—daphnids fed senescent non-dividing Chlorella grew slowly, failed to reproduce, and died after 11–13 days (Ryther, 1954). Gorham and Carmichael (1980) have also discussed unpredictable poisonings of livestock and wildlife, attributing these mortalities to waterblooms or scums of blue-green algae; (2) taste and odour problems; and (3) buildup of biochemical oxygen demand (BOD).

E. Algal Assay Costs

Considering the broad spectrum of aquatic tests available which represent several trophic levels, the investigator must accurately define his particular problem. The ecotoxicological tests routinely used to investigate perturbations of the aquatic ecosystem include a 96-hour fish LC_{50}, a 48-hour daphnid EC_{50}, and a variant of the algal assay bottle test, in which test material exposures ranged in duration from

96 hours to 14 days. These screening tests share the following characteristics: reproducibility, accuracy, simplicity, short duration, and cost effectiveness. Cost effectiveness will be explored in relation to aquatic testing costs and more specifically its relationship to algal assay costs.

A comparison of man-hours expended while performing each of these aquatic tests was determined (Table 2). The use of electronic cell-counting techniques of algal cells would reduce the man-hour expenditures required for algal testing when compared to time expenditures while performing cell counts using the haemocytometer method.

Table 2 Aquatic testing: man-hours

	Species		
Parameter	Algae	Daphnid	Fish
Test	EC_{50}	EC_{50}	LC_{50}
Duration (d)	4	2	4
End point	Biomass cells present	Immobilization	Mortality
Man-hours	15	8	15
Cost (\$)[a]	750	600	Static freshwater 600–700 Marine 650–800

[a] Approximate contract laboratory cost

In considering the cost of performing an algal assay, the basic assumption has been made that adequately trained individuals are performing the test. Although an exact value cannot be attributed to these tests, as overhead costs vary from locale to locale, the cost per algal test (96-hour) is comparable to that incurred during a 96-hour LC_{50} fish bioassay test. Fish and algal tests for detecting "harmful" chemicals have similar efficiencies; 56 and 62%, respectively (Pomeroy, 1980), and costs are similar. The 14-day algal test does possess the additional significant advantage that results of such experiments may provide a useful indication of the effects of chronic toxicant exposure (effects over several algal generations).

The comparative costs for performing multigeneration aquatic tests using either algae or daphnids as the test organism has been calculated (Table 3). Algae possess definite advantages over daphnids: (1) the test duration is shorter; (2) using a test chemical, equivalent sensitivities are demonstrated, as opposed to the greater variability seen in daphnid studies; (3) algal studies are more cost effective. Given the listed advantages for multigeneration algal tests, consideration should be

Table 3 Multigeneration aquatic tests: man-hours

	Species	
Parameter	Algae	Daphnids
Test	Multigeneration	3-brood study
Duration (d)	14	20
End-point	Biomass, cells present	Release of 3 broods
Man-hours	40	35
Cost ($)	1500[a]	4200[a]

[a] Contract laboratory, approximate cost

devoted to standardizing tests of this type for use in situations in which a greater environmental impact may be envisioned for a specific test material.

IV. Future Research

To facilitate the analysis and interpretation of data being presently gathered, the present scope of algal testing should be extended. To achieve this goal, it is suggested that:

(1) Testing techniques should be of practical value, producing data which are easily and identically interpreted by industry, academia, and governmental agencies;

(2) Test systems should undergo rigid evaluation by all sectors (industry, academia, and government) to develop a standard or consensus test method;

(3) Environmental scientists should continue to develop biomonitoring techniques;

(4) There should be definition of water quality factors, e.g. pH, hardness, etc., on an algal test;

(5) An attempt should be made to delineate the occurrence of synergistic, antagonistic, mixed responses occurring when toxicant is added to natural waters;

(6) An attempt should be made to define more precisely natural water systems and their relative impact on algal test procedures.

V. Conclusions

Algae are sensitive indicators of the state of their environment, and their responsiveness to changes in their environment can be readily and intelligently interpreted. In utilizing indicator algal species, background information relative to growth pattern, associated organisms, environmental conditions, and nutritional characteristics of the ecosystem must be recorded. Completeness of data will enhance the derivation of useful predictive conclusions.

A general review of the literature indicates that the utilization of algal cultures and assays by industry can be grouped thus: effects of (1) nutrients, (2) complex effluents, and (3) chemical substances. Test systems to evaluate these effects include the following:

(1) *Laboratory bioassay*, using single or multiple algal species.
(2) *Microcosm studies*, customarily involving simultaneous exposure of a number of species from different trophic levels having a controlled community structure less complex than those occurring in nature.
(3) *Field studies* are the ultimate test of the accuracy of a hazard evaluation because they examine the natural communities in their habitat.

Acknowledgment

The authors wish to express their appreciation to their co-workers, R. L. Bohon and D. L. Bacon, as well as to A. G. Payne and co-workers of the Environmental Safety Group at Procter and Gamble for their helpful comments and constructive criticism during the preparation of this work.

References

Allen, H. E., R. H. Hall and T. D. Brisbin (1980). *Enviro. Sci. Tech.* **14**, 441–443.
APHA-AWWA-WPCF (1975). "Standard Methods for the Examination of Water and Wastewater" (M. Franson, ed.), 14th edition, pp. 744–756.
Bothwell, M. L. and J. G. Stockner (1980). *Can. J. Fish Aquat. Sci.* **37**, 248–254.
Christensen, E. R. and J. Scherfig (1979). *Water Res.* **13**, 79–92.
Christensen, E. R. and P. A. Zielski (1980). *Bull. Environm. Contam. Toxicol.* **25**, 43–48.
Federal Register (1980). **45**, 79318–79379.
Friant, S. L., R. Patrick and L. A. Lyons (1980). *Water Poll. Control Fed.* **52**, 351–363.
Gorham, P. R. and W. W. Carmichael (1980). *Prog. Wat. Tech.* **12**, 184–198.
Greene, J. C., W. E. Miller, T. Shiroyama and T. E. Malone (1975). *Water, Air, and Soil Pollution* **4**, 415–434.
Greene, J. C., W. E. Miller, T. Shiroyama, R. A. Soltero and K. Putman (1976). *In* "Proc. Symp. Terr. Aqua. Ecol. Studies Northwest", pp. 327–336. EWSC Press Cheney, Washington.
Guy, R. D. and A. R. Kean (1980). *Water Res.* **14**, 891–899.
Heldal, M., S. Norland, T. Lien and G. Knutsen (1978). *Chemosphere* **7**, 247–255.
Honig, R. A. and A. L. Buikema Jr (1980). *Arch. Environm. Contam. Toxicol.* **9**, 607–618.
Hutchinson, T. C. and P. M. Stokes. "Heavy Metal Toxicity and Algal Bioassays", *Water*

Quality Parameters, American Society for Testing and Materials, 1973, pp. 320–343. Philadelphia, Pa.

James, A. (1979). *In* "Biological Indicators of Water Quality", (A. James and L. Evison, eds), pp. 1–3. J. Wiley, Chichester.

Joubert, G. (1980). *Water Res.* 14, 1759–1763.

Leischman, A. A., J. C. Greene and W. E. Miller (1979). "Bibliography of Literature Pertaining to the Genus *Selenastrum*" US EPA-600/9-79-021.

Miller, W. E., J. C. Greene, T. Shiroyama and E. Merwin (1975). The use of algal assays to determine effects of waste discharge in the Spokane River system. *In* "Proc. Stim. Nutr. Assess. Workshop", pp. 113–131. EPA-660/3-75-034.

Miller, W. E., J. C. Greene and T. Shiroyama (1976). Application of algal assays to define the effect of wastewater effluents upon algal growth in multiple use river systems. *In* "Biostimulation and Nutrient Assessment" (J. Middlebrooks, D. H. Falkenborg and T. E. Maloney, eds), pp. 77–92. Ann Arbor Science Publications, Ann Arbor, Michigan.

Miller, W. E., J. C. Greene and T. Shiroyama (1978). "The *Selenastrum capricornutum* Printz Algal Assay Bottle Test: Experimental Design, Application, and Data Interpretation Protocol." US EPA-600/9-78-018.

Mitchell, D. (1974). Algal bioassays for estimating the effect of added materials upon planktonic algae in surface waters. *In* "Bioassay Techniques and Environmental Chemistry" (G. Glass, ed.), pp. 153–158. Ann Arbor Science Publishers, Ann Arbor, Michigan.

OECD Guidelines for Testing of Chemicals (1981) Section 2, Effects on Biotic Systems Test 201 "Alga, Growth Inhibition Test" Adopted 12 May 1981.

Palmer, C. M. (1969). *J. Phycol.* 5, 78–82.

Patrick, R. (1973). *In* "Biological Methods for the Assessment of Water Quality", pp. 76–95. American Society for Testing and Materials, Philadelphia, Pa.

Payne, A. G. and R. H. Hall (1978). *Intl. Assoc. Theoret. Appl. Limnol.* 21, 507–520.

Payne, A. G. and R. H. Hall (1979). A method for measuring algal toxicity and its application to the safety assessment of new chemicals. *In* "Aquatic Toxicology" (L. L. Marking and R. A. Kimberle, eds), pp. 171–180. American Society for Testing and Materials, Philadelphia, Pa.

Pomeroy, S. E., S. E. Brauning, and G. H. Kidd (1980). *In* "Validation of the OECD Ecotoxicology Testing of a New Base Set." Final Report EPA Contract 68-01-5043.

Roline, R. A. and V. S. Miyahara (1979). "Evaluation of the Algal Assay Bottle Test", US Department of the Interior. REC-ERC-80-1.

Ryther, J. H. (1954). *Ecology* 35, 522–533.

Schlichting, H. E. and C. P. Mason (1980). *In* "Carolina Tips". 43, 25–28.

Schnoor, J. L. and D. J. O'Connor (1980). *Water Res.* 14, 1651–1665.

Simmons, M. S. and K. Sivaborvorn (1979). *Bull. Environm. Contam. Toxicol.* 23, 733–766.

US EPA (1971). "Algal Assay Procedures: Bottle Test", Corvallis, Oregon.

US EPA (1974). "Marine Algal Assay Procedure: Bottle Test", EPA-660/3-75-008.

US EPA (1976). "Bioassay Procedures for the Ocean Disposal Permit Program", EPA-600/9-76-010.

US EPA (1979). "Toxicology Handbook, Mammalian and Aquatic Data" (B. A. Schneider, ed.), EPA 540-9-79-003.

Vollenweider, R. A. and J. Kerekes (1980). *Prog. Wat. Tech.* 12, 5–38.

Walsh, G. E. and S. V. Alexander (1980). *Water, Air and Soil Pollution* 13, 45–55.

Walsh, G. E., L. H. Bahner and W. B. Horning (1980). *Environ. Pollut.* 21, 169–179.

12 Algal Bioassays of Industrial and Energy Process Effluents

G. E. Walsh

US Environmental Protection Agency
Environmental Research Laboratory
Gulf Breeze, Florida, USA

R. G. Merrill

US Environmental Agency
Industrial Environmental Research Laboratory
Research Triangle Park, North Carolina, USA

I. Introduction

Unicellular algae are ubiquitous in aquatic ecosystems, where they incorporate solar energy into biomass, produce oxygen that is dissolved in water and used by aquatic organisms, function in cycling and mineralization of chemical elements, and serve as food for herbivorous and omnivorous animals. When they die, they sink to the sediment, where their chemical constituents are transformed, solubilized, and recycled into the water. These functions depend on phytoplankton population dynamics, which in turn depend on seasonal variability in temperature, intensity of solar radiation, nutrient concentrations in the water, and grazing by animals. Natural and anthropogenic alterations of water quality can upset the balance of these controlling factors and bring about changes in species composition of the algal community, rates of production, biomass, and water chemistry. If water quality is altered by toxicants or growth stimulants from industrial, agricultural, or municipal sources, normal algal function may be upset, causing gross changes in structure and function of the receiving aquatic ecosystem.

A large amount of evidence indicates that industrial and energy process effluents contain algal and animal toxicants and algal growth stimulators. Algal laboratory tests are excellent indicators of pollution by these complex wastes. Like animals, algae in the laboratory respond to toxicants, but they also respond in an easily measurable way to substances that affect primary productivity. Algal population growth in the laboratory is rapid and may be measured directly in a variety of ways. In addition, rates of physiological processes within exposed populations may be estimated indirectly by measurement of evolution and uptake of oxygen.

This review presents laboratory data and their interpretation with regard to effects of complex wastes on marine and freshwater unicellular algae. The scientific literature contains relatively little data on algae and complex wastes, and specific methods need to be developed for field and laboratory studies. We have therefore included data from tests with organic compounds and heavy metals because both are present in complex wastes. Methods used in toxicity tests with these pollutants are discussed with regard to their application to studies on complex wastes. Inhibition and stimulation of algal population growth are considered to be equally undesirable, because any deviation of algal productivity or community composition from that which is usual for an aquatic ecosystem threatens the entire system.

II. Algae as Bioassay Organisms

Algae comprise a large and heterogeneous group of benthic and planktonic species that occur together in often complex communities. Members of the following classes are often prominent in natural assemblages: Chlorophyceae (green algae), Bacillariophyceae (diatoms), Cyanophyceae (blue-green algae), Rhodophyceae (red algae), Chrysophyceae (golden brown algae), Haptophyceae, Craspedophyceae, Xanthophyceae, Prasinophyceae, Pyrrophyceae (dinoflagellates), Cryptophyceae, and Euglenophyceae. Green algae, blue-green algae, and diatoms are commonly dominant in algal communities, but toxicity tests on species from the other classes are needed to gain a more comprehensive idea of effects of toxicants on algae.

Although there is a large number of algal species in fresh and salt water, many of which have been used in tests with pesticides, heavy metals, and other single toxicants, only a few have been tested for response to complex wastes. In freshwater toxicity tests, *Selenastrum capricornutum* Printz (now called *Monoraphidium capricornutum* Printz (Nygaard)) is used extensively to detect the presence of growth inhibitors and growth stimulators in complex effluents (Miller *et al.*, 1978; Walsh *et al.*, 1980). Other species commonly used and responsive to pollutants include *Chlorella homosphaera*, *Ankistrodesmus falcatus*, and *Oscillatoria agardhii* (Claesson, 1975), *Nitzschia linearis* (Patrick *et al.*, 1968), and *Euglena gracilis* (Falchuk *et al.*, 1975a,b).

Palmer (1969) listed 60 genera and 80 species of freshwater algae according to tolerance to pollutants. Although members of each group varied widely in response, green and blue-green algae generally were most tolerant and diatoms least tolerant to pollutants. The five most tolerant genera were *Euglena*, *Oscillatoria*, *Chlamydomonas*, *Scenedesmus*, and *Chlorella* and the five most tolerant species were *Euglena viridis*, *Nitzschia palea*, *Oscillatoria limosa*, *Scenedesmus quadricauda*, and *Oscillatoria tenuis*. The five most sensitive genera were *Crucigenia*, *Cymatopleura*,

Dictyosphaerium, Selenastrum, and *Stauroneis,* and the five most sensitive species were *Tetraedron muticum,* (now *Goniochloris mutica* (A.Br.) Fott), *Pyrobotrys gracilis, Euglena proxima, Gonium pectorale,* and *Cryptomonas ovata.* No species and only one genus, *Selenastrum,* listed above as pollution-sensitive is used in laboratory toxicity tests.

It is possible that *S. capricornutum* is not the most sensitive fresh-water algal species to pollutants in laboratory toxicity tests. Shiroyama *et al.* (1975) stated that *S. capricornutum,* although not indigenous to the United States, was chosen for use in the algal assay procedure bottle test because it is easy to culture and identify, displays minimum morphological changes during the population growth phase, it is single-celled and thus can be enumerated on an electronic particle counter, and it is widely distributed in eutrophic and oligotrophic waters. None of these characteristics addresses the question of sensitivity to pollutants. Rodhe (1978) stated that *Selenastrum* belongs to a group of ubiquitous algae that have a wide range of tolerance towards environmental conditions. Slobodkin and Sanders (1969) postulated that organisms adapted to stable environments would in general be more sensitive to environmental stress than organisms adapted to less stable conditions. Besides being more resistant to environmental change, species adapted to unstable conditions may adapt to culture conditions and thus their physiological responses may not be similar to those in the field. Highly adaptable species may adapt to the presence of bioactive substances from waste outfalls in laboratory tests. It is possible that only the most hardy species may be cultured in the laboratory and that species sensitive in natural systems cannot be evaluated in standard laboratory tests.

There is a need for screening studies to search for new, more sensitive freshwater species that respond to both growth inhibitors and growth stimulators. Many new algal species can be cultured and tested in the laboratory, and the more sensitive species might be isolated from geographical areas of interest. Care must be taken not to sample from contaminated waters because resistant forms may be collected.

Culture collections often have clones that can be tested. For example, the Culture Collection of Algae at the University of Texas at Austin lists as available: *Tetraedron* (5 clones), *Pyrobotrys* (1 clone), *Crucigenia* (2 clones), *Dictyosphaerium* (4 clones), *Selenastrum* (6 clones), *Stauroneis* (1 clone), *Euglena proxima* (1 clone), *Gonium pectorale* (6 clones), and *Cryptomonas ovata* (1 clone).

A large number of marine algal species have been tested with complex effluents. For example, Dunstan (1975) used 15 species to analyse effects of secondary treated sewage, Brack *et al.* (1976) used 4 species to describe synergistic and antagonistic effects of heavy metals, and Thomas *et al.* (1980a) reported effects of mixtures of 10 heavy metals on 4 species of marine algae. Rainville *et al.* (1975) used *Coccochloris elebans* because it is common in the marine phytoplankton, reacts rapidly to environmental change, responds to toxic substances, is genetically stable, and grows rapidly in suspension in laboratory cultures.

Before new species can be used in tests, their characteristics in laboratory culture must be defined to ensure that they do not depart greatly from those of natural populations. For example, original bioassay protocols for eutrophication studies (US Environmental Protection Agency (EPA), 1971) recommended *Microcystis aeruginosa* and *Anabaena flos-aquae* as test species. Wyatt (1973), however, pointed out that laboratory cultures of these cyanophytes differ greatly from natural populations. In culture, *M. aeruginosa* does not take the form of a gelatinous colony, and neither *M. aeruginosa* nor *A. flos-aquae* forms gas vesicles as in nature. Laboratory responses to toxicants with such atypical clones cannot be considered as estimates of responses in natural systems. On the other hand, morphological changes caused by pollutants in laboratory tests may be important detection tools and are discussed later.

Algal clones have been shown to respond differentially to pollutants. Fisher *et al.* (1973) showed that stenothermal and stenohaline clones of *Thalassiosira pseudonana*, *Fragilaria pinnata*, and *Bellerochia* sp. from the Sargasso Sea were more sensitive to PCBs than clones of the same species from estuaries and over the continental shelf. Jensen *et al.* (1976) demonstrated that two clones of *Skeletonema costatum* responded with equal sensitivity to copper ions, but differed markedly in response to zinc. It may therefore be necessary to test clones of many species to find those most sensitive in most tests.

Although algal tests are usually performed to detect toxicity, lack of toxicity is just as important to pollution-control workers. Thomas *et al.* (1980b) used an algal toxicity test with *Asterionella japonica*, *S. costatum*, *Chaetoceros affinis*, *Gymnodinium splendens*, *Gonyaulax polyhedra*, and *Dunaliella* sp. to demonstrate that ammonium from a sewage outfall was not toxic to the algae. However, growth was stimulated and the authors concluded that there is probably a balance between inhibitory and stimulating factors put into receiving waters by outfalls.

Most algal tests employ only one species, whereas in nature algae often occur in complex communities whose compositions change throughout the year. Although single pollutants are known to affect competition among algae, virtually nothing is known about effects of complex wastes on species composition or function of unicellular algal communities. Titman (1976) showed that the relative abundance of phosphate and silicate determined the relative abundance of *Asterionella formosa* and *Cyclotella meneghiniana*. Tilman (1977) demonstrated how the functional resource acquisition and/or utilization responses of *A. formosa* and *C. meneghiniana* in the presence of limiting amounts of phosphate and silicate regulate species numbers in mixed culture. Lewis (1977) showed that rare algal species are as likely to have an opportunistic strategy as common species and that filamentous algae are likely to be opportunistic (i.e. species that have marked power of increase under a narrow range of favourable conditions), whereas coenobial types are conservative (i.e. lack the ability to produce impressive growth surges but persist more effectively when conditions are unfavourable). It is not known how complex wastes may affect competition among algal species—whether by alteration of available resources, direct effects on the algae, or both. Research is needed on effects of pollutants on species composition of algal communities. Effects in culture must be related to field conditions such as temperature: Goldman and Ryther (1976) showed that results of competition are highly dependent on temperature.

It is probable that species in unialgal culture respond differently to pollutants than they would if exposed in multispecies tests. Gamulin-Brida *et al.* (1967) studied subtidal algal communities and concluded that single species in culture do not respond to physical and chemical conditions of the environment with the same sensitivity as when in competition with other species in natural situations. Fielding and Russell (1976) found this to be true *in vitro* when the marine macrophytic algae *Ectocarpus siliculosa*, *Erythrotricia carnae*, and *Ulothrix flacca* were exposed to copper singly and in combination. There were two types of algal reactions: competition and annidation (i.e. one species is stimulated by the presence of the other). For example, *U. flacca* grew better than *E. siliculosa* when grown together in untreated control culture. In the presence of 50 parts per billion (p.p.b.) (1 billion $= 10^9$) copper, growth was equal, but in 100 p.p.b., *E. siliculosa* grew better than *U. flacca*. At 100 p.p.b. *Ulothrix* enhanced the growth of *E. carnea*, which enhanced the growth of *E. siliculosa*. Fielding and Russell (1976) suggested that, in these cases, copper might have been complexed by extracellular products of algal metabolism or taken up by living or dead cells.

Fisher *et al.* (1974) demonstrated strong effect of 0·1 p.p.b. PCB on natural phytoplankton assemblages in a flow-through continuous culture system. Species diversity was reduced strongly after 146-hour exposure, and the authors suggested that the toxicant might alter the natural state of interspecific competition and decrease the diversity of natural communities. Goldman and Stanley (1974) studied growth interactions of *Dunaliella tertiolecta*, *Thalassiosira pseudonana*, and *Phaeodactylum tricornutum* in wastewater–seawater mixtures and suggested that a complex interaction of environmental factors is usually responsible for dominance of a species. The source of nitrogen was important in their studies. When the source was NO_3-N, the carbon:nitrogen ratio in the algae was high (7 to 8), when NH_4-N, the ratio was low (4 to 6).

It is clear that the species of alga and the conditions under which it is tested are important when estimating effects of pollutants on algae and algal communities. Most toxicological research with algae has been done with single pollutants such as pesticides and heavy metals, but it is not known whether species most sensitive to single pollutants are also most sensitive to mixtures. There is therefore a need for screening tests with algae and complex wastes to determine which species and conditions are best for detection of pollution from industrial outfalls. The following suggestions may be used as criteria for establishment of a testing program:

(1) Use ecologically important sensitive species;
(2) Measure the most important end-point for each species, e.g. inhibition and/or stimulation of growth, rates of photosynthesis and respiration;
(3) Use mixed-species tests to determine effects on communities, e.g. changes in dominance and rates of primary production.

(4) Use experimental conditions that relate to species survival and abundance and community structure under varying conditions found in natural systems, e.g. temperature, light, and salinity.

Use of indigenous species may not be protective of the environment because pollutants are transported easily to other areas. Use of sensitive species is much more important than use of indigenous species.

The above criteria imply knowledge of the population dynamics and physical and chemical requirements of sensitive algal species. Such knowledge can be used in multifactorial laboratory tests to produce algal toxicity data that relate to field conditions.

III. Algal Toxicity Tests

A variety of tests has been developed for measuring effects of toxicants and growth stimulators on algae. Although they vary according to composition of growth medium, species tested, size and shape of exposure vessel, intensity and cycle of lighting, etc., most are static tests with single species in enriched culture medium. Continuous culture techniques or flow-through tests have been used with monospecific and mixed-species cultures, but to a smaller degree than static batch tests. Both static and flow-through tests estimate growth-rate responses by measurement of change of biomass, chlorophyll content, cell numbers, or fluorescence. Rates of primary production have also been estimated by measurement of oxygen evolution or uptake of ^{14}C.

A. Field Tests

Algal tests have been used in the field and laboratory to determine biological potential of natural waters and of substances from industrial, municipal, and agricultural sources dissolved in them. They have also been used to directly estimate potential effects of waste streams. Schultz and Gerhardt (1969) described design and use of dialysis culture, and Trainor (1965), Jensen *et al.* (1972), Prakash *et al.* (1973), and Jensen and Rystad (1973) described methods of dialysis culture that can be used in the field and laboratory. Sakshaug and Jensen (1978) reviewed use of dialysis cultures for the study of marine phytoplankton and stated that such culture extended conventional bioassay methods because:

(1) Dialysis cultures can be easily grown *in situ*;
(2) Algal accumulation of trace compounds from large volumes of medium is possible;
(3) Dense populations can be obtained in oligotrophic waters;
(4) Control of one or more growth requirements is easily established;
(5) Growth rates and intrinsic algal properties can be monitored as functions of environmental factors, including pollutants.

In dialysis culture, algae are retained in dialysis bags or in vessels fitted with dialysis membranes and exposed to soluble pollutants or nutrients in natural waters that can pass through the pores of the bag or membrane. Growth of the algae can then be measured spectrophotometrically, by cell count, by direct gravimetric analysis of biomass, or by chlorophyll analysis. Jensen *et al.* (1972) demonstrated effects of non-enriched seawater, depth, and nutrients on growth of eleven species of marine unicellular algae by suspending dialysis bags with algae in flowing seawater in the laboratory or from a buoy in the open sea. Dor (1975) used dialysis culture to demonstrate stimulatory effects of soluble substances in domestic sewage on the freshwater species, *Scenedesmus obliquus*. Powers *et al.* (1976) exposed the marine dinoflagellate, *Exuviella baltica*, to low concentrations of DDE in dialysis bags and incubated them in a salt marsh. Population growth in bags in the field was inhibited as compared to laboratory and field untreated control cultures. Among control cultures, populations in dialysis bags suspended in the water of the salt marsh grew at a faster rate and attained greater biomass than did laboratory populations cultured in natural water or enriched seawater.

Although it has not been used extensively, and field data need to be compared to laboratory results, dialysis culture is potentially an excellent method for estimating effects of soluble components of industrial and energy process effluents on receiving water. Unialgal or mixed algal cultures could be placed at an outfall and at selected sites downstream to measure effects on growth, species composition of test communities, rates of photosynthesis and respiration, or chlorophyll concentration. Such field tests would be easy to perform and would relate more closely to natural conditions of light, temperature, and nutrient availability than laboratory tests.

B. Batch Culture

In freshwater, the most commonly used method for assaying the algal growth supporting potential of natural waters is the algal assay procedure: bottle test (EPA, 1971; Miller *et al.*, 1978). This method uses flasks instead of bottles and was developed to detect nutrient status of natural waters. One parameter used to measure effect in the bottle test is the maximum standing crop in each flask, which is determined gravimetrically or indirectly from cell counts and cell volume as determined with an electronic particle counter. Miller *et al.* (1978) recommended a growth period of 14 days or until the increase in biomass is less than 5% per day. Greene *et al.* (1975) used the method to detect effects of nitrogen and phosphorus from a primary sewage treatment plant on algal growth. The standard algal assay bottle test estimates effects of substances dissolved in natural waters by expression of population growth rate and dry weight biomass. Cain and Trainor (1973) modified the method to measure population growth by changes in optical density of algal cultures, transforming optical density data to cell number. They also added test water to their cultures each day to minimize effects of nutrient carryover after luxury uptake to determine growth rates under conditions of nutrient replenishment, as occurs in rivers. They demonstrated effects of a sewage treatment plant on the nutrient-carrying capacity of a river and stated that although final yield in cultures to which medium is not added each day is a function of available nutrients, such data may be of less value than growth rate data obtained under conditions of nutrient replenishment. Cain *et al.* (1979) compared results of freshwater algal tests to chemical analyses of effluent from a sewage treatment plant and concluded that their algal assay was highly efficient in detection of biological potential. It was especially efficient in estimating growth as it related to orthophosphate content of the effluent and receiving water.

Although the algal assay procedure: bottle test was developed to detect the nutrient status of natural waters, it has also been used to detect growth-inhibiting substances in natural waters and effluents. Shiroyama *et al.* (1976) detected inhibition of growth of *A. flos-aquae* by heavy metals in effluent from a primary sewage treatment plant, and Joubert (1980) demonstrated inhibition of algal growth by water from a leachate site. There is, however, a drawback inherent in this freshwater method. Some liquid waste effluents contain algae that have adapted to living in waste impoundments, so a sample taken at the outfall may contain algae that can grow in test medium. Miller *et al.* (1978) suggested filtration of all wastewater through a 0·45 µm filter to eliminate contaminating algae. Most complex wastes contain suspended particles, however, and pigments of coloured wastes are often particulate. Therefore filtered wastes are often grossly different from the originals, having lost all of the suspended matter, and highly coloured wastes may become colourless. We suggest that waste samples should be examined for the presence of algae before a test. If no algae are found, then filtration should not be done. Bacteria in unfiltered waste may flourish under algal culture conditions, and short-term tests of 48 to 72 hours (described below) may best avoid bacterial effects on algae and toxicants.

Lack of filtration precludes use of the gravimetric method and particle counters for estimation of biomass. However, the bottle test was developed for identification of the nutrient status of natural waters. The end-points for such a test must necessarily be different from those of a test to determine bioactivity of complex wastes that contain both growth inhibitors and growth stimulators. Whereas nutrients such as orthophosphate do not break down or attach permanently to flask walls and are thus available for algal growth at all times, toxic organic and inorganic components of complex wastes may be decomposed biologically, adsorb strongly to flask walls, break down in the presence of light required for algal growth, precipitate from solution, or form complexes with other substances. Algae may also develop resistance to the toxicants or the ratio of algal biomass to toxicant concentration may increase as the population grows slowly, ultimately resulting in a ratio favourable to rapid algal growth. All the factors that affect response of algae to complex wastes may operate in long-term tests, so short-term tests are best for estimation of effects on growth. Growth rates or doubling time may be estimated by cell counts with a microscope, optical density measurements, quantitative analysis of chlorophyll, or by fluorescence. Biomass after 14 days of growth is a poor estimate of effects of complex wastes, because toxicity is often lost after 5 or 6 days (Walsh *et al.*, 1982).

We suggest that an exponential phase plot of algal growth is not needed for toxicity tests. The purpose of such tests is to detect toxicity. Attempts to use them for any other purpose must consider fate of the toxicant, and this is usually impossible in standard tests. Our data show that suppression of growth after 48 or 72 hours exposure is the best indicator of toxicity.

Marine algal assays are similar to those with freshwater species because they are generally *in vitro* tests on selected, easily-cultured algae. Algal contaminants of liquid waste are seldom a problem in marine tests. Contaminant algae are usually low in number in waste streams, which are generally freshwater. Freshwater algae are killed by the salinity of the marine algal medium, and except for adding very small amounts of nutrients to the medium, have little effect on growth of test species.

Marine algal tests differ from one another mainly in the type of exposure vessel, growth medium, and algal species. Exposure vessels are usually Erlenmeyer flasks (Canterford, 1979; Williamson, 1980) or optically matched test tubes (Ferguson *et al.*, 1976; Walsh and Alexander, 1980). Optically matched test tubes that fit in the cuvette holder of a spectrophotometer are convenient because they take little space, and population size may be estimated by the absorbance value at any desired wavelength without removing samples from the tubes.

Freshwater algal growth medium is usually prepared by addition of nutrient salts, trace metals, and vitamins to well, distilled, or deionized water. Stein (1973) and Starr (1978) list a large number of liquid media, most of which support growth of many algal species. Few media have been tested for use in toxicity tests, but whatever medium is used, chelators should not be added. In tests with liquid industrial or energy process wastes, test media are prepared by adding nutrients directly to untreated effluent and then diluted with medium prepared with water. Medium prepared with water serves as the control medium. It may be necessary to use a nutritionally dilute medium that supports approximately one-half of the full growth in order to detect growth stimulants.

There are two basic types of marine algal growth medium: completely synthetic medium prepared with distilled or deionized water and salts, trace metals, and vitamins; and natural seawater enriched with nutrient salts, trace metals, and vitamins. Stein (1973) gives methods for preparation of marine algal media. Completely synthetic media may be prepared with commercial salt mix (Walsh and Alexander, 1980) or with reagent grade chemicals and distilled or deionized water (Harrison *et al.*, 1980). Harrison *et al.* (1980) tested 83 coastal and open-ocean species of algae with their medium and reported that only three did not grow in it.

The most commonly used enriched natural seawater medium is the f/2 formula of Guillard (Guillard and Ryther, 1962; Sears, 1967; Stein, 1973). Booth (1975) described growth of 26 species of phytoplankton in enriched seawater. Although algae grow well in it, enriched natural seawater, cannot be used in toxicity tests with algae and waste outfalls. Most outfalls are freshwater, and salts must be added to them in order to estimate potential effects when their constituents reach estuarine and oceanic systems.

Test medium for algae is prepared by adding salts, trace metals, and nutrients directly to liquid waste. Salinity may be between 5 and 35 parts per thousand (p.p.t.), depending upon data requirements and species tested. Euryhaline species need to be adapted by at least three serial transfers to each salinity. Waste with added salt is diluted to test concentrations with growth medium prepared with distilled or deionized water. Nutrient strength should be approximately one-half that which supports full growth so that growth stimulation can be detected.

Although most toxicity tests are performed on a single species of alga, it is best to test the most sensitive clones of at least three species in order to evaluate possible effects of toxicants on algae. Algae vary greatly in response to toxicants, for example a substance may be toxic to one species and non-toxic to another, whereas, the same two species may respond in an opposite manner to another substance (Hollister and Walsh, 1973).

A few tests have been reported on response of mixtures of algal species to toxicants in the laboratory. Dunstan (1975) stated that the most significant force exerted by sewage discharge is to change the pattern of phytoplankton production, distribution, and population dynamics and that ecological interpretation of results based on culture of one or two species can be misleading. Mosser *et al.* (1972) reported that the diatom *T. pseudonana* had a more rapid growth rate than the green alga *D. tertiolecta*, when each was grown singly and in mixed culture. However, in the presence of low concentrations of chlorinated hydrocarbon insecticides, dominance was reversed, with *D. tertiolecta* having the more rapid growth rate. Green algae are generally much more resistant to toxicants than diatoms (Fisher and Wurster, 1973), and presumably *D. tertiolecta* had a competitive advantage over *T. pseudonana* in the presence of chlorinated hydrocarbons. Walsh and Alexander (1980) reported that the diatom, *S. costatum*, had a more rapid growth rate than the rhodophyte, *Porphyridium cruentum*, in single-species and mixed cultures. Waste from a paper products plant had no effect on growth of *P. cruentum* but inhibited growth of *S. costatum* in single and mixed cultures, resulting in reversal of dominance in mixed culture. On the other hand, Walsh and Alexander (1980) reported that marine species (*P. cruentum*, *Neochloris* sp.) sensitive to the herbicide, neburon, were less sensitive when grown in mixed culture with resistant species (*Nitzschia angularum*, *Navicula inserta*). In the tests with neburon, it appeared that the less sensitive species protected the more sensitive species by initial rapid growth accompanied by adsorption of the toxicant, which reduced it to less than toxic concentrations. Sanders *et al.* (1981) exposed natural assemblages of marine phytoplankton to copper, chlorine, and thermal stress and concluded that chronic exposure to low concentrations of pollutants associated with power plant cooling systems can induce substantial shifts in species composition of the assemblages.

Laboratory tests with mixtures of species suggest that toxicants may affect algal population dynamics, resulting in dominance of resistant species. Harrison *et al.* (1977) showed dominance of copper-resistant species over copper-sensitive species when exposed to copper, and Jensen *et al.* (1974) showed that a clone of *S. costatum* from a zinc-polluted fjord was approximately two orders of magnitude more resistant to zinc than a clone from an unpolluted fjord. The mechanism of resistance, whether simply physiological or genetic, is unknown, but Fisher (1977) suggested that adaptation could involve changes in membrane structure and permeability, thus limiting uptake of the toxicants. Whatever the mechanism of adaptation, pollutants may cause changes in algal community structure with subsequent effects on herbivore populations. Also, resistant species may adsorb toxicants (Walsh, 1977) which, when passed upward in food chains, may accumulate to toxic concentrations. Ohkawa *et al.* (1980) used a simulated aquarium ecosystem that contained carp (*Cyprinus carpio*), daphnids (*Daphnia pulex*), snails (*Cipangopaludina japonica*), and a mixture of pond algae to test partitioning of the pesticides DDT and S-fenvalerate. After exposure for seven days, algae were efficient accumulators of the toxicants and the higher organisms obtained them either directly from the water or through food chains (algae→snails or algae→daphnids→fish). These concepts, however consistent with ecological theory, need to be investigated in the field because relationships between laboratory data and field theory have not been described.

C. Continuous Culture

Algal populations may be grown in continuous culture in which fresh growth medium is introduced at a measured rate and old medium washed from the system at the same rate. Since algal cells are removed from the system in the old medium, rate of flow can be regulated according to rate of growth of the algal population and any number of cells between zero and maximum standing crop may be maintained. Pavlou (1972) described methods for continuous culture of algae in chemostats, and numerous papers have reported methods for maintaining and testing of algae in chemostats.

Continuous culture techniques may be used to identify sublethal effects of toxicants in long-term tests and offer the opportunity to study algal responses to environmental change under conditions that resemble the natural environment more than batch culture. For example, continuous input of low concentrations of nutrients to a culture in the steady state resembles natural conditions more closely than batch culture, where exponentially growing cells draw upon a large reservoir of nutrients. Also, metabolites that may affect tests are removed from flow-through systems. Sharp *et al.* (1980) demonstrated that cellular chemistry and population physiology in batch culture are not synchronized, whereas they are in steady-state systems.

A very important advantage of continuous culture over batch culture is that many toxicants adsorb to the glass walls of exposure vessels: this could remove toxicants from solution and result in underestimation of toxicity. Continuous culture also allows study of physiological responses in relation to pollution, such as periodicity of metabolic processes (Eppley *et al.*, 1971), nutrient relationships (Davis *et al.*, 1973; Thomas *et al.*, 1980d), effects of temperature on growth, nutrient uptake and chemical composition of algae (Goldman, 1979), carbon budgets (Sharp *et al.*, 1980), and numerous others.

Continuous culture in chemostats may also be used to study effects of pollutants on algal communities, but the method is limited to use with species that are compatible with regard to growth rate. For example, the lowest acceptable flow rate would be determined by the doubling time of the slowest-growing species. This might confer considerable competitive advantage to fast-growing species and make toxicity tests useless. Research on development of flow-through systems with algal communities is needed.

Results of batch and continuous flow tests may be confounded if experimental conditions are not controlled carefully. Algal population growth and morphology are determined by many factors, such as temperature (Reynolds *et al.*, 1975); Nakamura and Miyachi, 1980; Terlizzi and Karlander, 1980), nutrients (Paasche, 1971; Turner, 1979; Terlizzi and Karlander, 1980), salinity (Terlizzi and Karlander, 1980), concentration of carbon dioxide (Nakamura and Miyachi, 1980), photoperiod (Chisholm and Costello, 1980), and average cell size (Banse, 1976). Temperature has profound effects on toxicity (Fisher and Wurster, 1973; Cairns *et al.*, 1975; Styron *et al.*, 1976). Although algae require heavy metals in their nutrition, they must be chelated or else they are toxic, causing slowed growth and abnormal morphology. Solvent carriers of toxicants, such as acetone or triethylene glycol, must be tested for toxicity before use with algae. For example, Stratton *et al.* (1980) demonstrated antagonistic, synergistic, and additive behaviour of acetone and the pesticide permethrin in three species of *Anabaena*. They recommended that solvent sensitivity of each organism be determined because of possible variation in species response and that every combination of solvent, algal species, and test compound be analysed to be sure that toxicities obtained are due to the compound and not to its interactions with solvent. In tests on complex industrial and energy process wastes, correct choice of solvent is very important in determining toxicities of chemical fractions as discussed later.

It is important, therefore, for maximal protection of aquatic systems, that sensitive algal species be tested under physical and chemical conditions optimal for detection of toxicity or stimulation.

IV. End-Points of Algal Toxicity Tests

There is a variety of ways by which algal response to pollutants can be measured. Growth response occurs either by enhanced growth or decreased growth relative to solvent control cultures. Complex wastes often contain both growth stimulants and growth inhibitors. Cells can be enumerated with counting chambers and microscopes, but not with electronic particle counters because wastes usually contain particulate matter that blocks the orifice. Particles in effluents also preclude accurate measurement of biomass by gravimetric methods. Total chlorophyll measurements give only an approximate estimate of the amount and physiological state of algal biomass, and by themselves do not give an accurate account of algal response to toxicants. Amount of chlorophyll per cell is a better estimate of effect than total chlorophyll alone. Traditional extractive methods for spectrophotometric quantitation of chlorophyll and their errors were discussed by Harjula (1979), but it should be recognized that occasionally a substance from the effluent may interfere with the tests. Jeffrey and Humphrey (1975) gave equations for calculating pigment concentrations in various types of algae. Generally, phaeopigment is not detected in appreciable amounts in healthy algal cultures. However, it should be quantified in all algal toxicity tests that use chlorophyll concentration as an end-point. It may be that phaeopigment concentration is as important as that of chlorophyll. Holm-Hansen (1978) described a fluorimetric technique for quantitation of extracted chlorophyll that also has the advantage of simple estimation of phaeopigments. Spectrophotometric determination of phaeopigment requires a neutralization procedure after acidification of chlorophyll, whereas the fluorimetric method does not, with a considerable saving of time.

Fluorescence of intact algal cells has been used extensively to estimate algal biomass. Application of this method was reviewed by Harris (1978). Although results with whole cells from natural systems are more variable than with extracts, *in vivo* techniques are used extensively in basic and applied research on algae. In the laboratory, *in vivo* techniques have the great advantage of ease of measurement, because time-consuming extraction is not required and environmental factors that cause variability in field populations do not operate under culture conditions. However, fluorometric measurement of chlorophyll *in vivo* in experimental studies on laboratory populations of algae may lead to erroneous conclusions. Fluorescence yield is a function of chlorophyll concentration and chloroplast shape. Complex wastes from industrial effluents often cause conformational changes in chloroplast structure, and Clement-Metral and Lefort-Tran (1974) showed that changes in thylakoid shape changed the fluorescence emission characteristics of the red alga, *P. cruentum*.

The rate of photosynthesis per unit biomass, per unit chlorophyll, or by a population has been used to estimate effects of pollutants on algae (Standyk *et al.*, 1971; Erickson and Hawkins, 1980). It is unclear whether the ^{14}C method measures net or gross primary production, because of possible losses of organic matter by excretion. Estimates of loss of organic matter range from undetectable (Williams and Yentsch, 1976) to more than 75% of photo-assimilated carbon (Sharp, 1977), and magnitude of excretion may vary with species and culture conditions. It is generally held that values from the ^{14}C method lie between the true net and gross photosynthesis (Strickland, 1960). The ^{14}C method may be used in toxicity tests if it is known that the toxicants do not affect excretion of photo-assimilated carbon. DeFilippis and Pallaghy (1976) showed that zinc and mercury inhibited excretion of glycolate from *Chlorella* sp., and this suggests that treated cultures could not be compared to control cultures in toxicity tests because their rates of excretion were different.

Two methods that may be used for measurement of effects of wastes on algae are the adenylate energy charge and adenosine triphosphate (ATP) assays. The adenylate energy charge is defined in molar concentrations of the adenine nucleotides as

$$\frac{([ATP] + \frac{1}{2}[ADP])}{([ATP] + [ADP] + [AMP])}$$

Since cellular growth and division utilize energy available in the adenylate pool, Falkowski (1977) suggested that the adenylate energy charge could be used as an indicator of physiological state of *S. costatum* and demonstrated differences in energy charge between nitrogen-limited continuous and batch cultures in relation to temperature.

Because ATP degrades rapidly after an organism dies, it can be used as an indicator of the amount of living biomass in a system (Holm-Hansen and Booth, 1966; Holm-Hansen and Pearl, 1972). ATP analyses have been used to estimate amount of living material in activated sludge (Patterson *et al.*, 1970; Chui *et al.*, 1973) and waste water (Levin *et al.*, 1975). Kennicutt (1980) demonstrated that ATP analysis can be used in routine toxicity tests and may be valuable in assaying impacts of pollutants on aquatic environments.

The adenylate energy charge may be useful in detecting sublethal effects of complex wastes on algae grown in them, and ATP analysis may be useful in defining effects of pollutants over selected time periods on algal populations. However, although both methods are potentially useful in algal toxicity tests, more research is needed to verify their utility and to adapt them to exposure techniques with complex wastes.

In static algal toxicity tests, toxicants are usually added to cultures with low numbers of algae, and population growth is measured over a period of time, often 96 hours. Walsh (unpublished) grew *S. costatum* for 48 hours in toxicant-free medium and then counted the number of living cells, dead cells, and those of abnormal morphology. Dilutions of the insecticide Dursban were made in triplicate cultures which were then incubated for 48 hours and again enumerated. Table 1 shows that the insecticide inhibited population growth and killed cells. It also caused deformations of cellular morphology at concentrations up to 2·5 parts per million (p.p.m.). Percentage of abnormal cells was greatest at 1·0 p.p.m. Low numbers of abnormal cells above that concentration were due to rapid death of cells after treatment, as indicated by cell numbers at 48 and 96 hours.

Studies of effects of complex waste effluents on survival and morphology of algae are important because they suggest mechanisms of action of the toxic factors. Death

Table 1 Effects of Dursban on survival and morphology of *S. costatum*.

	Before treatment			After treatment		
Dursban (ppm)	Average (Cells mm^{-3})	Dead (%)	Abnormal (%)	Average (Cells mm^{-3})	Dead (%)	Abnormal (%)
0	588	0·52	0	1806	0·54	0
0·1	604	0·58	0	1769	0·53	0·9
0·5	633	0·61	0	1482	0·96	5·9
1·0	600	0·61	0	1296	3·3	13·0
2·5	607	0·32	0	682	88	0·8
5·0	595	0·44	0	600	100	0

of cells is easily measured by use of mortal stains (Crippen and Perrier, 1974; Reynolds *et al.*, 1978), and the data for Dursban described above were obtained by use of the mortal stain Evans blue. Presumably, toxicants that kill algae (algicides) are more dangerous than those that only inhibit growth (algistats), but this may not be so if there is constant input of effluent into receiving water.

Fitzgerald and Faust (1963) recommended that toxicants, such as copper compounds, used to control algae should be studied for their algicidal and algistatic properties in order to evaluate optimal conditions for their use in swimming pools, ponds, and lakes. They suggested that algae whose growth was inhibited by toxicants be placed in fresh growth medium. If growth occurred, the toxicant was algistatic; if growth did not occur, it was algicidal. Payne and Hall (1979) gave a refined method of determining whether chemicals are algistatic or algicidal. The method works well, but is probably of little use with waste effluents because, as noted above, constant input of toxicants would cause continued reduced numbers unless the algae developed resistance.

A common response of algae to heavy metal and industrial waste pollution is change in cell size and morphology (Nuzzi, 1972; Azam *et al.*, 1973; Davies, 1974; Sunda and Guillard, 1976; Harrison *et al.*, 1977; Thomas *et al.*, 1980a). Falchuck *et al.* (1975a,b) described the role of zinc in cell size and morphology of *Euglena gracilis*. Zinc-deficient cells contained double the amount of DNA than did normal cells, and the authors concluded that zinc deficiency affects early mitotic events. Most studies report large and aberrant cells in the presence of above-normal concentrations of heavy metals. Davies (1976) suggested that large and aberrant cells are caused by the uncoupling of cell growth and cell division. Figure 1 shows aberrant cells of *S. costatum* produced by growth medium in which the metals were unchelated and by several industrial waste effluents. Abnormalities occurred in cell shape and internal structure. Elongated shape and the presence of indentations on the walls suggest that mitosis occurred but that cytokinesis was inhibited. When abnormal cells were placed in uncontaminated growth medium, only normal cells were seen after two days of growth (Walsh, unpublished). Figure 1 also shows

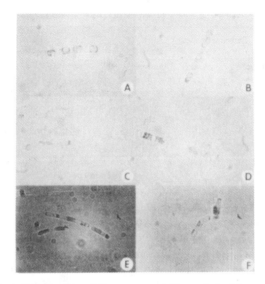

Fig. 1   Morphological abnormalities of *S. costatum* caused by unchelated heavy metals and industrial wastes. A: normal cells, B: cells from growth medium without EDTA, showing loss of chlorophyll, C: cells treated with waste from a textile mill, showing cell elongation and indentation of cell wall, D: cells treated with waste from a paint plant, showing round chloroplasts, E: cells treated with waste from a chemical plant, showing elongated cells with band-like chloroplasts, F: cells treated with waste from a textile mill, showing bending of elongated cells.

abnormal chloroplasts in exposed cells. Silverberg (1976) demonstrated ultrastructural changes in mitochondria of *Ankistrodesmus falcatus*, *Chlorella pyrenoidosa*, and *Scenedesmus quadricauda* grown in medium with cadmium chloride and concluded that the mitochondrion is the primary target for cadmium-associated toxicity.

Since heavy metals are often major constituents of complex industrial and energy process wastes, they may cause cellular abnormalities that affect the ability of cells to adjust to polluted conditions. Ibrahim (1970) found numerous aberrant cells living near outfalls of textile plants in Egypt. In tests with the mortal stain Evans blue, Walsh (unpublished) found that only apparently normal cells died in the presence of Dursban.

The gross morphological changes caused by complex wastes in the laboratory and field are possibly related to physiological and subcellular morphological adjustments to chemical stress. The small amount of data on the subject of adaptation of unicellular algae to chemical stress suggests that genetic selection need not be involved.

Choice of the end-points listed above depends upon requirements of each testing situation, and the strong and weak points of each test must be recognized so as not to draw unwarranted conclusions from the data.

V. Algal Bioassays with Whole Waste

Vocke *et al.* (1980) tested scrubber ash slurry from the settling pond of a coal-fired generating plant against *A. falcatus*, *S. obliquus*, *S. capricornutum* and *Microcoleus vaginatus*. The slurry was filtered through a 0·45 μm porosity filter, mixed with growth medium (EPA, 1971), and filter-sterilized through a 0·22 μm porosity filter. After two weeks of growth, population density was estimated by measurement of chlorophyll *a*. Calculated concentrations that would inhibit growth by 50% (EC_{50}) were: *A. falcatus*, 11·3%; *S. obliquus*, 8·8%; *S. capricornutum*, 15·4%; and *M. vaginatus*, 3·0%. The authors concluded that a standard species, such as *S. capricornutum*, the least responsive species, should be used only for the purpose of quality control and recommended use of several species in each test.

Walsh *et al.* (1980) tested effects of 23 textile mill wastes on growth of *S. capricornutum* and *S. costatum*. The freshwater species was tested with a modified algal assay bottle test (EPA, 1977) that recommended filtration and autoclaving or filtration only of the liquid waste. Growth medium for *S. costatum* was prepared by adding artificial sea salts and nutrients to raw waste. This medium was not filtered or sterilized so as not to change any of the dissolved or suspended substances. The method was described by Walsh and Alexander (1980).

All of the textile mill wastes affected growth of both species in one of three ways: stimulation, inhibition, or stimulation at low concentrations and inhibition at higher concentrations (Fig. 2). Autoclaving changed the bioactive properties of some wastes: stimulatory effects of two were reduced, inhibitory effects of three were reduced, and one was changed from highly inhibitory to highly stimulatory.

Walsh and Alexander (1980) showed that when two species of marine algae were grown together in the presence of textile mill waste, their relative numbers were changed drastically. *Skeletonema costatum* was sensitive to the waste at concentrations between 10 and 20%. *Porphyridium cruentum* did not respond to 20% waste. When grown together in untreated medium, cell number of *S. costatum* was approximately twice that of *P. cruentum*. In 20% waste, however, cell number of *P. cruentum* was approximately twice that of *S. costatum*, showing that the waste could potentially cause changes in relative sizes of the populations in algal communities in receiving waters.

Liquid waste effluents from industrial and energy process outfalls are complex mixtures of organic and inorganic substances, including heavy metals. Walsh *et al.* (1980), in a study of 23 effluents, detected no obvious relationship between bioactivity and colour, biological oxygen demand, chemical oxygen demand, or concentrations of sulphide, phenol, suspended solids, cyanide, organic carbon, total organics extracted with methylene chloride, metals, ammonia, nitrate, nitrite, Kjeldahl nitrogen, orthophosphate, or total phosphorus. They stated that bioactivity of a complex waste is probably related to interactions among the components, with no substance having a dominant effect. Growth stimulation was not correlated with high concentrations of nitrate or phosphate, and growth inhibition was not

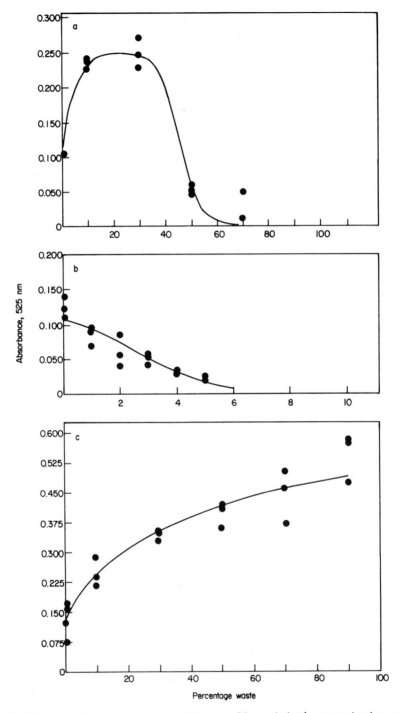

Fig. 2 Effects of textile wastes upon growth of *S. costatum*. SC_{20} = calculated concentration that would stimulate growth by 20%; EC_{50} = calculated concentration that would inhibit growth by 50% (from Walsh *et al.*, 1980). A, $SC_{20} = 1\cdot5\%$, $EC_{50} = 50\cdot0\%$. B, $EC_{50} = 2\cdot0\%$. C, $SC_{20} = 2\cdot3\%$.

correlated with concentrations of metals or organic substances. Some heavy metals were in very high concentrations in non-toxic wastes. It is probable that metals were chelated by the organic molecules or that antagonistic behaviour of some metals reduced their toxicity.

In the above tests (Walsh *et al.*, 1980), the procedure with *S. capricornutum* required incubation of the alga with wastes for 14 days, whereas the marine test required only four days. Only 35% of the wastes were toxic to *S. capricornutum*, and 71% were toxic to *S. costatum*. Walsh *et al.* (1982) tested *S. capricornutum* against ten unfiltered, unautoclaved industrial wastes and found that toxicity was lost in none by the ninth day of exposure, but was lost in nine of the ten after that time. Since toxicity and stimulation were apparent at four days, the authors recommended that tests be carried out for no more than four days. Eloranta and Laitinen (1981) also recommended a four-day incubation period for detection of algal growth stimulation by effluents from sulphite cellulose and paper factories.

Litton (1979a,b) reported the growth response of *S. capricornutum* to limestone scrubber samples and synthetic fuels. Figure 3b shows the response to extracts of scrubber filter cake. It is shown that there was an initial decrease in the number of cells on the first day in all but one treated culture. Growth of the control population was greater than that of all treated cultures on the fourth day of exposure, but the 11th day, there was little difference between treated and control cultures.

Figure 3a shows effects of synthetic fuel on growth of *S. capricornutum*. In this example, growth was inhibited by three concentrations and arrested by one. The three inhibited cultures recovered by the 16th day, when two of them were equal to

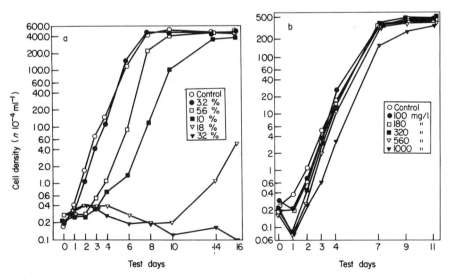

Fig. 3 Effects of energy process wastes on growth of *S. capricornutum*. 3A: extract from limestone scrubber cake. 3B: synthetic fuel.

the control in cell numbers. Growth rate of the third was rapid between days 10 and 16 and its curve was still rising on the 16th day.

These two examples are typical of the toxic response of *S. capricornutum* to liquid wastes: initial inhibition followed by recovery. Litton (1979a) concluded that the limestone scrubber filter cake extract was not toxic to algae despite the fact that growth was suppressed strongly by all concentrations in the early part of the test. The same is true for the synthetic fuel, where growth of algae exposed to 5·6% fuel was strongly suppressed. Litton concluded that growth was impaired only at 18% waste because only data from the 16-day exposure were used.

We believe these data show clearly the inaccuracies inherent in long-term tests with algae. The presence of growth inhibitors was demonstrated within four days in both tests. Algal assays are tests for growth inhibitors and growth stimulators, and they cannot address the problem of fate of toxicants in flasks. It was noted above that toxicants may degrade, adsorb to exposure vessel walls, attain non-toxic biomass to toxicant concentration ratios, etc. Algal toxicity tests cannot address these questions, which can be answered only by studies designed specifically to study fate. Long-term tests often do not identify toxic wastes and are so complicated by considerations of toxicant fate that they are nearly worthless.

Data from algal toxicity tests should be expressed in such a way as to allow comparison of responses to waste streams. For expression of growth stimulation, Walsh *et al.* (1980) recommended the term SC_{20} for the calculated concentration that would stimulate growth by 20%. In their tests, the mean upper 95% confidence interval for algal growth in control cultures was 17% above the control mean. Therefore, growth stimulation greater than 17% above control was a significant response, and 20% above control was a convenient value to calculate. The SC_{20} may be calculated by straight-line graphical interpolation (APHA, 1975) or preferably by more sophisticated methods, such as moving average (Stephan, 1977), which allow calculation of 95% confidence intervals.

The EC_{50} which is the calculated concentration that would inhibit population growth by 50%, is a standard expression that has been used for many years. It may be calculated by the same methods used for the SC_{20}, and 95% confidence intervals should be reported.

Payne and Hall (1979) recommended that the algistatic concentration that inhibits growth by 95% (AC_{95}) should be used to express effect. This value is based upon the assumption, that, if an algal population in the field is reduced by a waste to 5% of its density before exposure, it will recover to its original density after some unspecified period of time. We are not familiar with field data that support this assumption. Recovery of populations in the field is dependent upon many things, including grazing rates of herbivores, effects of wastes on herbivores and carnivores that eat them, nutrient content of the wastes and receiving water, wastes from other outfalls in the area, competitive ability of all algae in the immediate vicinity, and temperature. Unless recruitment occurs, populations reduced to small numbers tend to disappear.

It cannot be assumed that algae in the field are subjected to a single evanescent exposure to complex wastes as occurs in laboratory algistatic tests. It is most likely that they are exposed to wastes over protracted periods of time under varying environmental conditions. Given exposure over relatively long periods of time under field conditions, it is likely that algal populations and communities adapt either physiologically or genetically to toxicants. Neither the EC_{50} nor the AC_{95} addresses this aspect of algal population growth dynamics, and it may be an important omission of both approaches.

The AC_{95} concept is invalid in most tests because of statistical considerations. It is impossible to attain an accuracy of 5% in algal tests because of variability of population growth rates in cultures treated with complex wastes: 95% growth inhibition cannot be distinguished statistically from 90% or 100% inhibition. The method of Payne and Hall (1979) was described with single organic pollutants. Application of the AC_{95} to industrial waste toxicity is a misuse of the method.

It is sometimes argued (APHA, 1975) that EC_{50} values should not be used as an expression of effects of toxicants on algal growth because such data are quantitative, i.e. they inhibit population growth, and the number of organisms at the end of the test is greater than at the beginning. This is in contrast to quantal data obtained from tests in which organisms are killed, resulting in fewer organisms at the end of the test than at the beginning. It is argued that statistical evaluations, such as Duncan's Multiple Range Test or Dunnett's Test, should be used to identify significant differences between growth rates in concentrations of toxicants within a test. However, Hewlett and Plackett (1979) pointed out that quantal and quantitative changes are very closely related because "An individual organism responds quantally if an underlying quantitative change, that is, a gradual response, reaches a certain level of intensity characteristic of that individual organism." They then presented a hypothesis that unified the mathematical treatments of the two types of responses.

VI. Algal Bioassays with Waste Fractions

Algal bioassays with whole waste samples give an estimation of the potential hazard of such wastes to aquatic systems, but they do not aid in identification of the toxic or stimulatory substances. Walsh and Garnas (1979) described a method whereby complex liquid wastes may be fractionated chemically in order to identify the classes of toxicants and stimulants in them (Fig. 4). Fractionation and sub-fractionation were linked closely to bioassays in order to identify possible bioactive substances.

Table 2 shows growth responses of *S. costatum* to wastes and their fractions from an oil refinery and a sewage treatment plant. Toxicity and stimulatory ability of the petroleum plant resided in the organic fraction. The neutral subfraction was toxic and the acid subfraction was stimulatory. Analysis of sewage treatment plant waste

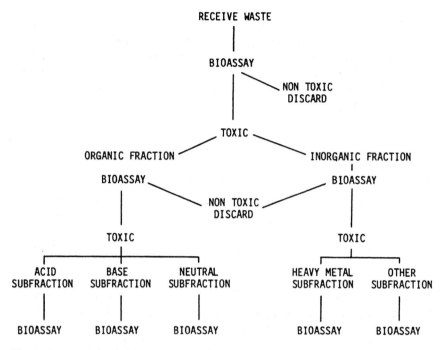

Fig. 4 Integrated chemical fractionation and biological toxicity tests for analysis of complex liquid wastes (Walsh and Garnas, 1979).

Table 2 Effects of whole wastes, fractions, and subfractions of an oil refinery and a sewage treatment plant on growth of *S. costatum*.

OIL REFINERY
  Whole waste: $EC_{50} = 45 \cdot 3\%$; $SC_{20} = 7 \cdot 5\%$
    Inorganic fraction: no effect
    Organic fraction: $EC_{50} = 43 \cdot 5\%$; $SC_{20} = 8 \cdot 1\%$
      Acid subfraction: $SC_{20} = 8 \cdot 6\%$
      Neutral subfraction: $EC_{50} = 84 \cdot 4\%$
      Base subfraction: no effect

SEWAGE TREATMENT PLANT
  Whole waste: $EC_{50} = 15 \cdot 4\%$; no SC_{20}
    Organic fraction: no effect
    Inorganic fraction: $EC_{50} = 15 \cdot 5\%$; $SC_{20} = 0 \cdot 4\%$
      Cation subfraction: $EC_{50} = 16 \cdot 5\%$
      Anion subfraction: $SC_{20} = 0 \cdot 9\%$

and its fractions showed that growth stimulants may be present but remain undetected in growth tests with whole waste. The EC_{50} of the whole waste was 15·4%, but no stimulation of growth was detected. The organic fraction has no effect on growth. However, the inorganic fraction had an EC_{50} of 15·5% and an SC_{20} of 0·4%. When subfractionated, toxicity was in the cation (heavy metals) subfraction and growth stimulation in the anion subfraction.

VII. Algal Tests Compared to Other Toxicity Tests

Algal data are seldom used for setting water quality standards because algae usually are less sensitive than invertebrates or fishes to individual organic pollutants. Recently, however, it has become clear that algae are usually much more sensitive to complex liquid effluents than animals and other organisms. Table 3 shows results of tests on leachate prepared by two methods from regenerator cyclone and second cyclone wastes of a fluidized bed combustion plant in health effects tests (Ames bacterial mutagenicity and cytotoxicity tests; chinese hamster ovary toxicity test) and ecological tests (*Daphnia magna* and *S. capricornutum*) (Eischen and Duke, 1980). It is shown that the leachates were more toxic to *S. capriconutum* than *D. magna* in three of the four samples. The Ames mutagenicity and Chinese hamster ovary tests did not detect toxicity, and the Ames cytotoxicity test detected toxicity in two of the four samples.

Litton (1979b) found *S. capricornutum*, *D. magna*, and *Pimephales promelus* (fathead minnow) to be more sensitive than bacteria to nine samples from a synthetic fuels production process in the Ames mutagenicity, Chinese hamster

Table 3 Effects of solid waste leachate from regenerator and second cyclone samples of a fluidized bed combustion plant (Eischen and Duke, 1980).

Sample/Test	Leachate method 1 (%)	Leachate method 2 (%)
REGENERATOR CYCLONE		
D. magna, LC_{50}	15·0	11·0
S. capricornutum, EC_{50}	4·6	0·6
Ames mutagenicity	Negative	Negative
Ames cytotoxicity	Negative	Toxic
Chinese hamster ovary	Not done	Negative
SECOND CYCLONE		
D. magna, LC_{50}	5·5	67·4
S. capricornutum, EC_{50}	6·3	0·4
Ames mutagenicity	Negative	Negative
Ames cytotoxicity	Negative	Toxic
Chinese hamster ovary	Not done	Negative

ovary, and rodent (rat) toxicity tests (Table 4). All wastes were highly toxic in all of the algal tests, but not in all of the others.

In other tests, Walsh and Alexander (1980) and Walsh *et al.* (1982) compared toxicity of numerous liquid wastes to freshwater and marine algae and to freshwater and marine crustaceans and fishes. Acute (96-hour) tests were done on 14 textile effluents and *S. capricornutum*, *Daphnia pulex*, and *P. promelus* of freshwater and *S. costatum*, *Palaemonetes pugio* (grass shrimp), and *Cyprinodon variegatus* (sheepshead minnow) of saltwater. Fishes were least sensitive to the wastes. *Skeletonema costatum* was the most sensitive species to 12 of the wastes, and *D. pulex* was most sensitive to two of them. However, 96-hour EC_{50}s and SC_{20}s of the algae were always lower than the EC_{50}s of *D. pulex*. Walsh *et al.* (1982) reported that wastes from a variety of industrial plant types were always more bioactive towards *S. capricornutum* or *S. costatum* than towards *D. magna* and *Mysidopsis bahia* (bay mysid). Of ten wastes tested, all affected the algae, but only three affected survival of *M. bahia* and two affected survival of *D. magna*. As examples of relative effect, a waste from a synthetic organic fibres plant had the effects shown in Table 5. The data are typical of comparative studies, and when a waste is stimulatory, the SC_{20} is usually much lower than the LC_{50} (calculated concentration that would kill 50% of exposed animals).

Walsh and Garnas (1979) reported on subfractions of nine industrial wastes that were toxic to both *S. costatum* and *M. bahia* (Table 6; Fig. 5). Toxicity due to heavy metals (cation subfraction) and neutral and acid extractable organics is shown. There was a small loss of toxicity in the fractionation step, but there was clear identification of toxic fractions and subfractions. Subfractions of the organic fraction of the paper products plant were recombined and the original toxicity was obtained.

All of the data we have gathered show that algae are more sensitive than animals to complex wastes in laboratory tests. Algal tests should be used for evaluation of pollution potential for such wastes because aquatic community structure is based upon the types of primary procedures present.

VIII. Practical Use of Algal Data

Algae have been shown to be sensitive to complex industrial and energy process wastes. They are generally better indicators of potential pollution than animals because they respond to both toxicants and growth stimulators. Tests on whole wastes allow estimation of possible effects in relation to volume of discharge (Walsh *et al.*, 1980). The expression:

$$\text{MEU d}^{-1} = \left(\frac{100}{\text{SC}_{20} \text{ or EC}_{50}} \right) (\text{discharge rate})$$

(where MEU = million effective units) may be used to estimate relative potential impact of a whole waste on a receiving stream. For example, a highly toxic or

Table 4 Effects of nine samples from a synthetic fuels production process in the Ames mutagenicity, Chinese hamster ovary, and rodent toxicity tests, on growth of *S. capricornutum*, and on survival of *D. magna* and *P. promelus*.[a]

Sample no.	Ames	Chinese hamster	Rodent toxicity	*P. promelus*	*D. magna*	*S. capricornutum*
1	NT	NT	NT	H	H	H
2	NT	H	NT	H	H	H
3	L	H	NT	H	H	H
4	M	—	—	H	H	H
5	M	H	NT	H	H	H
6	—	—	—	H	H	H
7	NT	NT	NT	M	H	H
8	—	—	—	M	H	H
9	NT	NT	NT	L	L	H

[a] H = high toxicity; M = moderate toxicity; L = low toxicity; NT = toxicity not detected.

Table 5 Effects of waste from a synthetic organic fibres plant and a textile mill on growth of *S. capricornutum* and *S. costatum* and on survival of *D. magna* and *M. bahia*.[a]

	S. costatum		*S. capricornutum*		*D. magna* LD_{50}	*M. bahia* LD_{50}
	SC_{20}	EC_{50}	SC_{20}	SC_{50}		
Synthetic organic fibres	1·0	21·2	2·0	45·6	44·9	NT
Textile mill	0·5	T	0·01	NT	NT	77·5

[a] T = toxicant present but not quantifiable; NT = toxicity not detected.

Table 6 Subfractions of industrial wastes toxic to *S. costatum* and *M. bahia* (Walsh and Garnas, 1979)

Industry	Toxic subfraction
Titanium oxide	Heavy metals
Oil refinery	Neutral organic
Tall oil products	Neutral organic
Phosphoric products	Heavy metals
Carpeting	Heavy metals
Amine products	Heavy metals
Textiles finishing plant	Heavy metals
Paper products	Neutral and acid extracted organics
Sewage treatment plant	Heavy metals

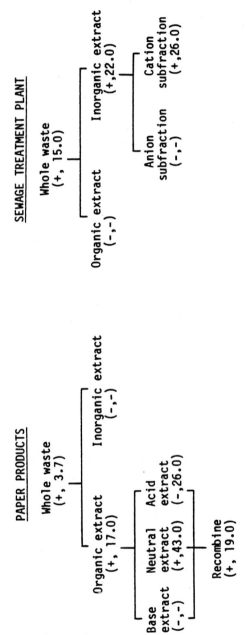

Fig. 5 Effects of industrial wastes and their organic and inorganic fractions and subfractions on growth of *S. costatum* and survival of *M. bahia*. + = toxic to *M. bahia*; — = non toxic to *M. bahia*; numbers are the EC_{50}s (%) for algal toxicity.

stimulatory waste emitted in large volume is more likely to have an adverse effect on aquatic systems than a weakly toxic waste emitted in low volume. Walsh *et al.* (1980) used this expression to rank textile mills according to their potential danger in screening tests (Table 7).

SC_{20} and EC_{50} values may also be used to calculate the dilution required for each waste to be rendered non-toxic or non-stimulatory. Walsh *et al.* (1982) reported their use in calculation of the instream waste concentration (IWC), which is an estimate of the amount of dilution that an effluent may be expected to undergo during continuous release into a receiving stream:

$$IWC(\%) = \frac{Qw}{Qr + Qw} \times 100\%$$

where Qw = volume of discharge, and Qr = the seven-day, ten-year low flow volume of the receiving water. The seven-day, ten-year low flow volume (Qr) required for safety can be calculated from the expression:

$$0.01 \times EC_{50} \text{ or } SC_{20} = \frac{Qw}{Qr + Qw} \times 100$$

where 0.01 = a safety factor currently used by EPA in IWC calculations for issuance of discharge permits.

Chemical fractionation tests with algae identify requirements for waste treatment. When the chemical nature of toxicants and growth stimulators is known, treatment systems that address specific problems can be devised and the considerable costs associated with treatment of whole waste can be reduced.

Table 7 Ranking of textile mill effluents according to relative impact on receiving water.[a]

Waste	Discharge $(m^3 \, d^{-1} \, 10^{-3})$	Response	Waste $(\%)$	MEU $d^{-1} \, 10^{-3}$
1	4·5	EC_{50}	9·0	50
2	1·0	SC_{20}	1·5	67
3	3·8	EC_{50}	2·0	190
4	9·5	SC_{20}	1·0	950
5	9·8	SC_{20}	0·5	1960

[a] Test organism = *Skeletonema costatum*, MEU = million effective units (Walsh *et al.*, 1980).

IX. Future Research

(1) The purpose of algal bioassays is to detect toxicity, and this is best done in tests conducted for between 48 and 96 hours after exposure. After that time,

interpretation of results is confounded by questions of toxicant fate, and such questions cannot be addressed in routine bioassays.

(2) Relatively few algal species have been screened for use in toxicity tests and those that are used are often chosen for convenience of culture rather than sensitivity to pollutants. There is a need for screening studies to search for new, more sensitive species that respond to both growth inhibitors and growth stimulators.

(3) Use of indigenous species in algal tests may not be protective of the environment because pollutants are transferred easily to other areas. Use of sensitive species is much more important than use of indigenous species.

(4) Algal cultures should be examined for number of dead cells to determine whether the toxicant exerts an algistatic or an algicidal effect. Also, exposed cells should be examined for morphological deformities.

(5) Because of the highly complex nature of industrial and energy-process effluents, effects upon algae cannot be predicted from chemical composition. It is suggested that effect is the result of additive, synergistic, and antagonistic behaviour of the chemicals in relation to physical properties of an effluent.

(6) Liquid effluents may be fractionated chemically to determine the nature of growth inhibitors and growth stimulators. Effects on growth may be overlooked if fractions are not tested.

(7) Algae are more sensitive to complex wastes than animals and should always be used for evaluation of potential hazard.

X. Conclusions

This review presents laboratory data and their interpretation with regard to effects of pollutants on marine and freshwater unicellular algae. Stimulation and inhibition of growth are considered to be equally undesirable. It is suggested that a search for new species for use in toxicity tests be made because the widely-used green alga, *Selenastrum capricornutum*, may not be one of the most sensitive freshwater species. Several algal species should be used to test each toxicant because there is not a single most-sensitive species and the conditions under which species are tested affects response. The batch method is the most commonly used exposure system for defining relative effects of pollutants, but mixed-species continuous culture presents a more realistic approach to estimation of effects in natural systems. Algae are more sensitive to industrial and energy-process wastes than animals. They respond to growth stimulants and growth inhibitors, and the stimulation response occurs at concentrations much lower than those that inhibit growth.

Note

This chapter is contribution No. 427 from the Gulf Breeze Laboratory.

References

APHA, AWWA, WPCF (1975). "Standard Methods for the Examination of Water and Wastewater" (14th ed.). American Public Health Association, Washington, DC. 1193 pp.

Azam, F., B. B. Hemmingsen and B. E. Volcani (1973). *Arch. Mikrobiol.* **92**, 11–20.

Banse, K. (1976). *J. Phycol.* **12**, 135–140.

Berland, B. R., D. J. Bonin, O. J. Guerin-Ancey, V. I. Kapove and D. P. Arlhac (1977). *Mar. Biol.* **42**, 17–30.

Booth, B. C. (1975). *Limnol. Oceanogr.* **20**, 865–869.

Brack, G. S., A. Jensen and A. Mohus (1976). *J. Exp. Mar. Biol. Ecol.* **25**, 37–50.

Cain, J. R. and F. R. Trainor (1973). *Phycologia* **12**, 227–232.

Cain, J. R., R. L. Klotz, F. R. Trainor and R. Costello (1979). *Environ. Pollut.* **19**, 215–224.

Cairns, J., Jr, A. G. Heath and B. C. Parker (1975). *Hydrobiologia* **47**, 135–171.

Canterford, G. S. (1979). *Aust. J. Mar. Freshwater Res.* **30**, 765–772.

Chisholm, S. W. and J. C. Costello (1980). *J. Phycol.* **16**, 375–383.

Chui, S. Y., I. C. Hoa, L. E. Erikson and F. T. Fan (1973). *J. Wat. Pollut. Control Fed.* **45**, 1746–1758.

Claesson, A. (1975). *Vatten* **4**, 333–338.

Clement-Metral, J. D. and M. LeFort-Tran (1974). *Biochim. Biophys. Acta* **333**, 560–569.

Crippen, R. W. and J. L. Perrier (1974). *Stain Technol.* **49**, 97–104.

Davies, A. G. (1974). *J. Mar. Biol. Assoc. U.K.* **54**, 157–169.

Davies, A. G. (1976). *J. Mar. Biol. Assoc. U.K.* **56**, 39–57.

Davis, C. O., P. J. Harrison and R. C. Dugdale (1973). *J. Phycol.* **9** 175–180.

DeFilippis, L. F. and Pallaghy (1976). *Z. Pflanzenphysiol.* **78**, 197–207.

Dor, I. (1975). *Water Res.* **9**, 251–254.

Dunstan, W. M. (1975). *Environ. Sci. Technol.* **9**, 635–638.

Eischen, M. A. and K. M. Duke (1980). Final Report on Comparison of Two Leachate Preparation Methods Using Selected Level 1 Bioassays. Contract No. 68-02-2686, Directive No. 120. Report from Battelle Columbus Laboratories to the EPA, Industrial Environmental Research Laboratory, Research Triangle Park, NC. 25 pp.

Eloranta, V. and O. Laitinen (1981). *Verh. Int. Verein. Limnol.* **21**, 738–743.

Eppley, R. W., J. N. Rogers, J. J. McCarthy and A. Sournia (1971). *J. Phycol.* **7**, 150–154.

Erickson, S. J. and C. E. Hawkins (1980). *Bull. Environ. Contam. Toxicol.* **24**, 910–915.

Falchuk, K. H., D. W. Fawcett and B. L. Vallee (1975a). *J. Submicrosc. Cytol.* **7**, 139–152.

Falchuk, K. H., D. W. Fawcett and B. L. Vallee (1975b). *J. Cell Sci.* **17**, 75–78.

Falkowski, P. G. (1977). *J. Exp. Mar. Biol. Ecol.* **27**, 37–45.

Ferguson, R. L., A. Collier and D. A. Meeter (1976). *Chesapeake Sci.* **17**, 148–158.

Fielding, A. H. and G. Russell (1976). *J. Appl. Ecol.* **13**, 871–876.

Fisher, N. S. (1977). *Am. Nat.* 111, 871–895.

Fisher, N. S., E. J. Carpenter, C. C. Remsen and C. F. Wurster (1974). *Microb. Ecol.* 1, 39–50.

Fisher, N. S., L. B. Graham and E. J. Carpenter (1973). *Nature* 241, 548–549.

Fisher, N. S. and C. F. Wurster (1973). *Environ. Pollut.* 5, 205–212.

Fitzgerald, G. P. and S. L. Faust (1963). *Appl. Microbiol.* 11, 345–351.

Gamulin-Brida, H., G. Giaconne and S. Golubic (1967). *Helgol. Wiss. Meeresunters.* 15, 429–444.

Goldman, J. C. (1979). *Microb. Ecol.* 5, 153–166.

Goldman, J. C. and J. H. Ryther (1976). *Biotechnol. Bioeng.* 18, 1125–1144.

Goldman, J. C. and H. I. Stanley (1974). *Mar. Biol.* 28, 17–25.

Greene, J. C., W. E. Miller, T. Shiroyama and T. E. Maloney (1975). *Water Air Soil Pollut.* 4, 415–434.

Guillard, R. R. L. and J. H. Ryther (1962). *Can. J. Microbiol.* 8, 229–239.

Harjula, H. (1979). *Ann. Bot. Fenn.* 16, 307–337.

Harris, G. P. (1978). *Arch. Hydrobiol.* 10, 1–171.

Harrison, P. J., H. L. Conway, R. W. Holmes and C. O. Davis (1977). *Mar. Biol.* 43, 16–27.

Harrison, P. J., R. E. Waters and F. J. R. Taylor (1980). *J. Phycol.* 16, 28–35.

Harrison, W. G., R. W. Eppley and E. H. Renger (1977). *Bull. Mar. Sci.* 27, 44–57.

Hewlett, P. S. and R. L. Plackett (1979). "The Interpretation of Quantal Responses in Biology." University Park Press, Baltimore. 82 pp.

Hollister, T. A. and G. E. Walsh (1973). *Bull. Environ. Contam. Toxicol.* 9, 291–295.

Holm-Hansen, O. (1978). *Oikos* 30, 438–447.

Holm-Hansen, O. and C. R. Booth (1966). *Limnol. Oceanogr.* 11, 510–519.

Holm-Hansen, O. and H. W. Pearl (1972). *Mem. Inst. Idrbiol. (suppl.)* 29, 149–168.

Ibrahim, A. M. (1970). Effect of Industrial By-products on *Chromulina pascheri*. M.Sc. thesis, Cairo University, Egypt. 139 pp.

Jeffrey, S. W. and G. F. Humphrey (1975). *Biochem. Physiol. Pflanz. (BPP)* 167, 191–194.

Jensen, A., B. Rystad and L. Skoglund (1972). *J. Exp. Mar. Biol. Ecol.* 8, 241–248.

Jensen, A. and B. Rystad (1973). *J. Exp. Mar. Biol. Ecol.* 11, 275–285.

Jensen, A., B. Rystad and S. Melsom (1974). *J. Exp. Mar. Biol. Ecol.* 15, 145–157.

Jensen, A., B. Rystad and S. Melsom (1976). *J. Exp. Mar. Biol. Ecol.* 22, 249–256.

Joubert, G. (1980). *Water Res.* 14, 1759–1763.

Kennicutt, M. C., II (1980). *Water Res.* 14, 325–328.

Levin, G. V., J. R. Schrot and W. C. Hess (1975). *Environ. Sci. Technol.* 9, 961–965.

Lewis, W. M., Jr (1977). *Ecology* 58, 850–859.

Litton Bionetics (1979a). Final Report: Level 1 Bioassays on Nine conventional Limestone Scrubbing Samples With and Without Adipic Acid Additive. EPA Contract No. 68-02-2681, Technical Directive No. 105. Industrial Environmental Research Laboratory, Research Triangle Park, North Carolina. 144 pp.

Litton Bionetics (1979b). Final Report: Level 1 Bioassays on Eight Synthetic Fuel Samples. EPA Contract No. 68-02-2681, Technical Directive No. 107. Industrial Environmental Research Laboratory, Research Triangle Park, North Carolina. 150 pp.

Ohkawa H., R. Kikuchi and J. Miyamoto (1980). *J. Pestic. Sci.* 5, 11–22.

Nuzzi, R. (1972). *Nature* 237, 38–39.

Miller, W. E., J. C. Greene and T. Shiroyama (1978). The *Selenastrum capricornutum* Printz Algal Assay Bottle Test: Experimental Design, Applications, and Data Interpretation

Protocol. US EPA, Environmental Research Laboratory, Corvallis, Oregon. EPA-60019-78-018. 126 pp.

Mosser, J. L., N. S. Fisher and C. F. Wurster (1972). *Science* **196**, 533–535.

Nakamura, Y. and S. Miyachi (1980). *Plant and Cell Physiol.* **21**, 765–774.

Overnell, J. (1975). *Pestic. Biochem. Physiol.* **5**, 19–26.

Paasche, E. (1971). *Physiol. Plant.* **25**, 294–299.

Palmer, C. M. (1969). *J. Phycol.* **5**, 78–82.

Patrick, R., J. Cairns, Jr and A. Scheier (1968). *Prog. Fish-Cult.* **30**, 137–140.

Patterson, J. W., P. L. Brezonik and N. D. Putnam (1970). *Environ. Sci. Technol.* **4**, 569–575.

Pavlou, S. P. (1972). Phytoplankton Growth Dynamics: Chemostat Methodology and Chemical Analyses. Technical Series 1, Special Report No. 52. Department of Oceanography, University of Washington, Seattle. 130 pp.

Payne, A. G. and R. H. Hall (1979). A method for measuring algal toxicity and its application to the safety assessment of new chemicals. *In* "Aquatic Toxicology" (L. L. Marking and R. N. Kimerle, eds), pp. 171–180. American Society for Testing and Materials, Philadelphia. ASTM 667.

Powers, C. D., R. G. Rowland and C. F. Wurster (1976). *Water Res.* **10**, 991–994.

Prakash, A., L. Skoglund, B. Rystad and A. Jensen (1973). *J. Fish. Res. Bd Canada* **30**, 143–155.

Rainville, R. P., B. J. Copeland and W. T. McKean (1975). *J. Wat. Pollut. Control Fed.* **47**, 487–503.

Reynolds, A. E., G. B. Mackiernan and S. D. Van Valkenburg (1978). *Estuaries* **1**, 192–196.

Reynolds, J. H., E. J. Middlebrooks, D. B. Porcella and W. J. Grenney (1975). *J. Wat. Pollut. Control Fed.* **47**, 2420–2436.

Rodhe W. (1978). *Mitt. Int. Ver. Limnol.* **21**, 7–20.

Sakshaug, E. and A. Jensen (1978). *Oceanogr. Mar. Biol. Annu. Rev.* **16**, 81–106.

Sanders, J. G., J. H. Ryther and J. H. Batchelder (1981). *J. Exp. Mar. Biol. Ecol.* **49**, 81–102.

Sargent, D. F. and C. P. S. Taylor (1972). *Can. J. Bot.* **50**, 905–907.

Schultz, J. S. and P. Gerhardt (1969). *Bacteriol. Rev.* **33**, 1–47.

Sears, J. R. (1967). *J. Phycol.* **3**, 136–139.

Sharp, J. H. (1977). *Limnol. Oceanogr.* **22**, 381–399.

Sharp, J. H., P. A. Underhill and A. C. Frake (1980). *J. Plankton Res.* **2**, 213–222.

Shiroyama T., W. E. Miller and J. C. Greene (1975). Comparison of the algal growth responses of *Selenastrum capricornutum* Printz and *Anabaena flos-aquae* (Lyngb.) DeBrebisson in waters collected from Shagawa Lake, Minnesota. *In* Proceedings: Biostimulation and Nutrient Assessment Workshop, University of Utah, Logan. Sept. 10–12, 1975. PRWG 168-1. pp. 127–147.

Shiroyama, T., W. E. Miller, J. C. Greene and C. Shigihara (1976). Growth response of *Anabaena flos-aquae* (Lyngb.) DeBrebisson in waters collected from Long Lake Reservoir, Washington. *In* Proceedings of the Symposium on Terrestrial and Aquatic Ecological Studies of the Northwest. EWSC Press, Eastern Washington State College, Cheney, Washington. pp. 267–275.

Silverberg, B. A. (1976). *Phycologia* **15**, 155–159.

Slobodkin, L. B. and H. L. Sanders (1969). *In* Brookhaven Symp. Biol. **22**, 82.

Standyk, L., R. S. Campbell and B. T. Johnson (1971). *Bull. Environ. Contam. Toxicol.* **6**, 1–8.

Starr, R. C. (1978). *J. Phycol.* **14** (suppl.), 47–100.

Stein, J. R. (1973). "Handbook of Phycological Methods: Culture Methods and Growth Measurements." Cambridge University Press, New York. 448 pp.

Stephan, D. E. (1977). Methods for calculating an LC_{50}. In "Aquatic Toxicology and Hazard Evaluation" (F. L. Mayer and J. L. Hamelink, eds), pp. 65–85. American Society for Testing and Materials, Philadelphia. ASTM 634.

Stratton, G. W., R. E. Burrell, M. L. Kurp and C. T. Corke (1980). *Bull. Environ. Contam. Toxicol.* **24**, 562–569.

Strickland, J. D. H. (1960). Measuring the Production of Marine Phytoplankton. Fish. Res. Board Canada, Bull. No. 122. Ottawa. 172 pp.

Styron, C. E. *et al.* (1976). *J. Mar. Biol. Assoc. U.K.* **56**, 13–20.

Sunda, W. and R. R. L. Guillard (1976). *J. Mar. Res.* **34**, 511–529.

Terlizzi, D. E., Jr and E. P. Karlander (1980). *J. Phycol.* **16**, 364–368.

Thomas, W. H., J. T. Hollibaugh, and D. L. R. Seibert (1980a). *Phycologia* **19**, 202–209.

Thomas, W. H., J. Hastings and M. Fujita (1980b). *Mar. Environ. Res.* **3**, 291–296.

Thomas, W. H., J. T. Hollibaugh, D. L. R. Seibert and G. T. Wallace, Jr (1980c). *Mar. Ecol. Prog. Ser.* **2**, 213–220.

Thomas, W. H., M. Pollock and D. L. R. Seibert (1980d). *J. Exp. Mar. Biol. Ecol.* **45**, 25–36.

Tilman, D. (1977). *Ecology* **58**, 338–348.

Titman, D. (1976). *Science* **192**, 463–465.

Trainor, F. R. (1965). *Can. J. Bot.* **43**, 701–706.

Turner, M. F. (1979). *J. Mar. Biol. Assoc. U.K.* **59**, 535–552.

US EPA (1971). Algal Assay Procedure: Bottle Test. US EPA, National Eutrophication Research Program, Corvallis, Oregon. 82 pp.

US EPA (1977). IERL/RTP Procedures Manual: Level 1 Environmental Assessment. Industrial Environmental Research Laboratory, Research Triangle Park, North Carolina. EPA-600/7-77-043. 106 pp.

US EPA (1978). Source Assessment: Textile Plant Wastewater Toxics Study, Phase 1. Industrial Environmental Research Laboratory, Research Triangle Park, North Carolina, Contract No. 68-02-1874. EPA-600/2-78-004h. 152 pp.

Vocke, R. W., K. L. Sears, J. J. O'Toole and R. B. Wildman (1980). *Water Res.* **14**, 141–150.

Walsh, G. E. (1977). *Chesapeake Sci.* **18**, 222–223.

Walsh, G. E. and S. V. Alexander (1980). *Water Air and Soil Pollut.* **13**, 45–55.

Walsh, G. E., K. M. Duke and R. B. Foster (1982). *Water Res.* **16**, 879–883.

Walsh, G. E. and R. L. Garnas (1979). Effects of liquid industrial wastes on estuarine algae, plants, crustaceans, and fishes. Proceedings of the Second US/USSR Symposium on Effects of Pollutants on Marine Organisms. Terskol, USSR.

Walsh, G. E., L. H. Bahner and W. B. Horning (1980). *Environ. Pollut. (Ser. A)* **21**, 169–179.

Williams, P. J. LeB. and C. S. Yentsch (1976). *Mar. Biol.* **35**, 31–40.

Williamson, C. E. (1980). *J. Exp. Mar. Biol. Ecol.* **43**, 271–279.

Wyatt, J. T. (1973). *Assoc. South. Biol. Bull.* **20**, 93.

Zuliei, N. and G. Beneche (1978). *Bull. Environ. Contamin. Toxicol.* **20**, 786–792.

Part Six

Modelling

13 Synthetic Microcosms as Biological Models of Algal Communities

F. B. Taub

School of Fisheries WH–10
College of Ocean and Fisheries Science
University of Washington
Seattle, Washington, USA

I. Introduction

Algal communities are too important to be ignored and yet too complex to be understood within complete aquatic ecosystems. Algae influence and are influenced by most aquatic processes. It is therefore not surprising that algae have been studied on many levels of organization. Within a single water body, so many processes are acting simultaneously that changes in the algal community can only be correlated with changes in other variables, and cause and effect can rarely be ascertained. Multiple enclosures of natural communities such as Controlled Ecosystem Pollution Experiment (CEPEX) and Marine Ecosystem Research Laboratory (MERL) allow different treatments to be made on subsamples of the same natural community, retaining small and medium-sized processes, but excluding large-scale processes (e.g. Thomas and Seibert, 1977; Elmgren *et al.*, 1980). Removal of samples from the natural environment to the laboratory allows greater physical control and often larger numbers of replicates, but with the loss of more processes and the risk that the community will lose some portion of its natural species assemblages. Nevertheless, such isolated portions of natural systems show many of the same ecological properties of their natural counterpart (Beyers, 1963). Many researchers have gone to unialgal cultures and to axenic (pure) cultures to further isolate processes for more detailed study. The study of purified compounds, such as chlorophyll, approach the ultimate in isolation of processes, although such studies can hardly be called simple.

An alternative to isolating processes is to combine those we believe to be important and study the behaviour of the resultant system. This can be done by mathematical models, or by biological models, that is, micrococosms. Microcosm research has been an active field, as the symposium, *Microcosms in Ecological Research* amply demonstrates (Giesy, 1980). An evaluation of microcosms can be found in *Microcosms as Potential Screening Tools for Evaluating Transport and Effects of Toxic Substances* (Oak Ridge National Laboratories and EPA, 1980). The early development of microcosm research was reviewed by Cooke (1977).

The ultimately defined microcosm not only has completely defined chemical and physical properties but is also gnotobiotic, that is, composed of known species. The feasibility of establishing gnotobiotic microcosms has been demonstrated. Nixon (1969) isolated several bacteria, an alga, and brine shrimp from a saline community, and synthesized microcosms by introducing these components into enriched seawater. Relatively stable populations of the bacteria and alga were established, but the *Artemia* populations did not persist. Taub (1969a) prepared gnotobiotic microcosms by inoculating different combinations of several bacteria, one alga, a protozoan, and a rotifer (all from standard laboratory stocks) in chemically defined medium. All of the organisms were sustained, with the exception of some of the bacteria. The presence of rotifers or protozoa delayed the maximal algal density from being reached, and the presence of the rotifers permanently decreased the protozoa densities. Continuous cultures of the alga, bacteria, and protozoa were established, and their responses to changes in light intensity and dilution rate were studied (Taub, 1969b). Gnotobiotic microcosms, while feasible, are very labour intensive, and apart from continuous cultures of bacteria combined with protozoa, have not been utilized extensively as investigative ecological tools.

Cultures of algae, grazers and fish have been used for foodchain studies. For example, Reinert (1972) studied the accumulation of dieldrin by the alga *Scenedesmus obliquus*, and sequentually by the grazer *Daphnia magna*, and the predator *Poecilia* (guppy). Although the study fulfilled its purpose in following the pathways of dieldrin accumulation, it did not yield any information on the ecological interactions of these organisms, or on modifications caused by the toxicant, because each trophic level was studied separately.

Simultaneous culture of algae, grazers and predators has been used in many studies to follow the fate of chemicals. For example, Metcalf and his colleagues have used microcosms with algae, *Daphnia*, snails, and mosquito larvae to study the bioaccumulation and degradation of a large number of radioactively labelled organic chemicals (e.g. Metcalf *et al.*, 1971). The organisms were selected for ease of availability and diversity of biochemical pathways, as well as representing typical ecological trophic levels. Ecological interactions and toxicity data have not been studied in these systems, although it has sometimes been noted that a species had to be reinoculated because the test chemical eliminated its population. Although a complete review of the published results is beyond the scope of this chapter, in general those laboratory communities that have been established to study chemical fate have rarely attempted to reproduce a naturally occurring assemblage of organisms. It has been assumed that the functionality of the communities would be similar to other laboratory and to natural systems.

Our laboratory has been involved in developing "synthetic microcosms" composed of a chemically defined medium and mixtures of easily cultured, and well-researched organisms that filled typical trophic niches. In addition to the inoculated organisms, they contain undefined microorganisms that are associated with the cultures of metazoa or that are introduced during sampling. Although the culture medium and containers are initially sterile, most of the cultures are not axenic or gnotobiotic, and therefore the final systems are not gnotobiotic. This represents a compromise between having maximal knowledge and control of the components and having practical handling techniques.

These synthetic microcosms were developed to allow us to study the effects of chemicals or other stresses among multiple algal species that were competing for a shared and limited supply of nutrients while being differentially impacted by several grazers with associated nutrient recycling. Our immediate goal was to develop a reproducible, non-site-specific, microcosm protocol that could be used to evaluate the effects of chemicals on algal–grazer–microbial communities. Although they are not "natural" assemblages, they appear to demonstrate characteristic ecosystem properties such as nutrient depletion associated with algal production, reduction of algal biomass associated with grazer production, and nutrient recycling and sustained algal production. They have the potential advantage that the same assemblage can be synthesized at a later date, or in another laboratory. It is also possible to evaluate the effects of chemicals on most of the component organisms in single-species tests, so that the researcher can distinguish between direct and indirect effects.

Theoretically, the microcosms should be better able than single-species bioassays to demonstrate the ecological effects of chemicals. The environmental effects of a chemical depend not only on the toxicology of the parent compound, but also its persistence, distribution, and degradation products (Cairns, 1978; Baughman and Lassiter, 1978). There is sound ecological reason to believe that the most dramatic effects may not be the reduced abundance of the most sensitive species, but the increased abundance of its food or competitor, or alternatively, extinction of its predator.

Like mathematical models, synthetic microcosms serve to simplify complex relationships and to encourage the testing of hypotheses. In fact, it is impossible not to test hypotheses in the selection of container, medium, and organism assemblage. Establishing the microcosm inevitably tests whether the conditions are adequate for the development and maintenance of a community. For example, my original hypothesis that a gnotobiotic community could be composed of a chemically defined algal medium (Knop's), a single species of alga, *Chlorella* or *Chlamydomonas*, and *Daphnia* was disproven; the medium had to be modified to have sodium rather than potassium as the major cation, and to include either bacteria or vitamins (Taub and Dollar, 1964, 1968; Taub and Crow, 1980). Once a satisfactory microcosm is developed, it can be used to test many other hypotheses.

Being live biological entities, synthetic microcosms differ from mathematical models in that the organisms bring with them far more attributes than those for which they were explicitly selected. For example, an alga may have been selected for its size and growth rate, but it necessarily brings additional morphological, chemical, and behavioural responses to environmental conditions. These additional attributes cannot be eliminated for the simplification of the experiment and the convenience of the experimenter. In contrast, a mathematical equation may bring with it certain mathematical relationships that are known only to sophisticated mathematicians, but it has no properties apart from its explicit relationships and parameters. For example, if a temperature relationship is not included in the model of growth rate as a function of nutrient concentration, the model is "blind" to temperature changes, and can not predict encystment or sexual responses to a particular temperature–light–nutrient condition. Because live organisms bring with them properties other than those explicitly defined, biological models are more specific and less general than most mathematical models.

The relationship of synthetic microcosms to single-species studies, or to interacting algal–grazer populations is obvious. Synthetic microcosms are merely the addition of additional species on each trophic level. This increased complexity allows competition between organisms within each trophic level and allows switching in feeding patterns of the grazers.

The relationship of synthetic microcosms to naturally derived microcosms, those derived from natural sources, is more complex; each type has its own relative advantages. If the researcher asks how a site-specific community responds to a stress, naturally derived microcosms offer the closest replicable subsamples. If the experiments are extended in time, the "naturalness" of the community can be compromised by the extinction of some species and immigration of others by laboratory contamination. Thus they may be considered subsets of the natural community for only a brief time. The researcher must accept the ever changing species composition of the natural community, and the likelihood that a similar experiment set up at another time may involve different taxa. Although it is often hypothesized that ecological functions may be independent of the taxa involved, and therefore that the results would be the same if other species are substituted, this has not been adequately tested to be universally accepted.

If the researcher asks how a chemical is likely to affect aquatic communities in general and further specifies that the results must be repeatable at other times and by other researchers at other sites, the synthetic microcosm offers advantages. Since all of the organisms are "laboratory weeds" the component organisms can be maintained indefinitely in laboratory cultures, and the same initial species composition, number, age, and physiological condition can be duplicated at a later date. The researchers must recognize that this exact assemblage of organisms has not evolved together, and while major feedback interactions have been designed into the system, unique control mechanisms found in a specific community may be absent. The ability to use the results of synthetic microcosms to predict the responses of natural ecosystems rests on the assumption that the same processes are analogous to those in natural ecosystems. The limitations of many microcosm and mathematical assumptions are discussed by Hill and Wiegert (1980).

The synthetic nature of these microcosms also make them suitable for comparing the effects of single-species toxicities on the component organisms with the effects of the same chemicals on the community. We have done this for several toxic substances, and have demonstrated community effects that were not predictable from the single-species bioassay results.

The studies reported here were performed in developing and testing the responses of the microcosms to a variety of toxic chemicals. The studies included studying the effects of diluting the medium, varying the initial algal density, increasing the silicate concentration, and altering the algal assemblage. The toxic chemicals were selected to have their direct effects on different trophic levels. Three experiments were performed with one of the chemicals, streptomycin, to determine the reproducibility of the microcosm responses. The experiments represent a series of "what if" games played with the microcosms. We asked, they answered.

II. Methodology

The methodology has evolved during the experiments reported here. The sequential order of the experiments can be determined by the experiment number; ME indicates Model Ecosystem and the number that follows is the experiment number; e.g. ME 14 was done before ME 31. Once an improved technique had been developed, it was incorporated into the following experiments. In general, the techniques for the full microcosm experiments have conformed to the description below. Many ME experiments were not full microcosm experiments, but smaller procedural tests; they will be so noted.

A. Procedures

The microcosms are synthesized from a chemically defined medium, sediment, and cultures of organisms. The containers are 1-gallon (3·8 litre), wide-mouth, glass bottles. The microcosm are prepared by adding sediment and most of the culture medium (minus the silicate, phosphate, trace metals, and vitamins) and autoclaving at 121 °C for 15 minutes (capped with aluminium foil). After cooling, the remaining sterile medium components are added to each bottle, the pH is adjusted to 7·0 with sterile 10% HCl, and the foil caps are replaced with transparent, sterile, Petri dish halves. With these few steps, the physical and chemical aspects of 30 microcosms are prepared and, within human error and the purity of the distilled water and chemicals, are identical replicates, and can be closely reproduced at a later time or in another laboratory.

The algae are inoculated on day 0 and allowed to grow in competition with each other. On day 4, the animals, including protozoa, are added and grazing begins. On day 7, the 30 microcosms are examined, 24 of the initial 30 are selected and randomly assigned to treatment groups. The reason for culling 6 of the initial microcosms is to allow the researcher to eliminate any microcosms whose bottles have cracked, or whose communities appear to be developing differently. Exclusion is based on statistical examination for outliers for a number of physical, chemical, and biological variables. Again, within experimental error and variability of the stock organisms and their initial reproduction, the technique provides microcosms with similar initial conditions at the beginning of the treatment, and with an initial species array that could be reestablished at a later time and in another laboratory.

The development of the microcosms is monitored twice weekly for some variables, and once weekly for others. The experiment is continued for at least 63 days; some have been continued longer. Prior to sampling, the bottle walls are scraped and the sediment well stirred to obtain mixed samples. A 100–600 ml sample is removed to count the macroinvertebrates; subsamples of these are taken for microscopic examination, primary productivity, and extracted chlorophyll; the remainder of the large sample is returned to the microcosm. Each microcosm has its own subsampling container, but some sampling devices and probes are shared by replicates within a treatment group.

The microcosms are incubated at approximately 20°C under fluorescent lights (General Electric Co. Power Groove warm white, F96PG17WW) at about 2100 ft c or $123.2\,\mu E\,m^{-2}\,s^{-1}$ PhAR. The light period is 12 hours long, from 9 a.m. to 9 p.m.

B. Medium

Studies in our laboratory had resulted in the formulation of a nutrient solution that was adequate for the growth of many algae, and nontoxic to *Daphnia*, designated T63 (Taub and Dollar, 1968). This was amended with the addition of vitamins and other growth factors which Murphy (1970) had found necessary for *Daphnia* reproduction (Taub and Crow, 1980). The amended medium, T63MV (Table 1) contained 0·5 mM nitrate as the limiting nutrient and could support algal populations of several million algal cells ml^{-1}, in the absence of grazers. Two additional modifications of the medium were used in the experiments reported here. A 10% T63MV, tested in ME 14, were merely diluted T63MV, thus having less nutrient, but also less osmotic concentration. A silicate enriched medium, T81MV (Table 1), was used to encourage diatom growth. The initial nutrient concentration of T63MV or T81MV is substantially higher than in most ponds or lakes and is capable of supporting potential maximum algal populations rarely seen outside of sewage ponds. Once the algae have grown, however, the nutrient concentrations are not unlike many mesotrophic bodies of water. The volume of medium was 2·5 l in ME 14, and 3·0 l in all subsequent experiments.

C. Sediment

The sediment is composed of 200 g silica sand amended with 0·5 g ground chitin

Table 1 Microcosm medium T63MV: a combination of Taub's medium 63[a] and Murphy's Vitamin Mixture[b]

Master solution	Compound	Molecular weight	Concentration mM	Element	mg l^{-1}
A	NaNO$_3$	85·0	0·5	N	7·0
B	MgSO$_4$·7H$_2$O	246·5	0·1	MG	2·43
C	KH$_2$PO$_4$	136·0	0·04	P	1·23
	NaOH	40·0	1·0	Na	0·32
D	CaCl$_2$·2H$_2$O	147·0	0·014	Ca	40·0
E	NaCl	58·5	1·5	Na	34·5
IA	Al$_2$(SO$_4$)$_3$·18H$_2$O	666·5	0·0048	Al	0·26
IB	Na$_2$SiO$_3$·9H$_2$0	284·0	0·080^c	Na	36·8
				Si	2·24
F[d]	FeSO$_4$·7H$_2$O	278·0	0·0224	Fe	1·25
	EDTA	292·0	0·0224	EDTA	6·5
	NaOH	40·0	0·067	Na	1·54
J[d]	EDTA	292·0	0·006	EDTA	1·75
	NaOH	40·0	0·018	Na	0·41
G[d]	H$_3$BO$_3$	61·8	0·015	B	0·16
	ZnSO$_4$·7H$_2$O	287·5	0·0005	Zn	0·03
	MnCl$_2$·4H$_2$O	287·9	0·005	Mn	0·27
	Na$_2$MoO$_4$·2H$_2$O	242·0	0·0005	Mo	0·048
	CuSO$_2$·5H$_2$O	249·7	0·0001	Cu	0·0064
	Co(NO$_3$)$_2$·6H$_2$O	291·0	0·00005	Co	0·0059
	Calcium pantothenate	476·5	0·00147		0·70
	Vitamin B$_{12}$	1355·4	0·000000022		0·00003
	Thiamin	337·3	0·00018		0·06
	Riboflavin	376·4	0·00011		0·04
	Nicotinamide	123·1	0·00106		0·13
	Folic acid	441·4	0·00035		0·33
	Biotin	244·3	0·00012		0·03
	Putrescine	161·1	0·00019		0·03
	Choline	181·7	0·00275		0·50
	Inositol	216·2	0·00509		1·10
	Pyridoxal	167·2	0·00299		0·50

[a] Taub and Dollar, 1968

[b] Murphy, 1970 (including an erratum by J. Murphy)

[c] Medium T81MV has a 10-fold increase in silicate (solution IB) so that Si concentration is 0·8 mM (22·4 mg l^{-1}); the complete medium is adjusted to pH 7·0 with sterile 10%, HCl. This silicate-enriched medium was compared with T63MV for diatom growth (Table 3) and microcosm use with the new assemblage (Table 2, Figs 1–9)

[d] Medium T82MV, similar to T82MV, but containing 1/20 the concentrations of the substances in solutions F, J and G (trace metals and ETDA) provides greater sensitivity to metal toxicity and is now recommended for use in the Standardized Aquatic Microcosm Protocol

Medium T63MV contains two corrections of errors contained in Taub and Crow, 1980; the molecular weight of folic acid is corrected, and the acid (not the sodium salt) of EDTA is indicated

and 0·5 g ground cellulose. The organics provide a carbon source for microbial growth until the algae grow and contribute organics.

D. Algae

Three assemblages of algae have been used (Table 2). The initial density of algal cells has varied in different experiments from 10^3 to 10^4 cells per ml of each species. In addition, small numbers of each species were reinoculated weekly, to allow repopulation if extinction should occur accidentally or as a result of temporarily toxic conditions. In each assemblage, the algae were selected to be capable of growing in unialgal culture in the medium, and to represent different size ranges of potential food for *Daphnia* or amphipods. The algae are each maintained in axenic or unialgal test tube cultures in T63-agar (later T81). Prior to use, test tube cultures are inoculated into one litre semicontinuous cultures. Subsequent to ME 22, the litre cultures were diluted with fresh medium daily so that the initial algal cells were not nutrient depleted. Standard aseptic technique is used in maintaining the algal stocks, and it is fairly easy to maintain test tube cultures without cross-contamination. If contamination occurs, older uncontaminated stocks can be used, the species reisolated through streaking on agar plates, or new cultures can be obtained from other laboratories or commercial sources.

The animal assemblages consisted of *Daphnia magna*, the ostrocod *Cyprinotus*, the amphipod *Hyalella* and a variety of rotifers and protozoa (Table 2). Rotifers, protozoa, and other microorganisms coexist in the animal cultures (except for the *Tetrahymena* cultures which were axenic, and the *Paramecium aurelia* which were monoxenic with a bacterium). All of the animals are available from commercial suppliers or other laboratories, and are easily reared in the laboratory. The animals were inoculated in small numbers on day 4 (*Daphnia*, 6 adults and 10 young; ostracod, 6 organisms per microcosm; amphipod 6 adults and 6 young per microcosm). Reinoculation of the animals was done each week only if a species was not observed at minimal density in a microcosm. This was done to allow repopulation if temporary toxic conditions or accidental extinction eliminated a population; this had been shown to cause divergence among replicates (Taub and Crow, 1978).

Monitoring includes twice weekly enumeration of all algal and animal species, *in vivo* fluorescence, midday pH, and oxygen concentrations prior to lights on and late in the afternoon (experimentation had shown these to be similar to values at lights out). Twice weekly for 3 or 4 weeks, and then once weekly, carbon uptake (^{14}C), alkalinity, extracted chlorophyll and phaeopigments, and algal nutrients have been measured in all microcosm experiments since ME 18.

We have chosen the term "variables" rather than "parameters" for the items we measure. The term "parameter" is used by some mathematical modellers to indicate a constant in a simulation equation, and by some statisticians to indicate the true estimate of the mean or variance.

Table 2

Original Assemblage in T63MV (ME14, ME28 (algal only), ME31)

ALGAE	ANIMALS
Anabaena	*Paramecium aurelia*
Ankistrodesmus	*Tetrahymena vorax*
Chlamydomonas	*Didinium nasutum*
Chlorella	*Daphnia magna*
Navicula	*Cyprinotus incongruens*
Pediastrum	*Hyalella*
Scenedesmus	
Selenastrum	
Stigeoclonium	
Ulothrix	

Simplified Assemblage in T63MV (ME40)

ALGAE	ANIMALS
Anabaena	*Daphnia magna*
Ankistrodesmus	*Hyalella*
Chlamydomonas	*Cyprinotus incongruens*
Chlorella	*Aeolosoma*
Scenedesmus	*Philodina*
Selenastrum	hypotrichs
Stigeoclonium	
Ulothrix	

New Assemblage in T81MV (ME40)

ALGAE	ANIMALS
Anabaena	*Daphnia magna*
Ankistrodesmus	*Cyprinotus incongruens*[b]
Chlamydomonas	*Hyalella*
Chlorella	*Philodina*
Gomphonema parvulum[a]	hypotrichs
Nitzschia kutzingiana	
Lyngbya	
Scenedesmus	
Selenastrum	
Stigeoclonium	
Ulothrix	

---

[a] In later studies *Gomphonema* has been eliminated from the standardized algal assemblage; 10 algal species are used in the Standardized Aquatic Microcosm Protocol

[b] In later studies the ostracod *Cypridopsis vidua* has been substituted because it is easier to culture and achieves measurable populations in control microcosms

The measurements of these variables are entered directly on coded data sheets for computer handling. Most of the data are keypunched as the experiment progresses. Data through day 28 are analysed statistically to examine the quality of the experiment and to test the statistical program. The statistical analyses consist of analyses of variance between the control and treatment groups as appropriate, depending on the experimental design; some experiments are 2×2 factorials, some are not. In practice, the statistical analyses can be completed in less than two weeks after the experiment, the graphics prepared a few days later, and the standard report completed 60 days after the experiment is over, or 123 days after it begins. This rapid processing of the data is possible only because the variables measured are standardized; it could be considerably faster if we had a committed computer. The program has provisions for noting additional species, should they be observed. The program also has provisions for entering "laboratory notes" to document any special events, such as power failures, non-functional instruments, or unusual observations; these notes are printed as the cover page whenever any data are printed by the computer.

In the first experiment, algal species abundances were estimated by twice weekly counts on haemocytometer slides. This choice of technique was unfortunate because larger cells and filamentous algae were unable to flow between the cover slip and slide (0·1 mm), and abundances below 10^4 cells ml^{-1} could not be observed. Many algal species, shown as absent by the haemocytomer counts, were observed on Sedgwick-Rafter slides. After experiment ME 14, Palmer cells with 0·4 mm depth (0·1 ml volume) have been used to avoid these problems.

The microcosm technique is described in greater detail (Taub and Crow, 1980), and a formal protocol entitled "Standardized Aquatic Microcosm Protocol" is now available from the author.

III. Effects of Medium Concentration

Because the initial T63MV medium was so concentrated relative to most lake and pond waters, it was decided to test whether a diluted medium of 1/10 the concentration would be as satisfactory (ME 14). Twelve microcosms were established with each of the two concentrations of media, T63MV and 10% T63MV; four microcosms of each type of medium served as controls, and four microcosms of each medium were subjected to 30% and 60% non-specific mortality on day four. The effects of the mortality treatments were transient and not of great magnitude (Taub et al., 1980), and the effects of the medium concentration were consistently shown in all 12 replicates of each medium.

The succession of algal species was similar in both types of microcosms. Most of the algal species achieved measurable populations at some point during the experiment; *Selenastrum*, *Chlorella*, and *Chlamydomonas* established early but non-persistent populations; *Ankistrodesmus* became the dominant population slightly later and persisted as the dominant during much of the experiment. *Scenedesmus* was present, but often at densities too low to be measured on the haemocytometer. *Anabaena* had a transient bloom; *Ulothrix* and *Stigeoclonium* became abundant late in the experiment. Only *Navicula* and *Pediastrum* failed to establish observed populations. Granted the limited ability to enumerate the larger and the less abundant algal cells with the haemocytometer, there were significantly lower populations of *Ankistrodesmus* in the microcosms with the more dilute medium for only a few days; after that period, the *Daphnia* in the T63MV microcosms had reduced the algal populations to densities similar to those in the dilute microcosms. It was rarely possible to document differences in the abundances of algal species after that. The differences in species number were not of great magnitude; also, in addition to the inadequacy of the counting method, slight differences in the timing of the algal species succession among replicate microcosms also contributed to the problems of demonstrating statistical differences between treatment groups when those differences were of small magnitude.

Functional attributes associated with primary production were much lower in the dilute medium, e.g. daytime oxygen change, *in vivo* fluorescence, and pH. The statistical differences of these variables could be demonstrated over much of the experiment. Exceptions occurred for a brief time while the algae were grazed down to low densities, and the attributes of primary production in the T63MV were lower than in the more dilute microcosms.

Daphnia populations barely persisted in the dilute medium microcosms; self-sustaining, but small *Daphnia* populations did persist in three of the 12 replicates, but *Daphnia* were never abundant in those three and rare in the others in spite of weekly reinoculation. In the concentrated medium, *Daphnia* populations rapidly became very dense, and after grazing down the algae, declined to lower, but generally self-sustained populations. The dilute medium may have excluded *Daphnia* either by providing an inadequate food base or by being osmotically too dilute.

Rotifers and other grazers persisted in the dilute medium, but not at higher populations in the absence of *Daphnia*, in spite of the standing crops of algal cells not being very different for most of the experiment. The relative absence of *Daphnia* did not allow an expansion of the other grazers.

The effects of the reduced nutrient and the reduced grazing by *Daphnia* did not allow any additional species of algae to develop. In both sets of microcosms, all of the algae tended to establish distinct peaks (except *Navicula* and *Pediastrum*), but most algal species became very rare over time, in spite of continued reinoculation, while *Ankistrodesmus*, and much later, the filamentous algae tended to dominate. This suggests that a few species were able to sequester most of the limited nutrients, and to prevent other species from becoming abundant. Contaminating diatoms did became established late in the experiment, indicating that successful competition can occur. Their source was probably Lake Washington water that was used to culture the amphipods. As mentioned, the larger animals are in "conventional culture", and coexist with undefined microorganisms. The two diatoms, morphologically similar but distinct from *Navicula*, were isolated in culture on T63MV medium, and identified as *Gomphonema parvulum* and *Nitzschia kutzingiana*.

It is interesting to note that the cell counts were less adequate in indicating the reduced primary production than were the chemical variables. The cell numbers represented a measure of standing crop, which might be expected to be less because less nutrient was available, but the algae were also subjected to less grazing because the *Daphnia* populations were low.

IV. Alterations in Algal Inoculum Condition

We also determined that the culture condition of the algal inocula and the initial number of algal cells had an effect on the development of the algal community, in the absence of grazers (ME 28, algae only, 14-day duration). *Anabaena*, *Ankistrodesmus* and *Selenastrum* developed larger populations in the microcosms when they had been grown in batch inoculum cultures, while *Chlamydomonas*, *Chlorella*, *Pediastrum* and *Scenedesmus* developed larger populations when grown in semicontinuous inoculum cultures. In the microcosms, phosphate was depleted more rapidly than nitrate, but both nutrients were essentially depleted by day 14, when this experiment was ended.

To investigate the effects of the initial concentration of algal cells, 12 microcosms each were initiated with either 10^3 or 10^4 cells ml^{-1} of each of the 10 species of algae (a few species had slightly fewer cells, but the proportions between the treatment groups were constant). Six of each type of microcosm received 32 mg l^{-1} of streptomycin. All microcosms had the same initial quantity of inorganic nutrients, and initial conditions of *Daphnia* and other grazers (ME 31).

Microcosms started with the greater initial algal densities had smaller peak algal blooms, in association with a more rapid increase in the *Daphnia* populations which cropped the algal populations down faster. The availability of nutrients per cell may also have contributed to differential behaviour of the treatment groups; nutrients were depleted more rapidly with the higher initial algal densities. Although *Daphnia* population densities in both treatment groups eventually converged, the timing of the algal–*Daphnia* cycles differed between treatment groups. These results suggest that very transient differences, initial algal densities and resultant differences in grazing pressure could affect the magnitude of an algal bloom even in the absence of any differences in the absolute nutrient concentrations. These data have been reported in greater detail (Taub *et al.*, 1981).

The two types of microcosms differed in the magnitude of their responses to streptomycin (32 mg l^{-1}), although the types of responses were similar. In both types of microcosms, there was a temporary reduction in primary production, *Scenedesmus* replaced *Ankistrodesmus* as the dominant *Daphnia* populations were less dense for a period after the reduced primary production, and ostracod populations were markedly increased.

V. Alterations in Silicate Concentration and Species Assemblage

The failure of the diatom *Navicula* to form measurable populations in the microcosms resulted in the absence of diatoms as a functional part of the algal community. Because diatoms form an important part of many freshwater communities and most salt water communities, their lack was undesirable. Since all of the algae, including *Navicula*, were able to grow as unialgal cultures in medium T63MV, the lack of *Navicula* to grow in the microcosms indicated that conditions were inadequate for growth in the presence of competition. This problem might be corrected by changing the medium or by using a more competitive diatom.

Cultures of 13 diatoms were tested singly with medium T63 (contains 0·08 mM silicate), and with increased silicate concentrations (0·8 mM), with and without the vitamins; four media in all (Table 3). The *Navicula* culture that had failed to compete in the microcosms was best able to grow in the original T63, with or without vitamins, but grew to greater densities with higher silicate concentrations. Most of the other unialgal cultures grew more rapidly and to higher densities with the higher silicate concentration but none was superior to *Navicula*.

The enriched silicate medium with vitamins, termed T81MV was tested on the full microcosm assemblage, with the addition of the diatoms *Cyclotella meneghiniana*, *Stephanodiscus minutus*, *Gomphonema parvulum* and *Nitzschia kutzingiana*. The copepod *Diaptomus clavipes* (from Oak Ridge National Laboratory) was also

Table 3 Responses of unialgal cultures of diatoms to silicate and vitamins[a]

	Species No.	Species	High Si		Low Si	
			T81	MV	T63	MV
Good	3	*Navicula* sp.	1·03	0·96	0·63	0·64
	8	*Nitzschia* sp.	—	0·93	0·10	0·04
	13	*Nitzschia kutzingiana*	0·49	0·52	0·12	0·11[c]
	11	*Stephanodiscus minutus*	0·41	0·37	0·17	0·19
Moderate	14	*Synedra delicatissima*	0·32	0·36	0·09	0·13
	12	*Gomphonema parvulum*	0·32	0·36	0·06	0·07
	10	*Cyclotella meneghiniana*	0·35	0·34	0·23	0·23
	7	*Cyclotella* sp.	0·34	0·32	0·09	0·16
	9	*Cylindrotheca fusiformis*	0·34[b]	—	0·03	0·05
	2	*Fragilaria* sp.	0·22	0·33	0·08	0·08
Poor	5	*Nitzschia punctata*	0·26	0·28	0·17	0·15
	1	*Synedra ulna*	0·26	0·22	0·09	0·07
	4	*Fragilaria crotonesis*	—[b]	—	0·06	0·09
	6	*Nitzschia curvilineata*	—	—	0·06	0·05

[a] Final optical density reached (day 39) ranked by descending OD

[b] Missing data, 3 of 5 replicate eliminated (contaminated, or sexual cells)

[c] Note highest OD before day 39

added to some of these microcosms. These were compared to microcosms having the original T63MV medium, with and without the new organisms (ME 37).

In the microcosms, *Navicula* was observed at low concentrations through day 21, and only rarely thereafter in the original medium; with additional diatoms, it was even rarer. With the silicate-enriched medium, *Navicula* still did poorly with and without the other diatoms. In brief, *Navicula* never attained appreciable populations when grown in the microcosms, in spite of its ability to grow well in the unialgal cultures. It is possible, though unlikely, that the *Navicula* may be growing attached to the glass or sand and not available to our sampling technique, but if that is so, it is not available to the *Daphnia* and other planktonic grazers, nor would we be able to demonstrate toxic effects.

Of the other diatoms growth improved in the enriched silicate medium. *Gomphonema* obtained populations of 10^4 and 10^5 cells ml^{-1} for a time; *Nitzschia* reached populations of 10^3 and 10^4 cells ml^{-1}; *Cyclotella meneghinia* had measurable populations from days 21–35. *Stephanodiscus* did not maintain measurable populations (it had been inoculated at a lower concentration than the others because its stock cultures grew slowly). The increase in silicate allowed normal development of the *Daphnia* populations and other organisms. Medium T81MV was therefore recommended for future use since it allowed the development of diatoms as well as the other organisms.

It is not surprising that increased silicate concentrations might allow diatoms to occupy a greater importance in the algal community. Kilham and Kilham (1978) have proposed that the algal dominance is determined by the relative amounts of nitrogen, phosphorus, and silicate, rather than by the absolute amounts of the limiting algal nutrients. Schelski (1975) has suggested that the change in diatom dominance in the Great Lakes is caused by the selective increase in phosphorus as compared to more modest increases in silicate. Titman (1976) demonstrated that algal dominance could be controlled by the relative ratio of algal nutrients.

The copepod *Diaptomus clavipes* survived in low numbers, but did not establish significant populations; in visual counts, 1–10 were usually noted as being present. Nauplii were not observed in subsamples that were examined microscopically. Although we can rear this copepod on algae and protozoa as its food, it apparently cannot compete in the microcosms, perhaps because of its long generation time. This is unfortunate, because it has been intensely studied at Oak Ridge, can be cultured, is amenable to single-species bioassays, and its population dynamics have been modelled (Maguire *et al.*, 1976a,b,c; Robertson *et al.*, 1974).

To further improve the design of the synthetic microcosms for testing the ecological responses to chemicals, additional studies were undertaken to identify new potential species for the microcosms, and the past studies were examined to identify species that rarely had measurable populations. *Lyngbya* was added to the microcosms as a second species of blue-green. The use of the single blue-green, *Anabaena*, had two disadvantages; it formed early blooms and was virtually absent late in the experiment, and we tended to assume that the single species could act as an indicator of all blue-greens, whereas in taxonomic groups that had several species, e.g. the green algae, it was clear that no single species was an indicator of the others. The oligocheate, *Aeolosoma*, that had been isolated from the *Diaptomus* cultures appeared to be a suitable addition. The rotifer, *Philodina*, and a large hypotrich protozoan that had established populations as contaminates were isolated from old microcosms and cultured. In summary, among the inoculated species that rarely had measurable populations in the microcosms were *Navicula*, *Pediastrum*, *Tetrahymena*, *Paramecium*, and *Didinium*, and there seemed to be little advantage in continuing to include them. All of these studies were used to design modified species assemblages for the next series of microcosm experiments.

VI. Responses of Two-Species Assemblages to Streptomycin

Streptomycin is an antibiotic and a selective algal toxicant. Single-species bioassays had shown it to be inhibitory to *Anabaena* and other blue-greens. Its toxicity to green algae was variable; *Selenastrum*, and *Ankistrodesmus* were quite sensitive, while *Scenedesmus* was less sensitive (Harrass *et al.*, in preparation). It did not appear to be toxic to the animals in our microcosms at 32 mg l^{-1} as judged by the rate of population increase.

Streptomycin was tested in two types of microcosms; "simplified" in which the older medium, T63MV was used with a simplified algal assemblage (8 species), and "new" in which the silicate enriched medium T81MV was used with an algal assemblage that included 2 blue-greens (*Lyngbya*, *Anabaena*), 2 diatoms (*Gomphonema parvulum* and *Nitzschia kutzingiana*) and 7 green algae. Both assemblages included 8 of the previously used algae, as well as *Daphnia*, amphipods, ostracods, *Aeolosoma*, *Philodina*, and hypotrichs (Table 2). This experiment (ME 40) demonstrated the effects of 32 mg l^{-1} streptomycin in both types of microcosms, but graphical results on only the "new" assemblage will be displayed here, and the comparison of two assemblages will be discussed. The experiment also served to compare these results with two earlier experiments (ME 22 and ME 31) in which the older assemblage of microcosms were tested for their responses to streptomycin.

The 32 mg l^{-1} streptomycin sulphate added on day 7 disappeared rapidly, and none was measured by day 28 (*Bacillus* bioassay). Without the use of radioactive tracers, we cannot state whether the molecule is chemically degraded, sorbed on to cell surfaces, or incorporated into the cells. We inferred that it had not been converted into ammonia or any other nitrogen source available to the algae, because it did not increase primary production as had glucosamine in a previous experiment. Also, the reduction in night-time respiration suggests that the sugars of strepto-mycin had not become available as energy sources to the bacteria.

The effects of streptomycin on the "new" synthetic microcosms are shown graphically for several variables (Figs 1–8) and the statistical differences between the control and treatment groups are summarized graphically (Fig. 9). The effects

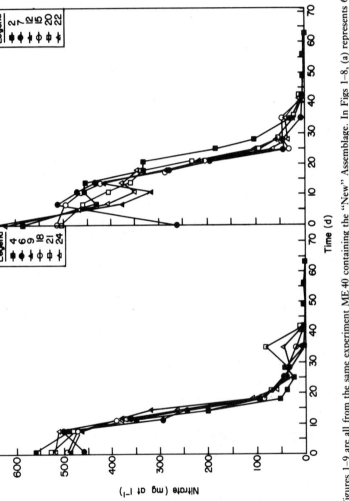

Figures 1–9 are all from the same experiment ME 40 containing the "New" Assemblage. In Figs 1–8, (a) represents 6 control replicates, and (b) represents 6 treatment replicates (32 mg l^{-1} streptomycin sulfate added on day 7). Each symbol and line represents an individual replicate microcosm. The spread of values among treatments indicates the degree of replication. The symbols are consistent among the figures, so that relationships among variables within individual microcosms can be identified.

Fig. 1 Nitrate concentrations. Note the delay in nitrate depletion in the streptomycin-treated microcosms. The single low value on day o for one of the treatment microcosms was almost certainly an analytical artefact, but the value has been included in all of the statistics.

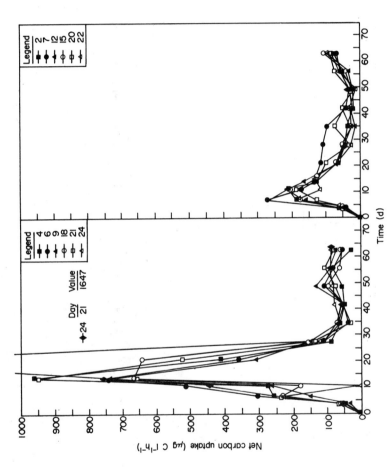

Fig. 2 Carbon uptake. The carbon uptake was calculated from the alkalinity and ^{14}C uptake. Carbon uptake was reduced markedly between days 7 and 14; differences were less marked during the remainder of the experiment, but were usually significantly lower than controls (see Fig. 9).

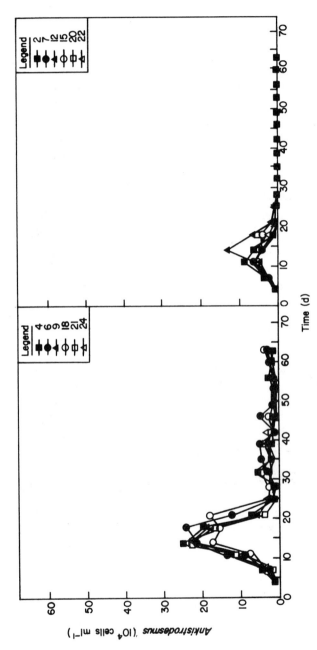

Fig. 3 *Ankistrodesmus* density was reduced by streptomycin; it was one of the more sensitive species.

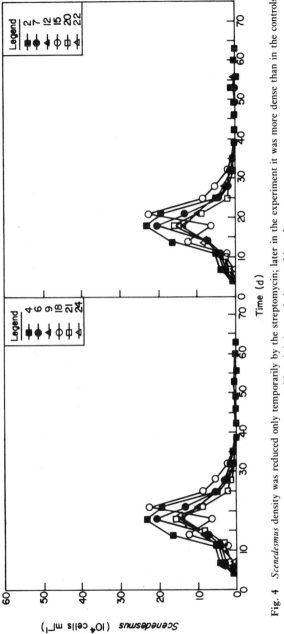

Fig. 4 *Scenedesmus* density was reduced only temporarily by the streptomycin; later in the experiment it was more dense than in the controls, although it is not obvious at this scale.

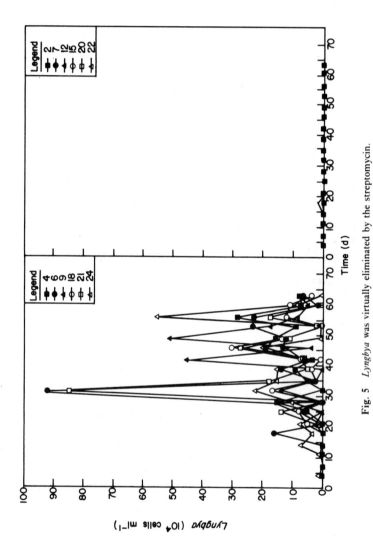

Fig. 5 *Lyngbya* was virtually eliminated by the streptomycin.

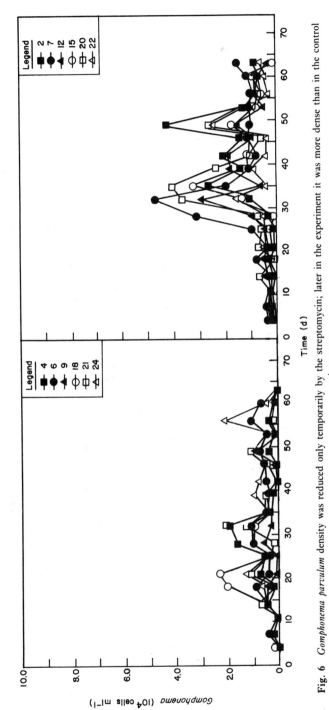

Fig. 6 *Gomphonema parvulum* density was reduced only temporarily by the streptomycin; later in the experiment it was more dense than in the control microcosms.

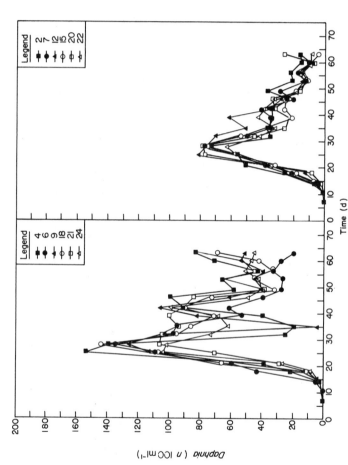

Fig. 7 *Daphnia* density (sum of small, medium and large). In the controls, the *Daphnia* population increased rapidly, grazed down the algae, and was reduced as small animals disappeared from the microcosms; increases in population indicated additional reproduction. The streptomycin reduced the magnitude of the *Daphnia* population peak, and there was scant reproduction later in the experiment.

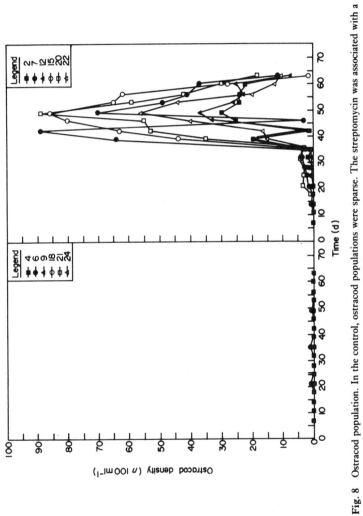

Fig. 8 Ostracod population. In the control, ostracod populations were sparse. The streptomycin was associated with a marked increase.

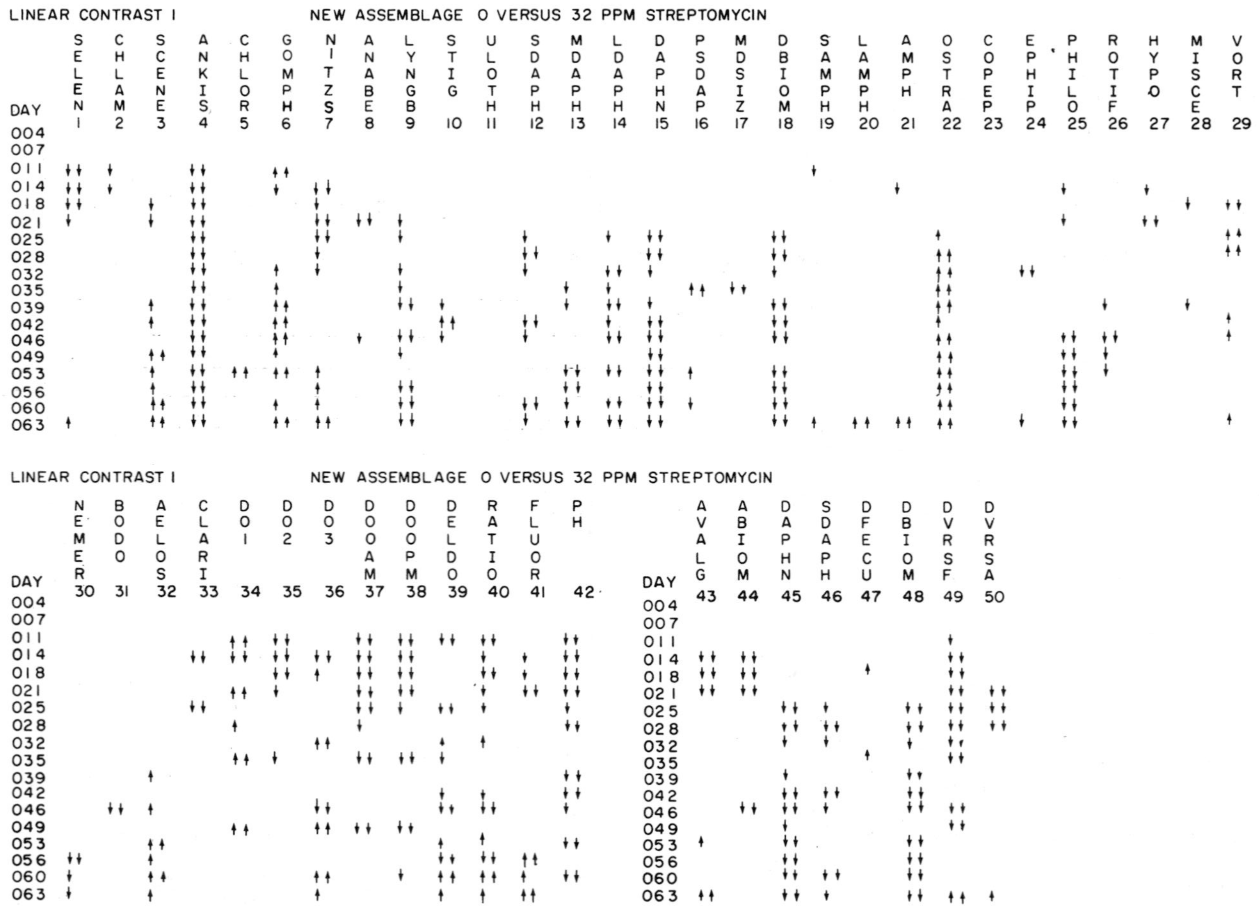

Fig. 9a (see page following)

Fig. 9a The measurements that are available on the day of sampling include: Columns 1–11 refer to the densities of the algal species; 12–18 refer to small, medium, large, total, percentage small, median size and biomass of *Daphnia*; 19–21 are small, large and total amphipods; 22 are ostracods; 23 copepods; 24 *Daphnia* ephippia; 25–26 rotifers; 27–32 Protozoa; 33 Clarity (Turbidity); 34–40 are oxygen concentrations at morning (pre–lights), afternoon, morning, increase over the lighted period (gross photosynthesis), decrease over the dark period (dark respiration), change over the 24 hours, ratio of gross photosynthesis/night respiration; 41 *in vivo* fluorescence; 42 pH. Some of the above data are used to calculate other variables: 43 an estimate of the volume of algae "available" to the *Daphnia*, 44 total algal biomass (volume); 45 total *Daphnia*; 46 small *Daphnia* (these are identical with numbers 15 and 12, but located here for easy comparison); 47 *Daphnia* fecundity; 48 *Daphnia* biomass (same as 18); 49 species biomass of "available" algae; 50 species diversity of all algal species.

Fig. 9b Other variables require time for analyses and are available after a brief delay; 51–55 are nutrients, phosphate (silicate values are compromised by freezing the samples), nitrate, nitrite, and ammonia; 56–58 are carbon uptake in the light, dark, and difference; 59 is the alkalinity; 60–63 are extracted pigments.

Fig. 9 Statistically significant differences between the 6 microcosms that received streptomycin and the 6 control microcosms, all of the "new" assemblage. A single arrow signifies a $P < 0.05$, and a double arrow signifies $P < 0.01$. If the arrow(s) are up, the variable was of greater magnitude in the treatment group than in the control, and the converse for down arrow(s). These probability displays are a convenient summary of where the effects of the treatment have occurred, but the computer tables must be consulted to see the magnitude of the differences. Figs 1–8, or the graphical display, Figs 1–8, or the computer tables must be consulted to see the magnitude of the differences. (The five letter abbreviation at the top of the column represents the variable name.)

LINEAR CONTRAST I — columns 51 (PO4), 52 (NO3), 53 (NO3), 54 (NO2), 55 (NH3); DAY rows: 000, 007, 011, 014, 018, 021, 025, 028, 035, 042, 049, 056, 063.

NEW ASSEMBLAGE, 0 VERSUS 32 PPM STREPTOMYCIN — columns 56 (CL4L), 57 (CL4D), 58 (CI44), 59 (ALK); DAY rows: 000, 004, 007, 011, 014, 021, 028, 035, 042, 049, 056, 063.

LINEAR CONTRAST I — columns 60, 61 (CHLLAFD), 62 (CHLLAFDL), 63 (PHAES); DAY rows: 000, 003, 004, 007, 011, 014, 018, 021, 025, 028, 032, 035, 039, 042, 049, 053, 056, 060, 063.

of the streptomycin was similar in the "simplified" assemblage. Depletion of nitrate was delayed in the streptomycin-treated communities (Fig. 1); phosphate depletion was also delayed. Carbon uptake was reduced, especially during the initial peak (Fig. 2). (Note that the rapid decrease in carbon uptake and many other algal attributes coincides with the rapid increase in *Daphnia* populations and their grazing.) *Ankistrodesmus*, which was the dominant alga in the controls during at least the first half of the experiment, was reduced throughout the entire experiment by the streptomycin (Fig. 3). *Scenedesmus*, which was the second most abundant in the controls during the first half of the experiment, was reduced during the initial peak, but reduced less than *Ankistrodesmus*, and thus replaced it as the dominant during the first half of the experiment. *Scenedesmus* remained more abundant in the streptomycin treated microcosms than in the controls during the later half of the experiment (Fig. 4). *Lyngbya* was virtually eliminated by the streptomycin (Fig. 5); as was *Anabaena*. *Gomphonema* became more abundant in microcosms that received streptomycin (Fig. 6). As a result of the combined changes among the algal species, the algal species diversity was reduced for a period, as was the "available algae"; i.e. those of a size available to *Daphnia*, and the total algal biomass.

The *Daphnia* populations were reduced after day 21, Fig. 7. Ostracods were notably increased in microcosms that had received streptomycin, Fig. 8, although the increase in ostracod populations did not become apparent until almost day 40, and the streptomycin had probably largely disappeared by day 14, and entirely by day 28.

Effects on many other variables, either measured or calculated, can be seen in Fig. 9 which displays the statistical differences between the control and streptomycin treated microcosms of this species assemblage. The addition of streptomycin altered many of the ecological processes. The patterns of effects are ecologically meaningful. Reductions in algal abundances are associated with delays in nutrient depletion, reduced *in vivo* fluorescence and extracted chlorophyll, reduced carbon uptake, and reduced oxygen production during the day. The resultant reduction in algae available to the *Daphnia* was associated with a reduced *Daphnia* population. The increases in the ostracod populations was entirely unexpected from the acute, single-species toxicity tests which can only indicate if a species is harmed. Presumably, the ostracods benefited from some change in the detrital processes; their increase was not consistently associated with decreases in *Daphnia* populations.

The nutrient–algal–grazer interactions among the controls are similar to the temperate aquatic ecosystems in the spring after nutrients have accumulated all winter while the light intensity and temperature were too low to support much algal activity. The available algal nutrients are depleted as the algae grows; their production is markedly decreased as the nutrients are depleted and the grazer population becomes abundant. Among the algal species, dominance relationships shift in a type of species succession, presumably influenced by competitive nutrient relationships and selective grazing. Primary production is sustained throughout the entire experiment by nutrient recycling; nitrate is depleted, nitrite is temporarily present, and ammonia—a product of recycling—is moderately abundant after the nitrate depletion. The *Daphnia* and other grazer populations respond to the available food in much the same way as in many natural lakes. Were these data presented as being from a eutrophic pond, rather than from a synthetic microcosm, they would be accepted as being ecologically reasonable.

Since the synthetic microcosms do not represent any specific natural community, the test of their ecological realism is their ability to predict the responses—in general—of aquatic communities impacted with the same or similar chemicals. Given the lack of predators in the synthetic microcosms, these represent nutrient-limited communities; predator-controlled communities might be expected to behave differently. (We have some suspicions that the amphipod may feed on juvenile *Daphnia*, although they are thought to be detritivores.) We are in the process of testing streptomycin on naturally derived communities from two lakes. We cannot expect species-specific predictions, although we may be able to predict taxonomic groups in the case where all members of a taxon are more sensitive than all members of another taxon. We may be able to predict the indirect and secondary effects in a general way; if primary production is reduced, it is likely that grazers will be reduced, at least for a time.

The "new" microcosm assemblages had slightly less primary and secondary production, but the standard deviations of the replicates were also slightly lower and the coefficients of variation were similar to the "simplified" microcosms. The diatoms, which established populations via contamination even in the simplified microcosms with lower silicate, developed larger populations more rapidly where they were initially inoculated and the silicate concentration was higher. *Lyngbya* established measurable populations. Nutrients were depleted more slowly by the 11 species of algae in the "new" assemblage than by the eight species in the "simplified" assemblage. Although the *Daphnia* population peaks were not as dense in the "new" assemblage, the pattern of a rapid increase and more gradual decline was retained. With the exception of the *Aeolosoma*, all the new organisms persisted and demonstrated measurable populations.

Our concern that the altered species assemblages or the modified medium might drastically affect the pattern of microcosm development, or might reduce the replicability within a treatment group proved to be unfounded. Both microcosm types seemed suitable for toxicity testing; the "new" assemblage had the advantage of having measurable diatom components and two blue-green species, thereby providing a degree of representation of these ecologically important groups not available in the previous assemblage.

These responses to streptomycin had all been observed in the two earlier microcosm experiments, ME 22, and ME 31. In all three sequential experiments, all variables associated with primary production were reduced temporarily, species shifts occurred usually benefiting *Scenedesmus* at the expense of other algal species, blue-greens were virtually eliminated, *Daphnia* populations were less abundant for a time following the reduced available algae, and ostracods were markedly increased. Minor changes had occurred in each of the experiments, e.g. ME 22 was generally less productive than previous and later microcosm experiments, but the trends and interactions between the algae and *Daphnia* were very similar. Species diversity among the algae was significantly reduced in ME 22, and in ME 40, but not in ME 31.

VII. Responses to Malathion

Malathion is an insecticide which is usually distributed on a wettable powder. There is some measurable (extractable) phosphate associated with the commercial Malathion, possibly associated with the clay of the wettable powder. When tested on unialgal cultures, Malathion tended to have a slight stimulatory effect which may have been caused by the associated phosphate.

This microcosm study (ME 44) was in progress at the writing, but even before the statistical analyses, it is obvious that the 10 µg l^{-1} of Malathion introduced on day seven caused massive mortalities of the *Daphnia*, ostracods, and amphipods, and allowed a massive accumulation of algae. Subsequent reinoculation of the macrofauna seven days after the Malathion addition resulted in survival and reproduction of the grazers which have now grazed down that surplus algal abundance. This implies that the Malathion had hydrolysed during the first week, a fact we had anticipated from the high pH that occurs when the algal population is dense and actively photosynthesizing. A similar insecticide, Dursban, was shown to cause increased algal standing crops as compared to control ponds (Hurlbert *et al.*, 1972). None of our single-species bioassays, nor any of the data summarized in a review of Malathion effects on isolated components of communities (Mulla and Mian, 1981), would have predicted the magnitude of the algal blooms obtained.

VIII. Conclusions

Results indicate that synthetic microcosms are feasible and display interactive behaviour between algal and grazer populations. Reproducibility among replicates was sufficient to allow the demonstration of statistically significant effects due to treatment. Results are reproducible between experiments, e.g. the three streptomycin experiments (ME 22, 31, and 40) which were set up at different times of the year. Synthetic microcosms show promise of providing tools for indicating the likely ecological effects of chemicals in aquatic communities.

Our laboratory is in the process of studying the responses of the microcosms to different types of chemicals. We hope to be able to compare some of our laboratory results to experimental ponds, enclosures or other more natural communities that have been stressed with similar chemicals. We are doing a limited amount of study on outdoor, naturally derived microcosms.

In many cases, the results demonstrated in the microcosms were not predictable from the single species results alone. The ability of an alga, such as *Navicula*, or *Selenastrum*, to grow well in unialgal culture does not assure that it will be able to compete in a more complete community that includes competing algae and grazers. From the acute toxicity tests on the algae, we knew that *Scenedesmus* was less sensitive to streptomycin, but no stimulation was demonstrated. We would not have been able to predict the increase in *Scenedesmus* abundance which we presume is caused by the reduced ability of the more streptomycin sensitive algae to compete for nutrients. Surely, we would not have predicted the nature of the *Daphnia* population decrease, nor the extreme increase in ostracod populations. With the Malathion, the single-species bioassays would have suggested a slight stimulation of the algae, not the dramatic increase that occurred while the grazers were eliminated.

The microcosms also appear to be more realistic than single-species bioassays in predicting the likely effects of a toxic material because they allow the recovery process to be demonstrated after a toxicant has degraded. Cairns *et al.* (1977) have documented the importance of the recovery process in damaged ecosystems. The synthetic microcosms also demonstrated delayed responses to toxicants, such as the permanent elimination of blue-greens and the increased abundances of ostracods long after the streptomycin could no longer be shown to be present. It is likely that the microcosms could be used to demonstrate the toxicity of a degradation product. These synthetic microcosms could be used to demonstrate the ecological effects of test chemicals in association with replicates treated with radioactively labelled test chemicals in which the fate could be determined, e.g. studies like those of Metcalf *et al.* (1971).

Although the species have not evolved together, and it can be argued that they lack specialized interrelationships that can co-evolve over geological time, they do display interactions. All of the algae can grow in the initial nutrient solution; that is one of the requirements of selection. When ten species (11 in later studies) are inoculated in the initial medium, all must compete for the available nutrients. Most of the species are able to deplete the initial nutrient supply in about two weeks and to form dense populations were no other species present. Some species become abundant quickly and become rare thereafter, e.g. *Chlamydomonas*, *Chlorella*, *Selenastrum*; others become more abundant and retain dominance, e.g. *Ankistrodesmus*. Given the limited amount of initial nutrient, competition for nutrients is forced by the initial conditions of the experiment.

Similarly, the reproductive potentials of the *Daphnia* and other grazers, combined with the lack of an alternative food supply, forces the strong interactions between the algae and grazers. The *Daphnia* can be expected to alter their reproductive rate in response to changes in algal density, e.g. Richman (1958). The grazers recycle nutrients in the very process of consuming algae and performing their own physiological functions. Undoubtedly bacteria and other microorganisms also play a role in nutrient recycling and competition. Microorganisms were probably responsible for many of the changes in community structure that were associated with streptomycin treatment.

Our results alerted us to the importance of the statistical power of the test. If variance among the replicates is as high as commonly found in biological data and if the sample size is small, the "acceptance region" may include practically all of the biologically possible range of values, thus rendering it impossible to demonstrate statistical differences between the control and treatment groups (Crow and Taub, 1979). The easiest way to increase the statistical power of the test is to increase the number of replicates. All effort should also be made to reduce biological causes of variance, such as starting all microcosms with the same numbers and stages of organisms. In practice, we have found a few cases in which errors have probably been made in microcosm inoculation, and we have allowed ourselves to exclude a few microcosms from statistical analyses (in experiments not reported here). A more troublesome source of variation among replicates has been the tendency for some microcosms to initiate the algal bloom slightly before or after the other replicates in its treatment group, and for the associated events to occur in a similar pattern but slightly out of phase with the others. This can present a very large variance when the data are analysed for each variable separately on each sample day. We are seeking alternative methods of analysing the data to allow the interactions between microcosm components to be compared (Taub *et al.*, 1981).

In the future, we plan to test the ability of mathematical models to describe the interactions between components. One of the questions that will be addressed is the interaction between grazing and nutrient limitation in the termination of the first algal bloom, and the role of food availability in structuring the *Daphnia* population. With that information, the single-species bioassays might be used to predict the effects of the toxic chemicals on microcosms, and the "mathematical" models compared to the "biological" models.

Other future plans include reanalysing the synthetic microcosm data to explore the concepts of stability and maturity (Orians, 1975), inertia and resilience (Westman, 1978), and constancy, resistance and resilience (Leffler, 1980). In results not described in this chapter, we have data on lower concentrations of toxic chemicals to which the microcosm responses were of lesser magnitude, and returned to the control conditions sooner. Another series of experiments is underway to test whether "young" or "mature" microcosms are better able to display the effects of toxicants. It would also be interesting to explore greater variation in the community structure, and its effect on toxic responses. Predator controlled microcosms would make interesting contrasts, and we have some efforts planned along these lines.

Acknowledgements

This chapter is School of Fisheries Contribution No. 558. Funding for this project was received from US Food and Drug Administration, Contract Number 223-80-2352. Many colleagues were involved in these studies: they included Dr L. L. Conquest, Michael E. Crow, Joan Hardy, Michael Harrass, Hans Hartmann, Andrew Kindig, Peter Munro, Patricia Read, and Sharon Roloff. Thanks are also owed to Dr John Leffler of Ferrum College for his comments and suggestions on an earlier draft of the manuscript.

References

Baughman, G. L. and R. R. Lassiter (1978). Prediction of environmental pollutant concentration. Ch. 3, Environmental concentration and fate. *In* "Estimating the hazard of chemical substances to aquatic life" (J. Cairns, Jr, K. L. Dickson and A. W. Maki, eds), pp. 35–77. STP 657, ASTM, Philadelphia, Pa.

Beyers, R. J. (1963). *Publ. Inst. Mar. Sci., Univ. of Texas* **9**, 19–27.

Cairns, J., Jr, K. L. Dickson and E. C. Herrick (eds) (1977). "Recovery and Restoration of Damaged Ecosystems". University Press Virginia, Charlottesville, Va.

Cairns, J., Jr, K. L. Dickson and A. W. Maki (1978). Ch. 1, Introduction. *In* "Estimating the Hazard of Chemical substances to Aquatic Life" (J. Cairns, Jr, K. L. Dickson and A. W. Maki, eds), pp. 3–11. STP 657. ASTM, Philadelphia, Pa.

Cooke, D. G. (1977). Aquatic laboratory microecosystems and communities. *In* "Aquatic Microbial Communities" (J. Cairns, Jr, ed.), pp. 59–103. Garland, New York.

Crow, M. E. and F. B. Taub (1979). *Intern. J. Environm. Studies* **13**, 141–147.

Elmgren, R., G. A. Vargo, J. F. Grassle, J. P. Grassle, D. R. Heinle, G. Langlois and S. L. Vargo (1980). Trophic interactions in experimental marine ecosystems perturbed by oil. *In* "Microcosms in Ecological Research" (J. P. Giesy, Jr, ed.), pp. 779–800. Technical Information Center, US Dept. of Energy, CONF-781101, Available; NTIS.

Giesy, J. P. Jr, (ed.) (1980). "Microcosms In Ecological Research" DOE Symposium Series, Augusta, Ga., Nov. 8–10–78, CONF-781101, NTIS. 1110 pp.

Harris, W. F. (ed.) (1980). Microcosms as potential screening tools for evaluating transport and effects of toxic substances. ORNL/EPA-4; EPA 600/3-80-042. US Dept. of Energy, Oak Ridge National Laboratory and EPA.

Harrass, M. C., A. C. Kindig and F. B. Taub (in preparation). Responses of blue-green and green algae to streptomycin in unialgal and paired culture. Presented at the Fall 1980 meeting of the American Society of Limnology and Oceanography, 30 December 1980.

Hill, James IV and R. G. Wiegert (1980). Microcosms in ecological modeling. *In* "Microcosms in ecological research" (J. P. Giesy, Jr, ed.), pp. 138–163. Technical Information Center, US Dept. of Energy, CONF-781101, Available; NTIS.

Hurlbert, S. H., M. S. Mulla and H. R. Willson (1972). *Ecol. Monogr.* **42** (3), 269–299.

Kilham, S. S. and P. Kilham (1978). *Verh. Internat. Verein. Limnol.* **20**, 68–74.

Leffler, J. W. (1978). Ecosystem responses to stress in aquatic microcosms. *In* "Energy and Environmental Stress in Aquatic Systems" (J. H. Thorp and J. W. Gibbons, eds), pp.

102–119. DOE Symposium Series, Augusta, Ga., Nov. 2–4, 1977. CONF-771114; Available; NTIS.

Maguire, L. A., C. W. Gehrs and W. Van Winkle (1976a). COPEPOD2: A Markov-type model for copepod population dynamics. Oak Ridge National Laboratory, Report No. ORNL/TM-4976, July 1976. 97 pp.

Maguire, L. A., S. R. Blum, C. W. Gehrs and W. Van Winkle (1976b). COPEPOD4: A discrete time-delay model of copepod population dynamics. Oak Ridge National Laboratory, Report No. ORNL/TM-4977, July 1976. 109 pp.

Maguire, L. A., S. R. Blum, C. W. Gehrs and W. Van Winkle (1976c). Submodels: Copepod population parameters as function of environmental factors. Oak Ridge National Laboratory, Report No. ORNL/TM-4978, July 1976. 61 pp.

Metcalf, R. L., G. K. Sangha and I. P. Kapoor (1971). *Environ. Sci. Technol.* 5, 709–713.

Mulla, M. S. and L. S. Mian (1981). *Res. Rev.* 78, 101–135.

Murphy, J. S. (1970). *Biol. Bull.* 139 (2), 321–332.

Nixon, S. W. (1969). *Limnol. Oceanogr.* 14 (1), 142–145.

Orians, G. H. (1975). Diversity, stability, and maturity in natural ecosystems. *In* "Unifying Concepts in Ecology" (W. H. Van Dobben and R. H. Lowe-McConnell, eds), pp. 139–150. W. Junk, The Hague.

Reinert, R. E. (1972). *J. Fish. Res. Bd Canada* 29, 1413–1418.

Richman, S. (1958). *Ecol. Monogr.* 28, 273–291.

Robertson, A., C. W. Gehrs, B. D. Hardin and G. W. Hunt (1974). Culturing and ecology of *Diaptomus clavipes* and *Cyclops vernalis*. EPA Report No. EPA-600/3-74-006, April 1974, 226 pp.

Schelski, C. L. (1975). Silica and nitrate depletion as related to rate of eutrophication in Lakes Michigan, Huron and Superior. *In* "Coupling of Land and Water systems" (A. D. Hasler, ed.), pp. 277–298. Vol. 10, Ecological Studies. Springer, New York.

Taub, F. B. and A. M. Dollar (1964). *Limnol. Oceanogr.* 9 (1), 61–74.

Taub, F. B. and A. M. Dollar (1968). *Limnol. Oceanogr.* 13, 607–617.

Taub, F. B. (1969a). *Verh. Internat. Verein. Limnol.* 17, 485–496.

Taub, F. B. (1969b). *Limnol. Oceanogr.* 14 (1), 136–142.

Taub, F. B. and M. E. Crow (1978). *Verh. Internat. Verein. Limnol.* 20, 1270–1276.

Taub, F. B. and M. E. Crow (1980). Synthesizing aquatic microcosms. *In* "Microcosms in ecological research" (J. P. Giesy, Jr, ed.), pp. 69–104. DOE Symposium Series 52, Nov. 8–10, 1978, Augusta, Ga. CONF-781101, Available; NTIS.

Taub, F. B., M. E. Crow and H. J. Hartmann (1980). Responses of aquatic microcosm to acute mortality. *In* "Microcosm in ecological research" (J. P. Giesy, Jr, ed.), pp. 513–535. DOE Symposium Series 52, Nov. 8–10, 1978, Augusta, Ga. CONF-781101, Available; NTIS.

Taub, F. B., M. C. Harrass, H. J. Hartmann, A. C. Kindig and P. L. Read (1981). Effects of initial algal density on community development in aquatic microcosms. *Verh. Internat. Verein. Limnol.* 21, 197–204.

Thomas, W. H. and D. L. R. Seibert (1977). Effects of copper on the dominance and the diversity of algae: controlled ecosystem pollution experiment. *Bull. Mar. Sci.* 27 (1), 34–43.

Titman, D. (1976). *Science* 192, 463–465.

Westman, W. E. (1978). *Bioscience* 28, 705–710.

14 Simulating the Effects of Increased Temperature in a Plankton Ecosystem: A Case Study

V. H. Dale and G. L. Swartzman

Center for Quantitative Science
University of Washington
Seattle, Washington, USA

I. Introduction

Because chemical reactions are temperature dependent, the thermal environment has a considerable effect on organisms and their competitive abilities. Increased water temperature of lakes is one of the by-products of many modern industries, especially of power plants, so thermal loading is important to environmental managers, who need an understanding of its impact. Perturbation studies are one of the classical methods of studying community dynamics, so thermal loading is of interest to ecologists trying to learn more about the interactions of natural systems. Temperature directly affects phytoplankton growth and zooplankton grazing and respiration, and one would expect that different patterns of responses would be more favourable to organisms living in warmer than normal waters. In this manner thermal loading may actually be a form of selection for certain patterns of population interactions or for those organisms having the types of responses to temperature changes favoured in the warmer water. Thus, this study of thermal loading in Lake Ontario has management, ecological and evolutionary implications.

The effects of increased temperature on plankton can be studied by field, laboratory or simulation experiments. By recording detailed biological and physical information, field and laboratory studies can detect sudden and/or slight changes in the ecosystem. Simulation experiments allow detailed consideration of the role of a given process or organism in model dynamics by examination of model behaviour under several alternative hypotheses (Swartzman, 1979a). Simulations are useful if it is not possible or advisable actually to manipulate the ecosystem. An example is the subjection of aquatic ecosystems to high water temperatures. Studies of the effect of previous high water temperatures carried out at the Savannah River Plant show that the return of the ecosystem to its natural state is largely determined by the presence

ALGAE AS ECOLOGICAL INDICATORS
ISBN 0 12 640620 0

of a natural habitat in close proximity (Gibbons and Sharitz, 1974). The nearer such an area is, the quicker the propagules or individuals can reach the disturbed site. In any case, there can be quite a long time before the disturbed system returns to its natural state. This indicates that care must be maintained in the initial perturbation of such systems if any value is placed on the undisturbed habitats. Computer models of ecosystems offer a means in lieu of manipulation of the natural ecosystem by which possible interactions of the system can be examined. Simulation study can provide an initial characterization of those types of processes which may be more advantageous in warmer than normal waters, and these will suggest laboratory experiments to provide further evidence for the types of organisms and interactions likely to exist in thermal plumes. The simulation experiments may also call attention to critical periods of the organisms and critical levels of nutrients, light or temperature. The importance and difficulty of knowing what details of the system to study in the field have been a major obstacle to biologists (Strickland, 1969). Understanding the nature of temperature-induced changes will lead to a greater accuracy in the prediction of the effects of thermal loading in lakes.

Both field and simulation studies can be used to explore the effects of increased temperature on plankton. These approaches are similar in that each must consider the physical and chemical environment in order to understand the biological interactions and each gives some measure of the seasonal dynamics. The primary difference between the two is that field experiments tend to focus on changes in the community structure whereas simulation studies concentrate on processes and their effects on the organisms. Furthermore, field studies are restricted by the manner in which the data is obtained (for instance, plankton net size), but collection allows for later re-examination of the material. A simulation study is relatively easy to run (once the program is created) and, in a deterministic model, is repeatable; however, it represents only the modeller's best understanding of the ecosystem. Because an ecosystem model is dependent upon field and laboratory studies for parameter estimation, its validity is a function of appropriate sampling of the physical and biological system.

The study of the effect of high temperature on plankton has received much attention in recent years (e.g. Brock and Brock, 1966; Coutant and Pfunderer, 1973, 1974; Gibbons and Sharitz, 1974). Some species may be more adept physiologically or competitively at the warmer than normal temperatures, and therefore species composition and diversity may change. The rate of metabolism is increased in warmer waters according to the van't Hoff principle that the rate of chemical reaction increases with rising temperatures. Thus one would expect a decrease in the average cell size, but an increase in biomass, primary production, respiration, chlorophyll a content and the cell division rate. A number of studies of the effect of increased temperature on phytoplankton are summarized in Table 1. Each of these studies had a control consisting of organisms in the normal environment to which characteristics of organisms in the warm waters were compared. In general, the effects of increased temperatures on plankton did not differ from those predicted.

Table 1 Summary of studies of the effect of increased temperature on phytoplankton

Reference	Phytoplankton studied	Site	Temperature increase causes changes in							
			Species composition	Average cell size	Biomass	Species diversity	Primary production	Respiration	Chlorophyll a	Cell division per day
Stockner (1967)	Diatoms	Runoff water from hotsprings, Wyoming	√							←
Cairnes (1965)	Community	Sabine River, Pennsylvania	√							
Daletzkaya and Chulanovskaya (1964)	Chlorella	Laboratory		→						
Margalef (1954)	Scenedesmus obliquus	Laboratory		→						
Garnier (1958, 1962)	Oscillatoria brevis	Laboratory							→[h] →[l]	
Patrick (1968)	Diatoms	Simulated Pennsylvania river			←	√[g]				
Patrick (1968)	Community	Potomac R.				←[d]				
Patrick (1968)	Community	Green R., Pennsylvania	√			→				←

Table 1 Summary of studies of the effect of increased temperature on phytoplankton

Reference	Phytoplankton studied	Site	Temperature increase causes changes in							
			Species composition	Average Cell size	Biomass	Species diversity	Primary production	Respiration	Chlorophyll a	Cell division per day
Stangenberg and Pawlaczyk (1961)	Community	Power station in Poland				↓ʃ				
Trembley (1960, 1965)	Community	Delaware R.				→				
Kreh (1973)	Community	W. Lake Erie	—		↑		↑s ↓w			
Sorokin (1959, 1967)	Chlorella						↑			
Steidinger and von Breedveld (1969, 1970)	Benthic marine algae	Crystal R. estuary, Florida			→	→				
Brooks (1972)	Community	Indian R., Delaware					→w			
Kelly (1971)	Community	Microcosms			↑		↑		←	
Werner (1971)	Coscinodiscus asteromphalus	Laboratory								←
Sobone (1974)	Thalassiosira rotula	Laboratory								←

Reference	Type/Species	Location				
Copeland and Davis (1972)	Community	South Creek estuary, N.C.	↑[s]	↑[s]	↑	
Sangar and Dugan (1972)	*Anacystis nidulans*	Culture				↑
Davis (1972)	Community	Pamlica R., N.C.	↑			
Specchi (1972)	Community	Gulf of Trieste, Italy			—	
Fairchild (1971)	*Achnanthes oxigua*	Hot springs		↓	↑	↑
Meeks and Castenholz (1971)	*Synechococcus lividus*	Laboratory		↑		↑

√ change observed
— no change
↓ decrease
↑ increase
d difficult to separate from nutrient increase
f for greater than 30°C
g greatest at 24°C
h at high light intensities
l at low light intensities
s summer
w winter

The interactions of the effects of temperature with those of light intensity and nutrients is apparent from the table.

II. The Lake Ontario Ecosystem

To ascertain the reasonableness of model output, it was compared to data collected on Lake Ontario during 1973 and 1974 (Stoermer *et al.*, 1975). Although Lake Ontario is not a typical lake, owing to the abundance of phytoplankton species either tolerant of or requiring eutrophic conditions, the availability of data for a number of years and the fact that our "default" model (that of Thomann *et al.*, 1975) was written as a description of this system made it appropriate to use Lake Ontario as the comparison ecosystem for this study. In all simulations discussed, environmental conditions and initial conditions for the state variable are those of Lake Ontario.

Lake Ontario (latitude 43°55′) is one of the five large northern lakes which make up the Great Lakes system in North America. Its surface area of 19 477 km² represents a watershed drainage area of 90 132 km². The volume is 1669 km³, and the lake has a maximum depth of 244 m. Situated at 74·01 m above sea level, the crytodepression has a depth of 170 m. The maximum length and width of Lake Ontario are 307 km and 87 km respectively, and the shoreline length is 1380 km. The major source of inflow into the lake is the Niagara River which has an average flow of 195 000 cfs (84% of the total discharge). Average annual precipitation in the area is 83·23 cm, and the average annual evaporation is 71 cm. The thermocline begins in late April to early May and dissipates in late September, with the average depth being 17 m. The hypolimnetic depth is 73·3 m. Because the lake is so large, the hypolimnetic retention time is 8·1 years. The principal zooplankton in the lake (91%) are *Cyclops bicuspidatus*, *Tropocyclops prasinus*, *Daphnia retrocurva* and *Bosmina longirostris* (Thomann *et al.*, 1975).

III. The Simulation Model

The plankton model simulator is a "library" of programs using different biological hypotheses translated into equations. To facilitate the reader's understanding of the simulator, this section includes a discussion of the library, its assumptions and restrictions, the comparison methodology, and the way thermal loading was simulated.

A. The Simulation Library

The simulation library consists of a number of models drawn from the plankton literature and incorporating the biological framework given in Fig. 1. The major state variables of the system represent phytoplankton, herbivorous and carnivorous zooplankton and detritus (mg l⁻¹), nitrogen (organic, ammonia, and nitrate),

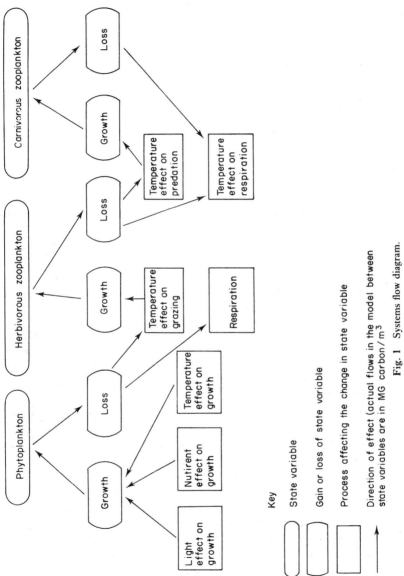

Fig. 1 Systems flow diagram.

Key

State variable

Gain or loss of state variable

Process affecting the change in state variable

Direction of effect (actual flows in the model between state variables are in MG carbon/m³

phosphorous (organic and phosphate) and silicon (mg 1^{-1}). The variables interact by means of the major processes considered by the model: photosynthesis, respiration, grazing and predation. Photosynthesis is broken down into the subprocesses of light, nutrient and temperature effects on growth. There are a variety of equation forms representing the hypothesized effects of each process (Table 2), and incorporating a number of these formulations in the simulator allows us to compare them and to determine how appropriate the different biological hypotheses are. The basis for the simulator is the model of Thomann *et al.* (1975), chosen because it includes most of the processes we wished to examine. The simulator output for this model was run under the environmental conditions of Lake Ontario (Fig. 2). Other process formulations for the effect of a physical parameter upon plankton were selected from the literature. All equations and parameter values used are listed in Table 2 and discussed in Swartzman and Bentley (1979). Examples of particular processes will be examined in a later section. The library of models is evaluated under a single set of environmental conditions so that a comparison of the effects of different processes assumptions can be made in terms of the total ecosystem.

B. *Assumptions and restrictions of the model*

By its very formation, the simulator is based on a number of assumptions about the ecosystem which can restrict its interpretation.

(1) In the model, phytoplankton growth is equal to the multiplicative effects of light, temperature, and nutrients from which grazing and respiration have been subtracted. Although the effects of light, nutrients and temperature are related, it is possible to separate their effects in the biological systems (Hutchinson, 1967).

(2) Because most of the models reviewed discuss phytoplankton fluctuations in terms of chlorophyll *a*, the simulator also uses this unit. Epply (1972) has pointed out the errors in assuming a constant chlorophyll *a* to carbon ratio, and the simulator is subject to these problems.

(3) The simulator includes the functional groups of phytoplankton, herbivorous and carnivorous zooplankton, and detritus. Physiological responses are frequently species specific, and some level of realism is lost by considering only large groups or organisms. Nevertheless, characteristics of general groups (e.g. blue-green algae) were considered to ascertain how that group might respond to specific conditions.

(4) A deterministic model such as the simulator can only be used to predict average year behaviour. Particularly seasonal dynamics are frequently due to environmental conditions specific to that year.

(5) Model interpretation must be wary of the time scale used. For this discussion, the model was run for one year with a time step of 12 hours.

(6) There are some effects associated with the increase in temperature which

Table 2 Equations in default model

Effect	Equation	Source
Light on phytoplankton growth	$PLIT_{1,m} = \dfrac{e}{K \cdot Z_m}\left(\exp\left[\dfrac{-IA_m}{IS}e^{-K \cdot Z_m}\right] - \exp\left[-\dfrac{IA_m}{IS}\right]\right)$	Steele (1962) Thomann et al. (1975)
Self shading by phytoplankton	$K = \alpha + \beta\, BP_m + \gamma BP_m{}^{2/3}$	Riley (1946)
Nutrients on Phyto growth	$PNUT_{1,m} = \dfrac{N1_m + N2_m}{KN + N1_m + N2_m} \cdot \dfrac{P_m}{KP + P_m}$	Thomann et al. (1975)
Temperature on phytoplankton growth	$PTMP_{1,m} = KM\,\theta^{T_m}$	Thomann et al. (1975) Fremer (1975) Thomann et al. (1975)
	$PTMP_{3,m} = \begin{cases} KM \cdot T_m & \text{if } T_m < TO \\ KM \cdot TO & \text{if } T_m \geq TO \end{cases}$	Bierman et al. (1974)
	$PTMP_{4,m} = \begin{cases} GM \cdot \exp\left[\dfrac{-2 \cdot 3(T_m - TO)^2}{(TL - TO)^2}\right] & \text{if } T_m < TO \\ GM \cdot \exp\left[\dfrac{-2 \cdot 3(T_m - TO)^2}{(TU - TO)^2}\right] & \text{if } T_m \geq TO \end{cases}$	Lehman et al. (1975)
Temperature on zooplankton respiration	$ZRSP_{1,m} = KR1 \cdot T_m$	Thomann et al. (1975)
	$ZRSP_{2,m} = KR2 \cdot \theta 4^{T_m}$	Kremer (1975)
	$ZRSP_{3,m} = KR3 \cdot \exp[KM(T_m - TO)]\left|\dfrac{TZ - T_m}{TZ - TO}\right|^{KM(TZ - TO)}$	MacCormick et al. (1972)
Herbivorous zooplankton density on carnivorous zooplankton grazing	$ZZRAT_{1,m} = BZ_m$	Thomann et al. (1975)
Food density on grazing preference for phytoplankton	$ZPREF_{1,m} = 1 \cdot 0$	Thomann et al. (1975)

Table 2—*cont.*

Effect	Equation	Source
Temperature on phyto respiration	$PRSP_m = RO \cdot (\theta 2)^{T_m}$	Thomann *et al.* (1975)
Temperature on zooplankton grazing	$ZTMP_{1,m} = KZ1 \cdot T_m$	Thomann *et al.* (1975)
Food density on herbivorous zooplankton grazing	$ZRAT_{1,m} = BP_m$	Thomann *et al.* (1975)
Food density on assimilation efficiency	$ZASS_{2,m} = KA \left[1 - \dfrac{BP_m}{KG + BP_m} \right]$	Thomann *et al.* (1975)

m = subscript for depth level ($m = 1$ denotes top layer—epilimnion)
$PLIT$ = light effect on phytoplankton growth
$PNUT$ = nutrient effect on phytoplankton growth
$PTMP$ = temperature effect on phytoplankton growth
$PRSP$ = phytoplankton respiration
$ZRAT$ = the effect of prey (phytoplankton) density on herbivorous zooplankton grazing
$ZTMP$ = temperature effect on zooplankton grazing
$ZASS$ = efficiency for assimilation of grazed material by zooplankton
$ZPREF$ = preference of zooplankton for eating phytoplankton relative to detritus
Similar variables exist with Z and ZZ substituted for P to denote analogous processes for zooplankton (herbivorous and carnivorous).

PHYTOPLANKTON
Light Effect
BP_m = phytoplankton biomass in depth level m (mg m^{-3})
GM = maximum growth rate all factors being optimum (d^{-1}) or maximum cell division rate
IA_m = solar radiation at surface of mth depth layer (ly d^{-1})
IS = optimal light intensity for growth (ly d^{-1})
K = light extinction coefficient (m^{-1} − m^{-1})
Z_m = depth of mth depth layer (m)

α = light extinction coefficient independent of phytoplankton (m^{-1})

β = effect of phytoplankton biomass on light extinction (self-shading)

$$\frac{m^3}{mg\ m} = \frac{m^2}{mg}$$

γ = effect of phytoplankton biomass on light extinction

$$\frac{m^{3\,2\,3}}{mg} \frac{1}{m} = \frac{m}{mg^{2/3}}.$$

Temperature Effect

T_m = Water temperature $(°C)$

KM = miscellaneous temperature exponent (dimensions)

TL = temperature below TO where growth rate is 10% of GM (°C)

TO = water temperature for growth $(°C)$

TU = temperature above TO where growth is 10% of GM $(°C)$

TZ = temperature above which no growth occurs $(°C)$

$\theta = Q_{10}$ value at 10 C

Nitrogen Effect

KN = Michaelis–Menten half-saturation constant for nitrogen $(mgN\ m^{-3})$

$N1_m$ = nitrate in depth layer m

$N2_m$ = ammonia in depth layer m

Phosphorus Effect

KP = Michaelis–Menten half-saturation constant for phosphorus $(mgP\ m^{-3})$

P_m = phosphorus concentration, depth layer m $(mgP\ m^{-3})$

Respiration

RO = constant which appears in linear or Q_{10} formulation of respiration (d^{-1})

$\theta2$ = the Q_{10} factor in respiration

ZOOPLANKTON GRAZING, GROWTH AND RESPIRATION

KG = half saturation constant of phytoplankton for grazing $(mg\ m^{-3})$

KA = assimilation efficiency

$KR1$ = respiration coefficient $(d^{-1}\ deg^{-1})$

$KR2$ = respiration coefficient (d^{-1})

$KR3$ = respiration coefficient (d^{-1})

$KZ1$ = increase in grazing rate with temperature $(d^{-1}\ deg^{-1})$.

TO = optimum grazing temperature (C)

TZ = temperature above which grazing ceases (C)

KZI = arbitrary zooplankton grazing coefficients

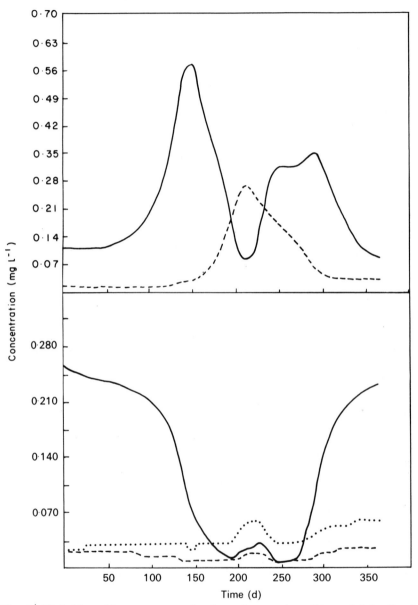

Fig. 2 "Default" model output run under the environmental conditions of Lake Ontario. (*Upper*) —— phytoplankton; – – – zooplankton. (*Lower*) ····· ammonia; —— nitrate; – – – phosphates.

have not been included in the computer models. In some cases, parasites or pathogens may do better at increased temperatures, but these organisms are not included in the model. Evidence has been found that the distribution of plankton throughout the water body determines the nature of temperature effects (Voronia, 1972), although Specchi (1969) found no such relationship, and the model allows us to examine this only in terms of a vertical column of water. In her review article, Patrick (1968) discussed the fact that an increase in temperature can cause an increase in the amount of bacteria mineralized organic material (a source of nutrients) and thus increase the nutritive value of the water. Increased temperaure and/or bacterial activity may also reduce oxygen levels, and this may cause the release of trace nutritive elements or toxic substances previously held by the oxidized microzone. The distance and velocity of vertical movement of diatoms may also be affected by the thermal environment (Hopkins, 1963), and the diatoms are very important for oxygen production. At least one organism— *Closterium leibleinii*, Munda (1960)—becomes more resistant to hypertonic solutions, such as higher chloride concentrations at increased temperatures. Although interesting, none of these effects of increased temperature can be examined with the simulator.

IV. The Simulated Effect of a Temperature Increase

The general model simulator was used to analyze the effects of temperature by comparing model output run under the thermally loaded and unloaded environments of Lake Ontario. The increased temperature of $8·33°C$ was obtained from the Ashbury–Frigo model (1971) which predicts areas affected by thermal plumes based on a correlation of surface area data from six power plants discharging into large northern lakes. The value of $8·33°C$ is found from studying 23 thermal plumes and is applicable to a surface area of $1·3$ km^2. Although this is a major increase in the water temperature, we believe that by changing the temperature so drastically yet within the bounds of past thermal plumes we shall be more likely to observe differences which might occur in the warmer than normal waters.

In the plankton models considered, temperature was hypothesized to have a direct effect on phytoplankton growth through photosynthesis, on zooplankton grazing through the maximum ration, and on zooplankton respiration. Although it is generally agreed that the process rates are positively correlated with increasing temperature, there is no agreement as to the exact nature of the curves.

A. Effect on Phytoplankton Growth

An increase in temperature causes an increase in phytoplankton growth up to a certain point in most plankton systems. There are three types of curve forms hypothesized for this effect that are used in the simulator (Figure 3 graphs these, and the equations are in Table 2). Thomann *et al.* (1975) used a linear effect of temperature on growth, although they cited no biological rationale for their choice. The "skewed normal" formulation proposed by Lehman *et al.* (1975) is identical to that presented by Lassiter and Kearns (1974) under certain conditions. The equations are derived from the general form of the curve for temperature effect on growth. Epply (1972) seems to have done the most perceptive study of the effect of temperature on phytoplankton growth. He points out that when a functional group is being considered, a Q10 formulation is the most biologically reasonable choice for the thermal effect, since this curve accounts for a change in species composition. Those species with a higher optimal temperature also tend to have a higher rate of photosynthesis at that temperature. Since the simulator lumps all phytoplankton into one group, the Q10 relation is probably the best equation for the effect of temperature upon growth in the simulator. Nevertheless, the lake may be dominated by a few species at a time, so we also tested the thermally loaded environment with the simulator using Lehman's "skewed normal" with an optimal temperature for growth of 25°C. The Q10 curve has additional advantages in its smaller number of parameter values and the fact that the change in the rate of response of organisms to each increase of 10°C is frequently measured.

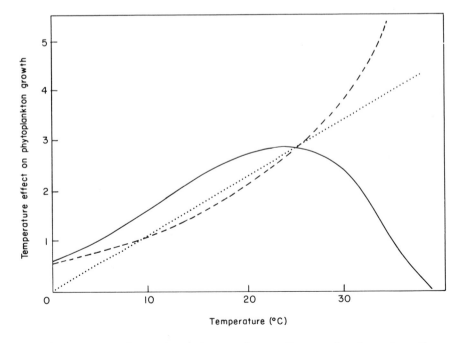

Fig. 3   Temperature effect on phytoplankton growth. – – – – Q_{10}; —— skewed normal; · · · · linear.

B. Effect on Grazing

Experimental evidence indicates that grazing is positively correlated with temperature. In the simulator, temperature affects grazing by altering the maximum ration available to the zooplankton for consumption (maximum grazing rate). Using a gamma function for the relationship between temperature and grazing, MacCormick *et al.* (1972) predict low grazing at most temperatures, particularly those lower than 15°C. Kremer's (1975) use of Q10 formulation for this interaction and the linear relationship of Thomann *et al.* (1975) are similar, although the linear equation predicts higher grazing at temperatures between 7 and 25°C. In the thermal loading studies, high, medium and low temperature effects on grazing were simulated by using the linear formulation with different slopes (Fig. 4). The lowest level of grazing resulted in an inability of the zooplankton to graze and thus their elimination. This low level is probably not representative of the true nature of the system, and it was not considered in detail.

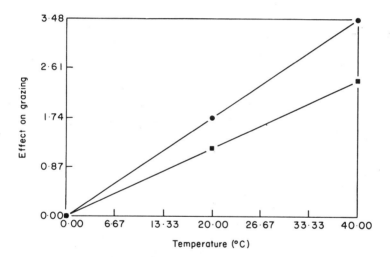

Fig. 4   Temperature effect on zooplankton grazing. Slope of temperature effect: ■ = 0·0600; ● = 0·0870.

C. Effect on zooplankton respiration

The predicted effects of temperature on respiration are similar to those for grazing. MacCormick *et al.* (1972) use a gamma function, Kremer (1975) proposed a Q10 relation and Thomann *et al.* (1975) a linear effect of temperature on zooplankton respiration. Although respiration rates are influenced by organism size, physiological state, feeding rate and seasonal acclimation, these factors were not considered in the models. A low, medium and high effect of temperature on respiration is modelled in the simulator by using each of the three predicted process equations (Fig. 5). The higher respiration curve (the Q_{10} formulation) usually resulted in rapid elimination of the zooplankton in the system and so was not considered as an appropriate model.

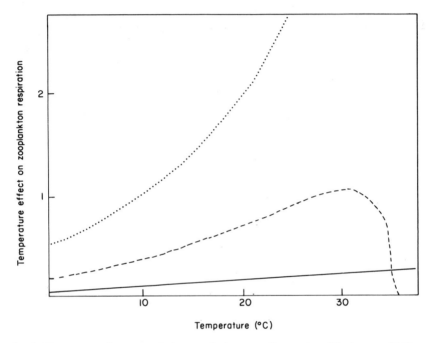

Fig. 5   Temperature effect on zooplankton respiration. · · · · · Kremer; — — — Wingra; ——— Di Toro.

V. The Simulation Comparison

In this study, graphical, tabular and statistical techniques are used to compare the model output to field data from Lake Ontario. Since seasonal dynamics of plankton and nutrients are the major areas of interest, the simulator output includes a graph of the fluctuations of these variables over one year (Fig. 2). To compare these time series, the computer produces a table of average and total production and the timing and values of the critical periods of each of the stable variables (Table 3). These criteria were chosen because they are believed to represent the major components of the dynamics of the Lake Ontario plankton system. Critical periods in the system frequently precede changes in the community structure, and these periods are usually correlated to a peak or trough of nutrients or biological activity. The timing and value of the critical level of each state variable are representative of seasonal dynamics. Total production over the year for each plankton group as well as average

Table 3a Temperature effect on phytoplankton growth increased

	Total production	Average standing crop	Maximum value	Day of maximum	Minimum value	Day of minimum
Phytoplankton	102·972	0·565	0·682	136	0·074	79
Herbivorous zooplankton	12·962	0·071	0·253	203	0·002	265
Carnivorous zooplankton	14·066	0·077	0·199	230	0·004	154
Organic nitrogen	—	0·291	0·247	190	0·090	1
Ammonia	—	0·045	0·044	365	0·011	187
Nitrate	—	0·219	0·249	1	0·001	189
Organic phosphorus	—	0·011	0·009	163	0·004	232
Phosphate	—	0·015	0·016	89	0·000	141

Table 3b Temperature effect on phytoplankton growth decreased

	Total production	Average standing crop	Maximum value	Day of maximum	Minimum value	Day of minimum
Phytoplankton	45·478	0·249	0·539	182	0·015	108
Herbivorous zooplankton	10·510	0·058	0·223	211	0·005	1
Carnivorous zooplankton	8·496	0·047	0·111	243	0·004	179
Organic nitrogen	—	0·203	0·213	203	0·053	365
Ammonia	—	0·082	0·087	233	0·015	1
Nitrate	—	0·379	0·279	365	0·028	204
Organic phosphorus	—	0·007	0·009	196	0·001	365
Phosphate	—	0·030	0·022	365	0·003	188

standing crop of each variable provides information on the importance of each of these compartments. These factors are most useful for comparisons between model runs since they represent the planktonic system behaviour.

The reasonableness of model graphical output is determined by placing overlays of the mean enveloped by one standard deviation of data of the state variable obtained from the Lake Ontario ecosystem (i.e. with identical hydrodynamic and initial conditions as the model). Model runs which went outside the bounds of this envelope were examined more closely to determine the cause of this aberrance. In most cases, the biological assumptions, the form of the process equations, the parameter values, or the combination of process equations used, were not representative of the Lake Ontario ecosystem. This comparison between the model, output and the timing of seasonal dynamics of the plankton allowed us to identify those types of processes characteristic of Lake Ontario.

Multivariate statistics were used to compare a number of model runs based on the attributes listed in Table 3. Clustering of the model runs indicates which are the most similar. Discriminant analysis provides a means of testing the relatedness of proposed groups based on the clustering results. Since discriminate analysis assumed the data are linear, additive and normally distributed, the statistical tests should be used with caution; the technique is appropriate as an advisory tool. Principal components analysis was used to elucidate the underlying gradient for different responses of the model and to illustrate similarities of attributes.

A. Graphical Comparisons

In runs of the simulator on Lake Ontario without thermal loading, phytoplankton usually decrease during April; a rapid drop follows, and a smaller peak in phytoplankton occurs in mid-September which is succeeded by a decrease (Fig. 2). Using the general model simulator, we determined that the height of the first peak is caused by phosphorus limitation and the subsequent decrease is due to zooplankton grazing. Predation by carnivorous zooplankton reduces herbivorous zooplankton in mid-summer and allows phytoplankton to have a second bloom in August. This phytoplankton bloom is enhanced by late summer turnover in the lake which causes an upward mixing of nutrients. Early winter reduction of plankton is due to falling temperature levels having constraints on respiration and growth.

As Epply (1972) points out, temperature is rarely a limiting factor in phytoplankton growth; instead it sets the upper limit for maximum growth. The simulation runs of the normal Lake Ontario ecosystem versus ones with increased temperature seem to demonstrate this. When reasonable sets of parameter values are used (for instance, cases with high respiration and low grazing are excluded since these are physiologically unrealistic), the thermally loaded runs as compared to the non-loaded cases result in an increase in average phytoplankton production. At the higher temperatures, phytoplankton can grow earlier in the year, and so the spring bloom occurs sooner (Fig. 6). Because more food is available during the first months of the year, herbivorous zooplankton are able to increase their biomass during this time. By increasing the temperature effect on grazing, zooplankton can track changes in their food abundance more quickly and thus can take advantage of the later bloom. This results in a greater production of zooplankton. Thermal loadings seem to cause oscillations in the plankton, but none is very extreme, eliminating the possibility of extinction of entire groups. This has an impact on the total ecosystem in terms of the type of food available. The phytoplankton surviving the warmer temperatures and the periods of heavy grazing are probably able to respond fairly quickly to these extreme conditions, since the heavy grazing does not last very long. Nutrients also play an important role in the plankton oscillations.

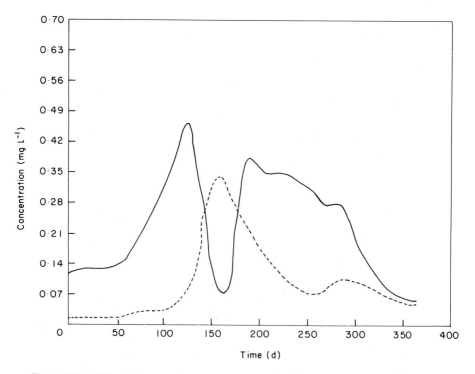

Fig. 6 "Default" model run under increased temperature. —— phytoplankton; --- zooplankton.

Less phytoplankton production results when the Q10 relationship between temperature is used for the temperature effect on grazing instead of the linear formulation. Since the Q10 function has higher grazing at high temperatures and lower grazing at medium temperatures, in the thermally loaded environment the grazing is increased as compared to cooler waters. When the "skewed normal" curve is used for the temperature effect on phytoplankton growth in the loaded run, the phytoplankton increase at a faster rate, the herbivorous zooplankton respond more quickly to the phytoplankton, and so the phytoplankton peak earlier in the year. These results are a function of the parameter values as well as the particular curve forms.

Nutrient limitation plays a different role in the model run under increased temperature as compared to normal runs. In contrast to the unloaded runs, phosphorus is not the factor limiting the first phytoplankton peak in the warmer waters. Since phytoplankton grow more quickly, the herbivorous zooplankton begin to graze them and thus decrease their biomass before phosphorus limitation occurs. The minimum nutrient effect on growth during this period of increase is twice the minimum during the similar period in the default run. This may result in more favourable conditions for those phytoplankton species which are normally phosphorus limited and so do not do well in ordinary conditions. The drop in phytoplankton biomass caused by grazing is followed by a period of growth as the herbivorous zooplankton are eaten by the rapidly increasing carnivorous zooplankton. The height of the second peak of phytoplankton is directly related to nitrogen availability, and the subsequent decrease is due to decreasing light and nutrients. The effect upon phytoplankton growth of temperature, light and nutrients is decreasing during July to mid-August. This interaction between light and phytoplankton growth does not play such a prominent role in the non-thermally loaded cases. Thus, those organisms which are not so limited by medium light intensities may have higher productivity in the thermally loaded waters of Lake Ontario.

The mid-August turnover in Lake Ontario causes a mixing of the water layers and results in an increase of nutrient supply in the euphotic zone. Incorporation of this phenomenon into the model allows for an increase in phytoplankton biomass shortly thereafter. Grazing characteristics of zooplankton determine whether phytoplankton can actually grow during this period of increasing nutrients. For example, the model with a medium temperature effect on herbivorous zooplankton grazing and a low respiration effect (Fig. 7a) has a higher grazing ration during this critical period than the model with a high temperature effect on grazing and a medium effect on respiration (Fig. 7b). Although both these models are run in the thermally loaded environment, only the second case has a significant phytoplankton bloom in mid-August.

B. Statistical Analysis

To ascertain the differences between the model runs for Lake Ontario under normal conditions and those in which the environment was increased in temperature, a number of model outputs were compared using multivariate statistics. The runs were made under varying temperature effects on phytoplankton growth and on zooplankton grazing and respiration, as indicated in the key to Fig. 8. The attributes used in this comparison were total production, average standing crop, maximum

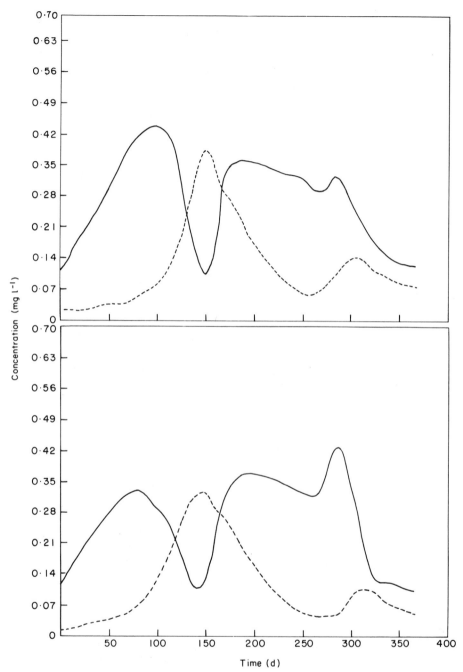

Fig. 7 Effect of grazing characteristics on phytoplankton dynamics. ———— phytoplankton, – – – – zooplankton. (*Upper*) Thermally loaded model with skewed normal temperature effect on phytoplankton growth and low effect on respiration. (*Lower*) Thermally loaded model with skewed normal temperature effect on phytoplankton growth, high effect on grazing and medium effect on respiration.

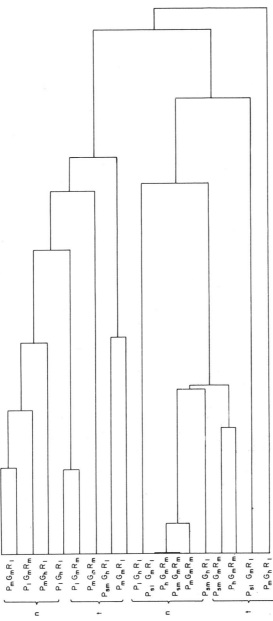

Fig. 8 Clustering of model runs. P = temperature effect on phytoplankton growth (see Fig. 3); G = temperature effect on grazing (see Fig. 4); R = temperature effect on respiration (see Fig. 5); m = medium; l = low; h = high; sm = medium skewed normal curve; sl = low skewed normal curve; n = normal temperature conditions. For instance run $P_m G_m R_l$ (the default run) has a medium effect of temperature of phytoplankton growth (P_m), a medium effect of temperature on grazing (G_m) and a low effect of temperature on zooplankton respiration (R_l).

and minimum value and timing of these events for each of the major compartments (Table 3). Since the phytoplankton usually have two blooms over the year, the value and timing of the second bloom was also included as an attribute.

A dendrogram (Fig. 8) was produced by MINFO (Goldstein and Grigel, 1971), an agglomerative, hierarchical, polythetic clustering method (Williams, 1976). Two major groupings seemed to occur, with the normal runs and those with increased temperature being subdivisions within the major groups. The coherence and meaning of these groups were examined by discriminant analysis and indirect ordination.

Discriminant analysis is dependent upon the number of subgroups defined and indicates the order of importance of the attributes as well as the percentage variance explained by the subgroups when one or more attribute is considered. When the two groups defined *a priori* were (1) normal and (2) thermally loaded runs of the model, total discrimination of the groupings was obtained on the basis of one variable, the timing of the herbivorous zooplankton maximum. The second variable to enter the discriminant equation was timing of the organic phosphorus maximum. When the four groups predicted by the dendrogram were considered by the discriminant analysis routine, 83·3% correct classification of the runs into groups was obtained by considering only timing of herbivorous zooplankton maximum and the organic nitrogen maximum value. All runs were correctly classified when timing of the ammonia maximum and of the detritus maximum were also included. Thus, this analysis gives greater confidence in the importance of nutrients and timing of herbivore to the plankton system of Lake Ontario under normal and stressed conditions.

Indirect ordination is a method by which characteristics are separated such that the location of each characteristic in two-dimensional space reflects some ecological gradient. By examining the position of each model run and attribute with respect to the axes, causal factors of the differences between groups of runs can be determined (Figs 9 and 10). The primary axis is related to the effect of thermal loading and explains 26% of the overall variation. The timing of the first phytoplankton peak and of the herbivorous zooplankton maximum are the major determinants of this axis. The second axis explains an additional 20% of the variation and is related to nutrients. Organic nitrogen, nitrate and phosphate average standing crop as well as the nitrate maximum value and the phytoplankton average standing crop contribute the most information to separation along this axis.

Examination of the ordination results indicates that in some cases thermal loading can produce greater changes in the timing and extreme values of critical events than in others. Pairs of runs (those under normal and increased temperature conditions) close together in the ordination space (Fig. 9) would be expected to represent conditions under which Lake Ontario would be less susceptible to increased temperature. For instance, a strong effect of increased temperature occurs in communities in which the zooplankton are characterized by a medium temperature effect on respiration and a high effect on grazing and the

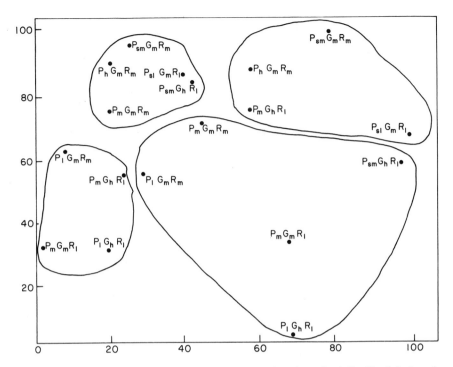

Fig. 9 Ordination of model runs. Time refers to time of maximum level. See Fig. 8 for legend.

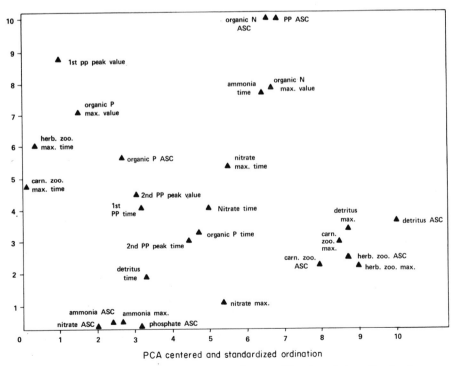

Fig. 10 Ordination of attributes. ASC = average standing crop; PP = phytoplankton; P = phosphorus; N = nitrogen.

phytoplankton have a temperature effect on growth which can be represented by the "skewed normal" curve (Fig. 11). Although the only difference in these two runs of the model is an increase in the water temperature, the simulation results are very different, as indicated by their separation in the dendrogram and the ordinations. Other pairs of runs exhibit relatively few differences (Fig. 12). Such analysis points out those types of plankton groups which are more susceptible to alterations in water temperature.

C. Summary of Differences

The three outstanding differences in limiting factors between the thermally loaded and unloaded runs observed from examining more than forty combinations of the simulator are summarized in Table 4. In the warmer waters of Lake Ontario, phosphorus limitation does not cause the first phytoplankton peak, nitrogen is in short supply in mid-June, and the light effect upon growth is a limiting factor in late summer. These differences suggest characterstics of organisms which may have higher productivity in the thermally loaded environment around a nuclear power plant or other discharge on Lake Ontario. These differences lead to hypotheses of

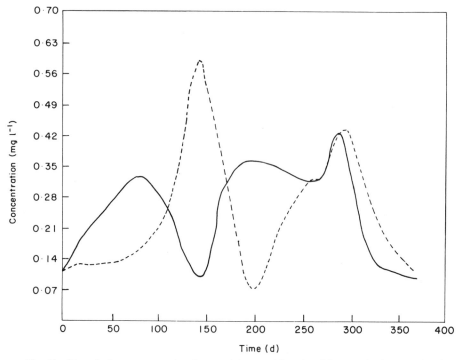

Fig. 11 Phytoplankton concentration from model runs with major differences in the normal and thermally loaded conditions. —— thermally loaded; – – – – normal conditions.

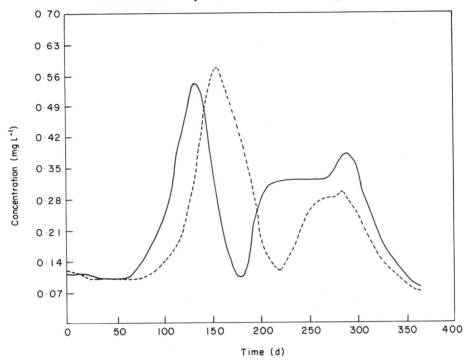

Fig. 12 Phytoplankton concentration from model runs which are similar under different temperature conditions. ——— thermally loaded; ———— normal conditions.

characteristics of organisms which can survive better in the warm waters. These proposed relations were tested by means of the general model simulator, and those that support predictions lead to hypotheses to be examined in a laboratory or field experiments.

The major differences between model outputs from the thermally loaded and the unloaded runs are an increase in phytoplankton and zooplankton over the year and a change in the timing of phytoplankton blooms. The first is a measure of total productivity for the plankton system. Experimental and field results from thermal plumes usually show an increase in chlorophyll *a*, primary production and biomass (see Table 2) giving some confidence in the model's ability to predict successfully the outcome of thermal perturbations. These changes may in part be due to alterations in species diversity and/or composition. To determine the degree to which different species groups cause a change in productivity, the nature of environmental pressures at the different times of the year in the thermally loaded case was explored.

Table 4   Comparison of the causes and timing of threshold values in a nonloaded versus thermally loaded simulation run

Occurrence	Normal temperature		Temperature increased	
	Cause	Day	Cause	Day
First peak	Light and phosphorus limitations	143	Grazing	122
subsequent decrease	Grazing	144–208	Grazing	123
Second peak	High respiration and decreasing temperature effect	250	Nutrient limitations	190
subsequent decrease	on growth	250–265		
Third peak	Nutrient mixing	285	Nutrients not limiting	225
subsequent decrease	Decreasing temperature	290–365	High respiration and low light effect on growth	230–280
Fourth peak			Nutrient mixing	285
subsequent decrease			Decreasing temperature	290–365

VI. Causes of Differences of Plankton Dynamics under Increased Temperature

Organisms which survive and reproduce in the thermally loaded waters of Lake Ontario will be a subset of those species which exist in the lake. Due to genetic variability these are different species from those which are found in the normally warmer waters of a more southern lake. Freshwater species are often able to avoid adverse conditions by ceasing divisions or forming resting stages. Cairns (1956) demonstrated this in an experiment in which he showed that the temperature range for diatoms is 20–30°C, for green algae is 30–35°C and for blue greens is greater than 35°C. Outside the temperature ranges for each group, those algae were not found in the water, although readjustment to the range of that group allowed the species to reappear and grow. This implies that species are not able to compete successfully at some temperatures. The simulator permits examination of the temperature effect in terms of the whole Lake Ontario ecosystem. The comparisons of the thermally loaded versus nonloaded runs point out some of the differences in physiological limitations the organisms may experience. Thus, one would predict that those species which have a lower phosphorus tolerance and a higher nitrogren tolerance would have a greater productivity in the thermal plumes. This was tested by running the general model simulator in the thermally loaded environment with the half-saturation constant for nitrogen increased and for phosphorus decreased. In all combinations considered, phytoplankton and zooplankton productivity increased, supporting the prediction.

Similarly, the thermal loading comparisons indicate that in the warmer than normal waters light becomes limiting in August, whereas it is never the most limiting factor in the runs under normal environmental conditions. This leads to the prediction that those algae species with a lower light limitation would produce more. To test this prediction, the light effect on growth was changed, as indicated in Fig. 13. The optimal light intensity was decreased and the coefficient was lowered so that growth is greater up to 500 langleys per day of surface light intensity. Because this is the intensity at which the second bloom occurs in the thermally loaded case and because light was shown to be limiting during this time, this is equivalent to relaxing the light limitation. Although average productivity does not increase, the amount of phytoplankton in the second bloom does. This is because the first bloom

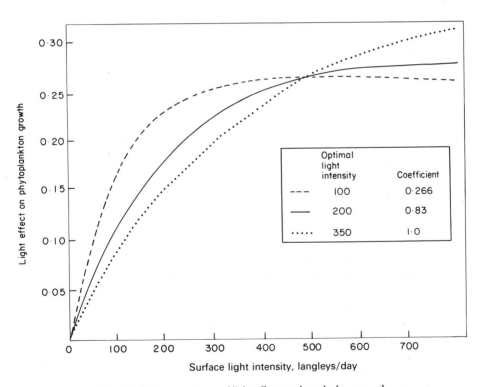

Fig. 13   Different patterns of light effect on phytoplankton growth.

is not related to light limitation, is not significantly altered by the parameter change and has the greater contribution to average productivity of the two blooms.

The simulator has demonstrated that those species with a lower phosphorus tolerance, a greater tolerance for nitrogen, and a lower optimal light intensity for growth will be relatively more favoured by warmer temperatures. These characteristics are typical of blue-green algae (Garnier, 1962; Patrick, 1968). Thus, based on physiological considerations, the prediction of the simulation model is that blue greens will have a better ability to survive thermally loaded conditions in Lake Ontario. Numerous experiments have demonstrated that this is the group of algae usually found in warm waters (e.g. Cairns, 1956; Patrick, 1974).

Identification of limiting conditions in thermally loaded waters and of physiological characteristics of phytoplankton able to live in warm waters is a preliminary step for further experimentation in the laboratory and field. The simulation study indicates that laboratory experiments would be helpful to explore such questions as:

(1) What particular species have these physiological characteristics?
(2) What blue-green algae do not?
(3) Are there any costs associated with increased ability to live in the warmer waters?
(4) If warm temperatures create nuisance plankton blooms, are there manageable ways to avoid these outbreaks of algae?

Furthermore, being aware of conditions allowing particular algae to occur gives us a better understanding of the aquatic community structure. Species diversity and trophic structure are directly related to the composition of algae species.

VII. Future Research

The results of this study have a variety of implications for future research on algae as ecological indicators. The ramifications are applicable to phytoplankton research using models and involving field work.

Future research with simulation models could involve further investigation of particular processes within one lake system or expansion of the model to other lakes. Since particular groups of phytoplankton are more abundant under certain environmental conditions, the future use of functional groups within the model framework is justified. For instance, Lehman et al. (1975) successfully modelled plankton dynamics by considering blue-greens, chrysophytes, diatoms and dinoflagellates. Furthermore the role of patchiness within the lake system can be explored via a model such as that of Vlymen (1977) for fish larvae feeding in a patchy environment.

The case study to which this particular model is applied could be extended to other lakes (the model is also parameterized for Lake Washington in Seattle,

Washington). Also the phytoplankton can be examined in the framework of a model for the entire lake ecosystem. The model discussed in this chapter has been extended to include fish and is running for Lake Keowee (South Carolina). Finally, by assessing the applicability of the model to other lakes within a region, one can determine whether a regional approach to thermal impact is justified. Do lakes within one region behave similarly in response to thermal loading?

Model implications for field work pertain to the manner in which data is collected. Phytoplankton groups could be classifed according to their mechanistic response to impact rather than taxonomic affinities. Such a functional rather than structural classification is similar to the range in which nitrogen limitation and light limitation occur. Most blue-green algae have these characteristics. Thus, this study demonstrates that physiological tolerances to the limiting conditions plays a major part in determining species composition in Lake Ontario.

The value of simulation studies of perturbation has also been considered. Computer models allow for examination of changes to the physical components of the ecosystem without actual (and permanent) manipulation of that system. Processes can be explored with a simulation model which is not always the case with laboratory or field experiments. Also, changes of a dynamic seasonal nature can be investigated rather than just averaged. Behaviour during critical periods of the year can be examined closely. Finally, formulation of hypotheses for laboratory experiments can be made, thus reducing their costs and frustrations.

VIII. Conclusions

The effects of increased temperature on phytoplankton and zooplankton seasonal dynamics in Lake Ontario were examined using a simulation model. A review of published results indicated possible changes to be explored with the model. A simulation "library" of hypotheses from various models was used to examine state variables and process effects on a daily basis, and this output was compared to data collected in Lake Ontario. Comparisons of plankton dynamics in thermally unloaded and loaded runs of the model showed differences between the two situations. Under thermal loading the spring phytoplankton peak occurred earlier and was limited by grazing whereas phosphorus limited the bloom in the unloaded environment. Also, more phytoplankton blooms occurred in the model with increased temperature, and light became more limiting during the late summer. These comparisons led to hypotheses about adaptations favourable to phytoplankton in a thermally loaded environment including (1) a lower phosphorus and a greater nitrogen tolerance, and (2) lower optimum light intensity. Tests of these hypotheses with the simulator indicated that phytoplankton with these theoretically favoured adaptations had increased productivity in the warmer waters. As a group, blue-green algae are known for the physiological characteristics we found to be favourable in the thermally loaded environment.

Acknowledgement

Funding to support this research was supplied in part by the US Nuclear Regulatory Commission under Contract No. NRC 04-75-222.

References

Ashbury, J. C. and A. A. Frigo (1971). A phenomenological relationship for predicting the surface areas of thermal plumes in lakes. Argonne National Lab., Argonne, Ill.

Brock, T. D. and M. L. Brock (1966). *Nature* **209** (5024): 733–734.

Brooks, A. S. (1972). The influence of a thermal effluent on the phytoplankton ecology of the Indian River Estuary, Delaware. Ph.D. dissertation. John Hopkins Univ., Baltimore, Md.

Cairns, J., Jr (1965). *Indus. Wastes* **1** (4), 150–152.

Copeland, B. J. and H. L. Davis (1972). Estuarine ecosystems and high temperatures. Water Resources Res. Inst., Univ. of North Carolina, Chapel Hill, UNC-WRR1-72-68, PN211-808.

Coutant, C. C. and H. A. Pfuderer (1974). *Journal WPCF* **46** (6), 1476-1540.

Daletzkaya, I. A. and V. Chulanovskaya (1964). *Bot. Zhurn.* **49** (8), 1147–1159.

Davis, J. S. (1972). *Biologist* **54** (2): 52.

Eppley, R. W. (1972). *Fish. Bull.* **70** (4): 1063–1085.

Fairchild, E. D. (1971). A physiological investigation of the hot springs diatom, *Achnanthes exigua* Grun. Ph.D. dissertation, Univ. of Montana, Missoula.

Garnier, J. (1958). *C.R. Acad. Sci., Paris* **246** (4), 630–631.

Garnier, J. (1962). *C.R. Acad. Sci., Paris* **254** (12), 2218–2220.

Gibbons, J. W. and R. R. Sharitz (1974). *Amer. Sci.* **62**, 660–670.

Goldstein, R. A. and D. F. Grigal (1971). Computer programs for the ordination and classification of ecosystems. USAEC ORNL, Publ. 417. 125 pp.

Hopkins, J. T. (1963). *J. Mar. Biol. Assoc. U.K.* **43** (3), 653–663.

Hutchinson, G. E. (1967). "A Treatise on Limnology. Vol. II. Introduction to Lake Biology and the Limnoplankton." J. Wiley & Sons, Inc. New York. 1115 pp.

Kelly, R. A. (1971). The effects of fluctuating temperature on the metabolism of freshwater microcosms. Ph.D. dissertation, Univ. of North Carolina, Chapel Hill.

Kreh, T. V. (1973). An ecological evaluation of a thermal discharge. Part IV. Postoperational effects of a power plant on phytoplankton and community metabolism in Western Lake Erie. National Tech. Info. Service. Springfield, Va. 92 pp.

Kremer, J. N. (1975). Analysis of a plankton-based temperate ecosystem: an ecological simulation model of Nanagansett Bay. Ph.D. dissertation, Univ. of Rhode Island, Kingston.

Lassiter, R. R. and D. K. Kearns (1974). "Modeling the Eutrophication Process." (Middlebrooks, Falkenborg, and Makney, ed.). Ann Arbor Sci. Publ., Ann Arbor, Mich. pp. 131–138.

Lehman, J. T., D. B. Botkin and G. E. Likens (1975). *Limnol. Oceanogr.* **20** (3) 305–496.

MacCormick, A. J. A., O. L. Loucks, J. F. Koonce, J. F. Kitchell and P. R. Weiler (1972). An ecosystem model for the pelagic zone of Lake Wingra. Eastern Deciduous Forest Biome, Memo Report No. 72-122. 103 pp.

Margalef, R. (1954). *Hydrobiologia* 6 (1–2), 83–94.

Meeks, J. C. and R. W. Castenholz (1971). *Arch. Mikrobiol. (Ger.)* 78 (1), 25.

Munda, I. (1960). *Bioloski Vestnik Ljubljana* 7, 3–9.

Patrick, R. (1969). Some effects of temperature on freshwater algae. *In* Biological Aspects of Thermal Pollution, Krenkel, P. A. and F. L. Peter (eds). Vanderbilt Univ. Press. pp. 161–185.

Sangar, V. K. and P. R. Dugan (1972). *Appl. Microbiol.* 24, 732.

Schone, H. K. (1972). *Marine Biol. (W. Ger.)* 13, 284.

Sorokin, C. (1959). *Nature* 184: 613–614.

Sorokin, C. (1967). *Science* 158 (3805): 1204–1205.

Speechi, M. (1969). The influence of temperature on the superficial microdistribution of plankton in the Gulf of Trieste. Pubbl. Sta. Zool. Napoli (It.) 37 (2): 338; Aquatic Sci. Fish. Abs., Aq. 795 7M (1972).

Strangenberg, M. and M. Z. Pawlaczyk (1961). Nauk. Pol. Wr. Wroclaw No. 40, Inzyn. Sanit. Water Poll. Abst. 1: 67–106.

Steidinger, K. A. and J. F. van Breedveld (1971). Benthic marine algae from waters adjacent to the Crystal River Electric Power Plant (1969 and 1970). Fla. Dept. Natural Resources, Prof. Paper Ser. 16.

Stockner, J. F. (1967). *Limnol. Oceanogr.* 12 (1), 13–17.

Stoermer, E. F., M. M. Bowman, J. C. Kingston and A. K. Schaedel (1975). Phytoplankton composition and abundance in Lake Ontario during 1F4GL. National Environmental Research Center Pub. No. EPA-660/3-75-004.

Strickland, J. B. (1969). Remarks on the effects of heated discharges on marine zooplankton. *In* "Biological Aspects of Thermal Pollution", pp. 73–79. US Dept. Education and Welfare, 999WP-25. US Govt. Printing Office, Washington, D.C.

Swartzman, G. L. (1979a). Evaluation of ecological simulation models. *In* "Contemporary Quantitative Ecology and Related Ecometrics" (G. P. Patil and M. L. Rosenzweig, ed.). International Co-operative Publishing House, Fairland, Maryland. pp. 295–318.

Swartzman, G. L. (1979b). *J. Environ. Manage.* 9 (2), 145–164.

Swartzman, G. and R. Bentley (1979). *Internat. Soc. Ecol. Modeling J.* 1, 30–81.

Thomann, R. V., D. M. Di Toro, R. P. Winfield and P. J. O'Connor (1975). Mathematical modeling of phytoplankton in Lake Ontario. I. Model development and verification. EPA-660/3-75-005.

Trembley, F. J. (1960). Research project on effects of condenser discharge water on aquatic life. Progress Report, 1969–1970. Lehigh University Institute of Research, Bethlehem, Pa.

Trembley, F. J. (1965). Effects of cooling water from steam-electric power plants on stream biota. *In* "Biological Problems in Water Pollution" pp. 334–335. US Dept. Health, Education, and Welfare, 999WP-25. US Govt. Printing Office, Washington, D.C.

Voronina, N. M. (1972). *Mar. Biol.* 15, 336.

Vlgmen, W. J. (1977). *Envir. Biol. Fish.* 2 (3), 211–233.

Werner, D. (1971). *Arch. Mikrobiol. (Ger.)* 80 (1): 43; Aquatic Sct. Fish. Abs. Aq. 834M (1972).

Williams, W. T. (ed.) 1976. "Pattern Analysis in Agricultural Science." CSIRO, Melbourne, Elsevier. 331 pp.

Subject Index